NANOSAFE 2008: International Conference on Safe Production and Use of Nanomaterials

Journal of Physics: Conference Series Volume 170

Grenoble, France
3-7 November 2008

ISBN: 978-1-61738-313-7

Printed from e-media with permission by:

Curran Associates, Inc.
57 Morehouse Lane
Red Hook, NY 12571
www.proceedings.com

Some format issues inherent in the e-media version may also appear in this print version.

Printed by Curran Associates, Inc. (2010)

Institute of Physics
Dirac House, Temple Back
Bristol BS1 6BE UK

Phone: 44 1 17 929 7481
Fax: 44 1 17 920 0979

techtracking@iop.org

Accepted papers received: 6 May 2009
Published online: 12 June 2009

PREFACE

EXPOSURE ISSUES

CHARACTERIZATION, DETECTION AND MONITORING

NANOMATERIALS LIFE CYCLE

TOXICOLOGY

COMMERCIAL EQUIPMENTS RELATED TO SAFE PRODUCTION AND USE OF NANOMATERIALS

PERSONAL PROTECTION AND SECURE INDUSTRIAL PRODUCTION

SAFETY PARAMETERS EVALUATION

3 - 7 NOVEMBER / MINATEC - FRANCE

Preface

The first international conference on "*Safe production and use of nanomaterials*": NanoSafe2008 was held on 3-7 November at Minatec congress center located in Grenoble, France. As recognised, the rapidly developing field of nanotechnologies present many opportunities and benefits for new materials with significantly improved properties as well as revolutionary applications in the fields of energy, environment, medicine, *etc.* However, the potential impact of these new materials on human health and the environment is viewed with apprehension. All the security aspects have to be solved in order to reach an acceptable level of risk. The new nano industry can only develop dynamically if these legitimate fears are satisfactory allayed.

This event was organized by the French Commissariat à l'Energie Atomique (CEA) in the frame of the Integrated European project NanoSafe2 funded by the European Commission under FP6 programs. NanoSafe2008 had broad international participation from 25 countries representing the 5 continents, with registered delegates numbering over 250. The scientific program involved the presentation and discussion of 112 papers, classified as 8 plenary lectures, 65 oral and 39 posters.

The conference attracted many of the best known leading scientists and specialists in almost all the different fields of expertise covering nanosafety issues: Exposure, Characterisation, Detection, Monitoring, Life cycle, Toxicology, Personal protection, Secure industrial production, Safety parameter evaluation, Standardization, Regulation and Education.

In parallel with the conference an exhibition took place where 10 exhibitors presented the commercial equipment available today in relation to safety during the production or the use of nanomaterials.

We believe that this conference, with the above topics, has provided a forum where the many matters of interest to the NanoSafety community have been debated. The success of the conference was such that attendees asked the organizers to perpetuate this event on a two year basis in order to become a global manifestation with greater success in bridging the gap between the best scientists, engineers, exhibitors and participants from countries concerned with safe and responsible nanomaterial activities. Therefore, the second International NanoSafe Conference "NanoSafe2010" will be held in Grenoble at Minatec, 15-19 November 2010.

The meeting was financially supported by generous contributions from the European Commission, La Région Rhône Alpes, la Ville de Grenoble, la Metro, and partners such as CEA, INERIS, Cilas, Niro A/S, Nanosight, Arkema, Faure Ingénierie, Philips, Sperian, TSI, Ecomesure, Mecachrome, Nanogate, INRS and Cordouan Technologies.

The conference Organizers
Carole Sentein
Frédéric Schuster
Francois Tardif

Journal of Physics: Conference Series **170** (2009) 011001 doi:10.1088/1742-6596/170/1/011001

Conference committees:

Local Organising Committee

Carole SENTEIN (CEA)
Dominique BAGUET (CEA)

Jacques BOUILLARD (INERIS)
Eric CHARIKANE (ECRIN)
Audrey SCARINGELLA (MINATEC)
Frédéric SCHUSTER (CEA)
François TARDIF (CEA)

International Advisory Committee

Chair: Frédéric SCHUSTER (CEA, FR)
François TARDIF (CEA, FR)
Co-chair: Georgios KATALAGARIANAKIS (EC, BE)

Daniel BERNARD (ARKEMA, FR)
Jorge BOCZKOWSKI (INSERM, FR)
Jacques BOUILLARD (INERIS, FR)
Jean-Philippe BOURGOIN (CEA, FR)
Peter HATTO (ISO, UK)
Peter HOET (KUL, BE)
Martin KLENKE (NANOGATE, DE)
Carole SENTEIN (CEA, FR)
Louis TREPIED (MINEFE, FR)
Jacques VENDEL (IRSN, FR)

Participants

AGUY S.
AITKEN R.
ALEXANDRE S.
ALVES O.L.
AMOABEDINY G.
ANDAR A.
ANGELONI S.
ANTIPOV S.
APPERT-COLLIN J.-C.
ARMAND L.
ASIMAKOPOULOU A.
ASTIER O.
AUGER A.
AUGIER S.

BACKMAN U.
BAGUET D.
BALAS F.
BARD D.
BARDINI P.
BARET G.
BARTHES V.
BAU S.
BAUMANN B.
BERNARD D.
BERNE P.
BERRET J.-F.
BLANC MANTOY C.
BLAZKOVA A.
BLOCH D.
BOCZKOWSKI J.
BOGDANOVIC-GUILLON A.
BOIZE M.
BOMBARDIER P.
BORDÈRE S.
BORDES S.
BOTTA A.
BOUILLARD J.
BOULANGER P.
BOURGEON M.
BRAYNER R.
BREDEAU S.
BROUARD C.
BROUWER D.
BRÜSKE-HOHLFELD I.

CALOGINE D.
CARRIERE M.
CARRIQUIRY M.
CASTELLANO P.
CHABLI A.
CHARTON F.
CHAZELET S.
CHERNS P.
CHIVAS C.
COMTE M.
CONCHESO A.

DE MONTLUC B.
DELTEDESCO M.
DISSETTE V.
DOBASHI R.
DUFOUR J.-P.
DUSINSKA M.
DUTOUQUET C.
DUVAL-ARNOULD G.

ELGRABLI D.

FAROUZ T.
FAURE L.-P.
FAVIER A.-V.
FEDERZONI L.
FERREIRA I.
FILIMUNDI E.
FJELLSBØ L.
FONG A.
FRESNAY C.

GAILLARD P.
GALLET S.
GALVEZ PEREZ V.
GATTI A.
GEYS A.
GIACOBBE F.
GIRAUD S.
GITZHOFER F.
GOEPFERT C.
GOLANSKI L.
GOOSSENS H.
GOUGEON O.
GROSSEAU P.
GUIOT A.
GUIZARD B.

HÄNNINEN O.
HATTO P.
HAYASHI T.
HAYWOOD G.
HOET P.
HOLE P.
HONNERT B.
HOUSIADAS C.
HUBERT L.

ICHIHARA G.
ILKOV V.

JANKA K.
JEANSON H.
JOKINIEMI J.

KAISER J.-P.
KASPER M.
KASPER G.
KASTRINAKI G.
KATALAGARIANAKIS G.
KESSELS S.
KINCAID L.
KLENKE M.
KUGA H.

LABILLE J.
LACOURSE A.
LACROIX G.
LANDSIEDEL R.
LANONE S.
LE BARNY P.
LE BIHAN O.
LE DUR D.
LE GALLAIS H.
LECERF P.
LILEY M.
LINDELOEV J. S.
LUCHININ V.
LUIZI F.
LUTHER W.
LYNCH I.

MACHUEV A.
MAISONNEUVE C.
MALAKOUTI KHAH J.
MALARD S.

REVIRAND P.
RICAUD M.
RIEDIKER M.
RIEHLE M.

VANDRIESSCHE S.
VAQUERO C.
VEDAM H.
VENDEL J.

MALLOY A.
MANODORI L.
MARCHE P.
MARRA J.
MATSUI Y.
MAXIT B.
MEISSNER T.
MELIQUE T.-M.
MIHAILOVICI M.-S.
MOEHLMANN C.
MONTIGNY G.
MONTIGNY Y.
MOTELLIER S.
MOUGIN B.
MOUTIERS G.
MUIR R.
MÜLHOPT S.

NADEJDE C.
NADERI A.
NAU K.
NEUMEISTER L.
NORMAND A.

OGURA I.
OZAWA A.
OZIL S.

PANZARU C.
PARISOT P.
PASTRE S.
PATRICK S.
PERY A.
PIMENOFF J.
POIRIER C.
POURCHEZ J.
POZZI MUCELLI S.
PRAOLINI F.
PUI D.

RACCURT O.
RADOSAVKIC D.
RAINIERI S.
RATOI M.
REINWALT J.-M.
RIPPERT C.
ROGERIEUX F.
RUIZ J.-C.

SAINT-JALMES S.
SAMUEL J.
SANDRIN M.
SAOUT C.
SAVOLAINEN K.
SCHMID K.
SCHMITT F.
SCHUSTER F.
SENTEIN C.
SHAKESHEFF A.
SICARD Y.
SIKHWIVHILU L.
SIRVEN J.-B.
SIVONEN T.
SKAIKY H.
SOM C.
SOSNOV A.
SPIELVOGEL J.
SPRUIT O.
STUURMAN B.
SUTER G.
SUZUKI T.
SZENKNECT S.

TABET L.
TANARRO GOZALO C.
TARDIF F.
TEILLET A.
TENEGAL F.
TERRAY M.
THIERET N.
TOFAN A.
TOMA L.
TREPIED L.
TRUFFER M.

UCHIYAMA T.
USTACHE A.

VALLES J.-L.
VAN DEN BRINK W.
VIGNES A.
VILLIERS C.
VINCENT J.
VITORGE E.
VOETZ M.

WAGNER S.
WANG Y.F.
WEIL M.
WITSCHGER O.
WUEST I.

YAMADA H.
YANAGIUCHI M.

ZHAO Y.
ZIMMERMANN E.
ZITOUNI L.

All papers published in this volume of *Journal of Physics: Conference Series* have been peer reviewed through processes administered by the proceedings Editors. Reviews were conducted by expert referees to the professional and scientific standards expected of a proceedings journal published by IOP Publishing.

Nanoparticles monitoring in workplaces devoted to nanotechnologies

Laura Manodori[1] and Alvise Benedetti[1,2]

[1]Associazione CIVEN, via delle Industrie 5, 30175 Venice, Italy

[2]Department of Physical-Chemistry, Ca' Foscari University, Calle Larga S. Marta 2137, 30173 Venice, Italy

E-mail: manodori@civen.org

Abstract. The impact of engineered nanoparticles (ENP) on human and environmental health is largely unknown at the moment, although their applications involve a wide range of fields from electronics, medicine, environmental remediation, packaging, etc. and their use are becoming very common. Currently the best way to estimate the human exposure to ENP is by monitoring the workplaces, which are the first environment involved in their involuntary release.

This study has been carried out in the research laboratories of the CIVEN Association, devoted to the promotion of research and training activities in different fields of nanotechnology. Its aim is to evaluate the emission of particles in the full range from 5 nm to 20 µm from two different deposition processes of thin film (plasma enhanced chemical and physical vapour deposition, PECVD and PVD respectively) and from the compounding of polymers with nanofillers by means of twin screw extruder.

The results show very different scenarios of emissions. In particular a significant increase of particles concentration has been detected at the opening of the PECVD vacuum chamber and during nanocomposite compounding. In the first case it rises to 3 000 particles cm^{-3}, while in the second one the emissions are two orders of magnitude larger, reaching more than 1.5×10^5 particles cm^{-3}.

1. Introduction

Nanotechnologies are emerging processes that involve materials at the nanoscale dimension, which applications include a wide range of fields and are becoming very common and of ordinary usage. The debate is open about the impact of engineered nanoparticles on human and environmental health; currently workplaces are the most likely places of exposure, and recently Boccuni *et al.* [1] have estimated that workers potentially exposed to engineered nanoparticles in Italy are more than 350 000. However, exposure can not be easily evaluated in workplaces because ultrafine particles are emitted by many industrial activities, such as hot processes (melting and refining of metals, welding), combustion and fumes [2], and actually the real-time sampling instruments are not able to differentiate between the sources of emission. Moreover, the indoor concentrations of particles can vary largely due to the other activities occurring in the background, the number of people present, the differences in workers techniques, the changes in ventilation and the variation in outdoor concentration [3].

The aim of this study is to assess the number concentration of particles in the fine and ultrafine ranges in research labs devoted to nanotechnology, and to evaluate their sources as first approach to estimate the potential exposure of involved workers to engineered nanoparticles. Samplings have been carried out at the CIVEN laboratories, an association devoted to the promotion of research and training activities in different fields of nanotechnology, belonging to the Italian Cluster of Nanotechnology. The production techniques investigated in this study are widely used in industrial processes and concern surface coating via vapour deposition (both Plasma Enhanced Chemical Vapour Deposition PECVD and Physical Vapour Deposition PVD) and polymer nanocomposite production. The former processes allow deposition of thin film on substrates to improve its surface characteristics (tribologic, protecting and functional coatings), while the second one improves the bulk properties of polymers, such as their permeability to gases or flame resistance.

2. Experimental

Indoor aerosol was sampled in the different lab rooms, both during the process run and in non-working conditions, with a Condensation Particle Counter (CPC model 5403 Grimm Aerosol Technik) to characterize the temporal variation of the concentration of particles lower than 370 nm. Successively, samplings were done with a Wide Range Aerosol Sampler (WRAS, Grimm Aerosol Technik) composed by the CPC equipped with the Classifier "Vienna" Type DMA (model 5.500) and the portable aerosol spectrometer (model 1.108) to assess the dimensional distribution of particles in the full range from 5 nm to 20 μm.

Instruments were put on a trolley, 115 cm high above the floor, as near as possible to the operator (about 120-130 cm) to estimate his potential exposition to emitted particles but avoiding the interferences with the worker movements.

3. Results and discussion

The results show very different scenarios of contamination regarding both the background concentrations and the emission of particles.

PECVD lab is a clean room class 10 000, so the background concentration (particles with aerodynamic diameter less than 370 nm) is very low and stable: it is about 500 particles cm^{-3} when the instrument pumps are switched on and less than 50 particles cm^{-3} when they are off. The particles concentration sharply increases when the operator opens the door of the vacuum chamber, growing up to 800-2 900 particles cm^{-3}. Afterwards it rises more than one order of magnitude, reaching the value of 10 000 – 17 000 particles cm^{-3}, because of the vacuum cleaner used for removing the waste products from the chamber. That great increase could be due to nanoparticles emitted from the vacuum cleaner or may be recycled from it. In any case, the great influence of the vacuum cleaner on the air concentration has been previously reported by Maynard [4].

Samplings performed by WRAS do not evidence the peak emission due to the chamber opening, because of the limited time of the event (about 4 minutes), however, they clearly show that the particles emitted by the vacuum cleaner belongs to the $<$ 100 nm range.

The importance of the background level is confirmed by the results obtained from the monitoring of the PVD process (Figure 1), which instrumentation is located in a laboratory with conventional air conditioning and filtering system, and where the concentration in non working conditions is quite high and not stable (about 1 000-3 000 particles cm^{-3}). The chamber opening, occurring at the end of different deposition processes, does not result in concentration peaks, but it could not be stated that it is due to the absence of particles emission or because of the too high level of the room baseline.

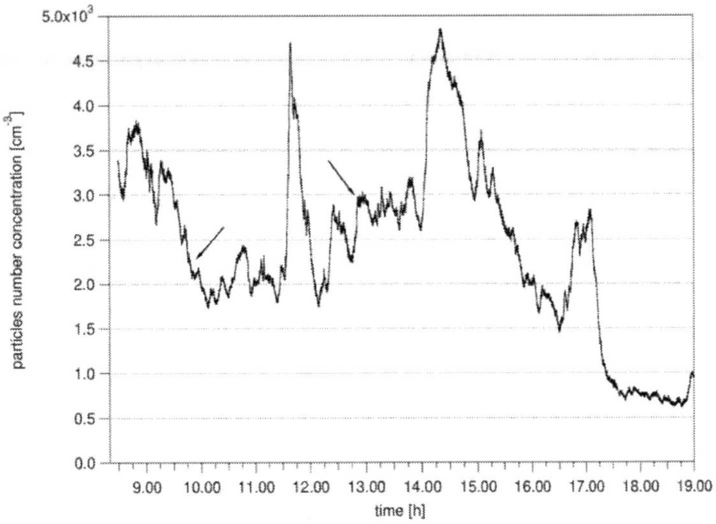

Figure 1. Temporal trend of the concentration of particles (5 – 370 nm) detected in the PVD lab. Arrows point the moment of the chamber opening.

The largest particles concentration has been detected during the compounding of polymers with nanofillers by means of a twin screw extruder heated up to 180-240°C. In fact during this activity the room concentration grows up to 2 x 10^5 particles.cm^{-3}, and it is mainly due to particles smaller than 100 nm, as depicted in Figure 2. These emissions occur both when the operator introduces the polymer and the nanofillers into the mixing chamber and during the cleaning by polyethylene (without any nanoparticles).

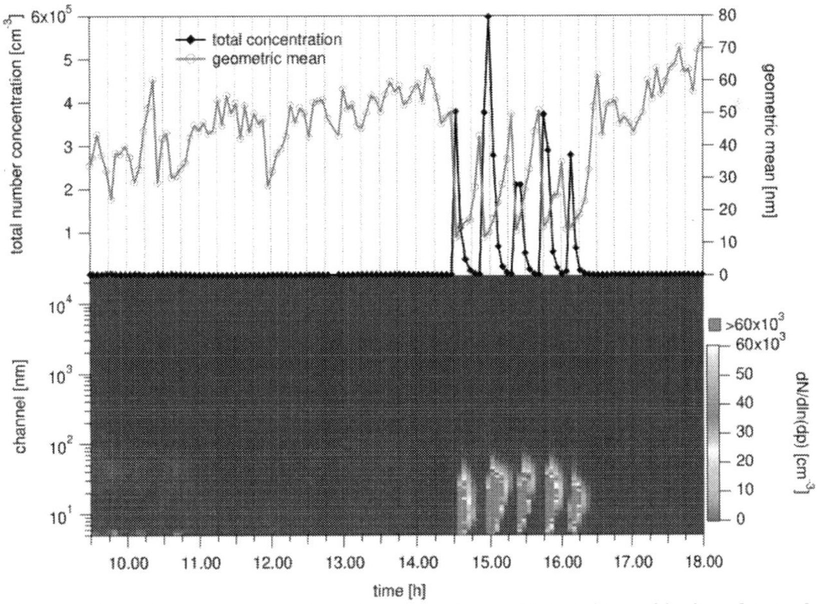

Figure 2. Temporal trend of the concentration of particles (5 nm – 20 μm) detected in the polymers lab by WRAS: total concentration (up) and dimensional distribution (down).

As claimed by Tsai *et al.* [5], who monitored a similar process into an industrial plant, detecting concentration larger than 10^5 particles.cm^{-3} (size range: 5.6 - 560 nm) during the production of nanofilled polymers, it is reasonable to suppose that the nature of the detected nanoparticles is not engineered, but composed by condensed polymer fumes and degraded carbonaceous materials.

4. Conclusions

This study discusses the temporal concentration of fine and ultrafine particles in research laboratories involved with nanotechnological processes, namely vapour deposition in vacuum conditions (both plasma enhanced chemical VD and physical VD) and compounding of polymers with nanofillers. The results obtained are very different in respect to the processes involved and the background conditions, and are briefly resumed in Table 1.

Table 1. Particles number concentration and likely sources.

	PECVD	PVD	nanocomposite compounding
background concentration* [particles cm^{-3}]	5×10^2	1×10^3 - 3×10^3	3×10^3
maximum concentration* [particles cm^{-3}]	3×10^3 - 3.5×10^4	7×10^3	7×10^6
geometric mean § [nm]	10	-	30-80 $^\#$
	120	-	15 $^+$
likely source	silica nanoparticles from deposition chamber and nanoparticles from vacuum cleaner respectively	unknown	polymers fumes and degraded materials

* measured by CPC; § measured by WRAS; $^\#$ non working conditions; $^+$ mixing

The results agree with those reported by other authors [6], and confirm the difficulties to detect the engineered nanoparticles emissions due to the production processes, because of the presence of interferences from other sources. However, to the best of our knowledge this study should be the first attempt to estimate the potential operator's exposure to nanoparticles produced by the involved processes, in particular from the PECVD technique.

References:

[1] Boccuni F, Rondinone B, Petyx C and Javicoli S 2008 *Journal of Cleaner Production* **16** 949
[2] Vincent J H and Clement C F 2000 *Phil. Trans. Royal Society London* **358** 2673
[3] Yeganeh B, Kull C M, Hull M S and Marr L C 2008 *Environmental Science and Technology* **42** 4600
[4] Maynard A D, Baron P A, Foley M, Shvedova A A, Kisin E R and Castranova V 2004 *Journal of Toxicology and Environmental Health, Part A* **67** 87
[5] Tsai S-J (Candace), Ashter A Ada E, Mead J L, Barry C F and Ellenbecker M J 2008 *Aerosol and Air Quality Research* **8** 160
[6] Kuhlbusch T A J and Fissan H 2006 *Journal of Occupational and Environmental Hygiene* **3** 558

Experimental investigation on blood magnetic contamination in the presence of drug molecules

D.E. Creangă[1], Gh. Iacob[2] and C. Nădejde[1]

[1]"Al. I. Cuza" University, Faculty of Physics, 11A Blvd. Carol I, RO-700506, Iasi, Romania

[2]"Gr. T. Popa" University, Faculty of Biomedical Engineering, Kogalniceanu Street, No. 9-13, RO-700454, Iasi, Romania

E-mail: nadej_dia@yahoo.com

Abstract. The purpose of the present project was to study the interference of magnetic nanoparticles with drug molecules – rifampicin, used in lung infectious disease and respectively, sodium diclofenac, an antiinflammatory steroid. The controlled magnetic contamination was accomplished using colloidal nanoparticles supplied from diluted magnetic fluids. Various concentrations of diluted aqueous magnetic fluids, based on magnetite cores coated with citric acid and respectively sodium oleate, were tested. The experiment was focused on the capacity of the magnetic nanoparticles to form reversible complexes with the drug molecules, as well as on the monitoring of the nanoparticle-drug complex dynamics, under the action of external magnetic field. The level of released rifampicin ranged between 4 mg/100 ml and 7 mg/100 ml for the magnetic exposure of 20 mT, while the sodium diclofenac decomplexation level was not higher than 2.5 mg/100 ml under magnetic exposure of 60 mT. The experimental arrangement was proved to be an adequate model for the dynamical study of magnetite reversible complexation with drug molecules, evidencing certain specific values of drug concentration and magnetic field induction that favour such interactions.

1. Introduction

The interaction magnetite-biological molecules could occur following accidental magnetic contamination of human body tissues or blood, since very fine magnetic nanoparticles seem to be present in most biotic and abiotic elements of the environment – the magnetic loading of water, air and soil increasing continuously in the modern industrial era. In the same time, environmental magnetic fields enhance the risk of chronic magnetic exposure. The present study was motivated by the issue of magnetic contamination of both the environment and living bodies, that shaped more and more during latest decades, based on the iron abundance – the fourth element on the Earth, on the detection of ferromagnetic particles in aerosols, foods and sterile laboratory liquids, as well as on the existence of magnetosomes. According to [1], ferromagnetic particles can be present in the body tissues mainly as contaminants of the lungs and other organs. The magnetic domains of these particles can be aligned by the application of an external magnetic field; the magnitude and time course of the resultant field depend on the quantity of magnetic material and the degree of particle motion. In this respect, other authors [2-6] formulated a possible explanation of the living body sensitivity to electromagnetic fields, based on the hypothesis of magnetic particle contamination. According to this theory, small polluting magnetic particles from the environment are present not solely in the air dust, but they are also deposited on the surface of laboratory devices and penetrate into plastics and glass, into chemicals and purified water [7]. Their mean size is about 10^{-5} cm and they consist of ferro- and ferrimagnetic compounds (they possess spontaneous magnetization). The authors have shown that routine laboratory procedures like pouring or/and rinsing give rise to enrichment of the cell cultures by pollutants. The number of particles in cultures could be ten times more than that of cells. The energy of such a particle is approximately three orders of magnitude greater than kT. In the authors [8-9] opinion, the magnetic particle absorbed by a cell membrane can transfer its energy to the adjacent biophysical structures, for example, mechanically activated ion channels. On the other hand, internalized magnetic nanoparticles are shown to be an endogenous source of chronic magnetic exposure that increases the local concentration of free radicals. Experimentally, ultrafine particles (12–14 nm in size) were shown to be internalized by human monocyte cells and significantly increase, by 40–45%, the release of free radicals [10]. An enhanced level of leukaemia in early childhood is assumed to originate from magnetic nanoparticles located in hematopoietic stem cells [11], while progressive neurodegeneration is supposed to be initiated by metal crystalline compounds [12-14].

The main objective of the experiment presented below was to investigate the putative side effects of residual magnetic nanoparticles, supplied by environmental sources or following insufficiently controllable medical procedures (such as Magnetic Resonance Imagery or cancer therapy with magnetic fluids) in the case of patient consecutive treatment with usual pharmaceutics. Rifampicin antibiotic was chosen due to its ability to interact with iron ions, but sodium diclofenac was also tested in the frame of physical model proposed below.

2. Materials and methods

Magnetic nanoparticles sources were consistent with magnetite stabilized with citric acid (CA) and respectively sodium oleate (SO), that were supplied as adequate volumes from various dilutions of the corresponding aqueous magnetic fluids (CA-MF and SO-MF) [15-16]. The dilutions (in distilled water), that favored the magnetite reversible complexation with the drug molecules, were: C1 = 1 ml CA-MF + 9 ml water; C2 = 2 ml CA-MF + 8 ml water; C3 = 4 ml CA-MF + 6 ml water for magnetite-citric acid complexation with rifampicin; C4 = 5.0 ml SO-MF + 5.0 ml water; C5 = 2.5 ml SO-MF + 7.5 ml water; C6 = 1.25 ml SO-MF + 8.75 ml water for magnetite-sodium oleate complexation with sodium diclofenac. The stock solutions of the tested drugs were prepared from pharmaceutical powders in the concentrations of 30 mg/100 ml rifampicin and 40 mg/100 ml diclofenac. Magnetic field (rare earth permanent magnets) induction was of 20 mT for rifampicin and 60 mT for diclofenac as measured with a Hall probe teslameter. Spectral device (Shimadzu UV-VIS 1700 spectrophotometer) was utilized for quantitative monitoring of drug molecules decomplexation in aqueous medium.

The level of released sodium diclofenac was assessed in the ultraviolet range (UV) at 275 nm, while the absorption bands of rifampicin were recorded in the visible range (VIS) at 474 nm and ultraviolet domain at 334 nm. For each magnetic fluid concentration, the same percentage of drug solution volume was added: 50 % (v/v) - by careful stirring, followed by 24 hours incubation at the temperature of 35.0°C in the INCUCELL thermostatic room. Magnetic exposure was applied by placing all samples onto permanent magnets (Figure 1); distilled water columns of 10 ml were added very slowly onto the mixture of fluids - simplified physical model of blood serum, suitable for spectral assay based on light absorption. Aliquots of 0.5 ml were withdrawn to measure light extinction at t = 0, t = 0.5 h, t = 1.0 h, t = 1.5 h, t = 2.0 h, t = 2.5 h and t = 24.0 h – temperature monitoring being carried out continuously.

Three replicates of every experimental variant were considered in the experimental project in order to get the statistical insurance. Average values and standard deviations were taken into account for discussion; statistical *t*-test was applied to compare the rifampicin levels measured by using two absorption bands, as well as to discuss diclofenac behavior for the three tested concentrations.

3. Results and discussion

The dynamics of reversible interaction CA-MF - rifampicin was monitored by spectral assay, the rifampicin concentration being provided by light extinction measurement and calibration curve – in mg/100 ml or mg%. The data obtained for the three combinations mentioned above (noted with C1, C2 and C3: C1<C2<C3) are presented in Figures 2a, 2b and 2c. The concentration of rifampicin reached over 4 mg % in the case of C1 – as shown by graphical plots of both UV and VIS data (Figure 2a). In the case of C2, over 6 mg % were detected by spectral investigation (Figure 2b), while in the case of C3 (Figure 2c) the highest level of rifampicin decomplexation was evidenced – about 7 mg %. The data provided by the measurements carried out in the maximum of the rifampicin VIS absorption band are similar to those corresponding to the measurements in the UV band maximum, the standard deviation being of about 5 %. The differences between the two data series obtained for each spectral band and for each of the concentrations C1, C2, C3 are negligible - according to t-test - except for two time values, i.e. 25.5 and 26.0 h in the case of C1 (Figure 2a) and respectively, for 24.0 and 24.5 h in the case of C2 (Figure 2b) ($p<0.05$). The statistic comparison of any pair of data series (each pair of Ci, Cj concentrations, i,j=1,2,3) corresponding to each spectral band, resulted in statistically significant differences ($p<0.05$).

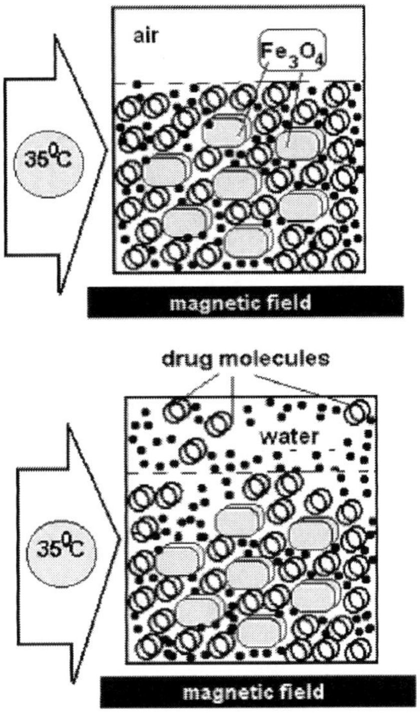

Figure 1. Reversible interaction magnetite-drug molecules in the presence of the magnetic field

The interpretation of the rifampicin dynamics during its decomplexation is related to the specific conditions characteristic for the experimental project described above. So, the experiment was designed to model the situation when the magnetic contamination occurs during a patient treatment with rifampicin – it could be the case of magnetite nanoparticles up-taken by lung tissue simultaneously with antibiotic administration against infectious lung disease. This could happen actually in the case of professional exposure of mine workers; but in this case, we have to do with natural stores of iron and iron-oxidization compounds – namely magnetite – where magnetic field exposure is also unavoidable since local magnetic field is considerable increased around iron mines. Consequently, the rifampicin molecules could be temporarily retained by the magnetite nanoparticles present in the lung tissues, under the supplementary influence of the relative intense magnetic field. Then gradually – as it was shown by the data presented above – the water from the interstitial liquid as well as that from the blood serum can extract rifampicin molecules from the complexes formed with magnetite nanoparticles. In nature, as well in the artificially arranged spaces, free magnetic nanoparticles seem to be dispersed in significant levels, but when they are internalized within the human body fluids, colloidal forms could develop as result of their interaction with various biomolecules.

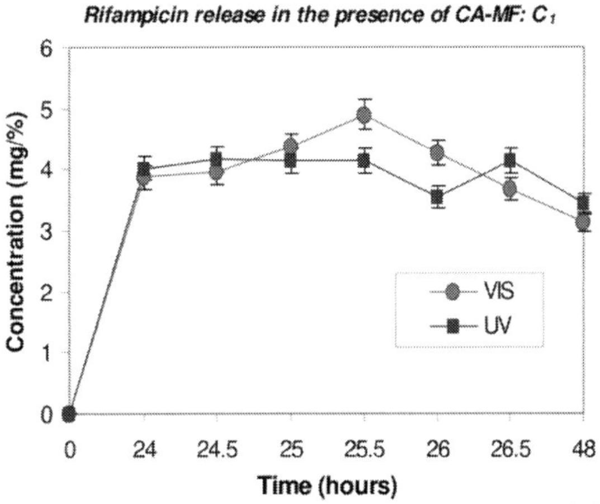

Figure 2a. The dynamics of rifampicin - magnetite reversible complexation as resulted from measurements on the UV-VIS spectrum – relative low concentration (C1) - the time axis was scaled from 0 to 48 h by including the 24 hours incubation

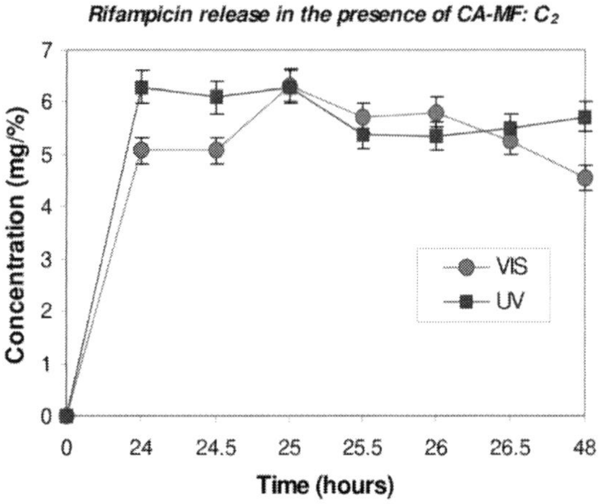

Figure 2b. The dynamics of rifampicin-magnetite reversible interaction as resulted from UV-VIS spectral assay – medium concentration (C2) - the time axis was scaled from 0 to 48 h by including the 24 hours incubation

In the frame of the experiment discussed inhere, the utilization of magnetite powder was found to be inadequate since it does not disperse spontaneously in aqueous media, except when colloidal suspensions are used – this is why the supply of magnetite nanoparticles was carried out by means of diluted magnetic fluids. In this situation, the involving of the coating molecules could play a role in the interaction between the iron oxides and the drug molecules.

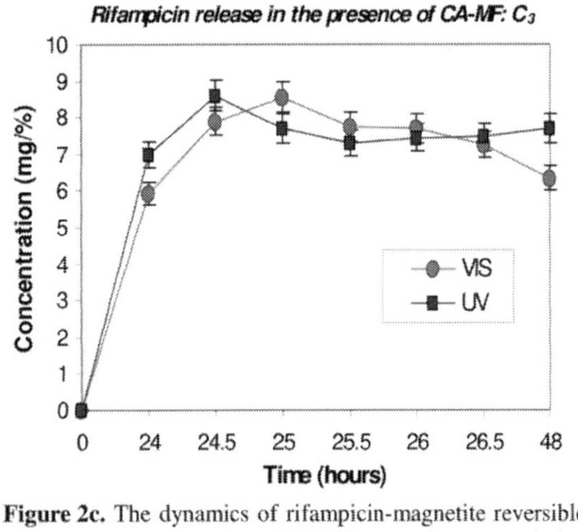

Figure 2c. The dynamics of rifampicin-magnetite reversible interaction as resulted from UV-VIS spectral assay – relative high concentration (C3) - the time axis was scaled from 0 to 48h by including the 24 hours incubation

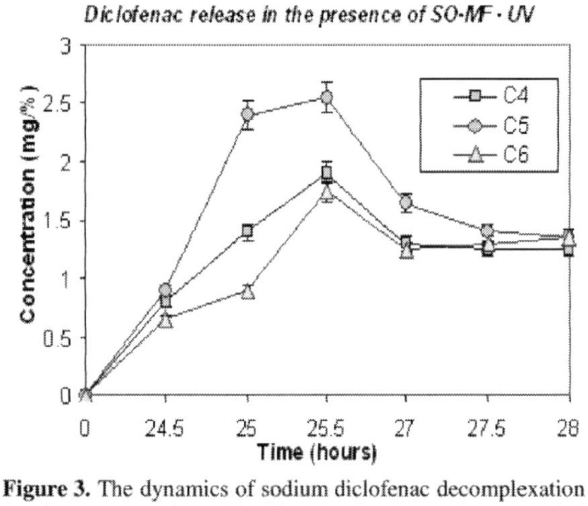

Figure 3. The dynamics of sodium diclofenac decomplexation for the concentrations C4, C5, C6 - the time axis was scaled from 0 to 48h by including the 24 hours incubation

The rifampicin is a large molecule that was shown to be able to interact with iron ions like those composing magnetite, but when citrate ions are coupled to the magnetite particles, the complexation phenomenon could be more complicated. Though the measurements performed in the frame of this research could not provide details regarding the intimate molecular arrangements into the studied magnetic systems, however it was evident that the level of rifampicin released following decomplexation - favored in aqueous environment - is increasing with the concentration of magnetite (Figures 2a, 2b, 2c). We need to mention that for higher magnetic induction (60 mT) also tested within this work, the decomplexation was rather slightly visible; on the other side, the mixture between rifampicin and SO-MF was not homogeneous enough to enable optimal quality spectral readings.

In Figure 3 the data obtained for the interaction SO-MF-diclofenac are shown. Since diclofenac absorbs only ultraviolet radiation, the measurements were carried out on a single spectral band. The dynamics of diclofenac decomplexation is described by similar curves as in the case of rifampicin, except the diminution tendency revealed for longer observation times. One could say that the precipitation of recently released drug molecules could occur in time, so that light extinction gradually diminishes. The maximum levels of drug molecules in the supernatant ranged between 1.6 and 2.5 mg/100 ml for the three mixtures of diclofenac solution and magnetic fluid (C4>C5>C6). It seems that for similar concentrations of the two drugs used in the present experiment, the decomplexation of diclofenac is several times weaker than that of rifampicin, which could be possibly related to the particular chemical features of diclofenac (sodium salt) and the magnetite shell (sodium oleate). The two sodium ion sources could generate competitive processes which could underlie the non-monotone variation of released diclofenac level to the monotone variation of magnetite concentration. Indeed, the highest diclofenac level was obtained for the medium SO-MF concentration – C5, the differences recorded between all three data series (considered at the same time moments) being statistically significant ($p<0.05$) for the time moments when the maxima of the three curves were reached: 24.5, 25.0 and 25.5 h. Neither the mixing of sodium diclofenac with CA-MF nor the application of the relatively low magnetic field (20 mT) did not lead to convenient homogeneous systems.

4. Conclusion

The magnetite nanoparticles in colloidal form were let to interact with two drug molecules in the presence of static magnetic field aiming to provide an experimental model of blood magnetic contamination and its putative side effects. The reversibility of magnetite complexation with rifampicin and sodium diclofenac was studied. Optimal experimental conditions for the observation of the decomplexation phenomenon were found testing two magnetic fluids – as magnetite sources – and two magnetic field inductions. The increased decomplexation of rifampicin was evidenced to the increase of magnetite ratio within the studied magnetic mixtures, the optimal experimental conditions involving a 20 mT magnetic field. Comparatively, the diclofenac reversible interaction with magnetite was found to release lower drug concentration in the surrounding aqueous medium while higher magnetic field was also needed. The study of these interactions could be of interest for the situations of people magnetic contamination – as in professional exposure - during simultaneous treatment with pharmaceutical products.

References

[1] Valberg P A, Burler J P 1987 *Biophys. J.* **52** 537
[2] Kirschvink J L, Kirschvink A K, Diaz-Ricci J C, Kirschvink S J 1992 *Bioelectromagnetics* **1** 101
[3] Bazylinski D A, Frankel R B 2004 *Nat. Rev.Microbiol.* **2(3)** 217
[4] Johnsen S, Lohmann K J 2005 *Nature Rev. Neurosci.* **6** 703
[5] Binhi V N, Chernavskii D S *Europhys. Lett.* **70(6)** 850
[6] Binhi V N 2006 *Bioelectromagnetics* **27(1)** 58
[7] Kobayashi A K, Kirschvink J L, Nesson M H 1995 *Nature* **374** 123
[8] Binhi V N 1997 *Biophysics* **42(2)** 317
[9] Binhi V N 1998 *Bioelectrochem. Bioenerget* **45** 73
[10] Simko M, Hartwig C, Lantow M, Lupke M, Mattsson M O, Rahman Q, Rollwitz J 2006 *Toxicol.Lett.* **161(1)** 73
[11] Huntly B J, Gilliland D G 2005 *Nat. Rev. Cancer* **5(4)** 311
[12] Purdey M 2004 *Med.* **63(5)** 793

[13] Hautot D, Pankhurst Q A, Kahn N, Dobson J 2003 *Proc. Roy. Soc. Bio.* **270(1)** S62
[14] Quintana C, Bellefqih S, Laval J Y, Guerquin-Kern J L, Wu T D, Avila J, Ferrer I, Arranz R, Patino C 2006 *J. Struct. Biol.* **153(1)** 42
[15] Racuciu M, Creanga D E, Calugaru Gh 2005 *J. Optoel. Adv. Mater.* **7(6)** 2859
[16] Bredetean O, Camarut St, Ciochina A D, Camarut N, Iacob Gh 2005 *Rev. Med. Chir.* **109(2)** 445

Dustiness testing of engineered nanomaterials

Isamu Ogura, Hiromu Sakurai and Masashi Gamo

National Institute of Advanced Industrial Science and Technology (AIST), 16-1, Onogawa, Tsukuba 305-8569, Japan

E-mail: i-ogura@aist.go.jp

Abstract. We investigated the dustiness (the propensity of a material to generate airborne dust during its handling) of various nanomaterials, including carbon nanotubes and metal oxides, by the vortex shaker method. The number concentrations and size distributions (~10->10 000 nm) of aerosol particles released during agitation were measured. It was found that the modal diameter was greater than 100 nm for all tested nanomaterials, and for most of them some sub-100 nm particles were observed. The dustiness differed by two (or three) orders of magnitude among the test nanomaterials.

1. Introduction

There is growing concern about the potential health impact of engineered nanomaterials. Inhalation of airborne nanomaterials is potentially a significant exposure route, but there is little information currently available on the nature and magnitude of aerosol releases during manufacturing and handling nanopowders.

In the field of occupational hygiene in the UK, the concept of 'dustiness' has been applied to powders. Dustiness is defined as the propensity of a material to generate airborne dust during its handling, and provides a basis for estimating the potential health risk due to inhalation exposure [1, 2]. Various dustiness test methods have been developed; however, they cannot be applied to nanomaterial powders directly because a test which needs smaller quantities of test materials is more suitable for costly nanomaterials or potentially toxic nanomaterials and the dustiness evaluation based on weight concentration is not necessarily adequate to nanosized aerosols.

We built a 'dustiness tester' for nanomaterials based on the technique by Maynard *et al.* [3], which needs only small quantities of test materials for testing (<~1 cm^3 of test materials) and simple/compact instrumentation. We have been investigating the dustiness of various nanomaterials, including carbon materials and metal oxides.

We are particularly interested in: (1) the nature of aerosols generated by the technique, (2) the amounts, size distributions, and morphologies of particles released from nanomaterial powders, (3) relative emission/exposure potentials among nanomaterial powders, and (4) the factors that affect the dustiness.

2. Experimental methods

The list of test nanomaterials is shown in Table 1. A schematic diagram of the experimental setup is shown in Figure 1. A spoon (0.25, 0.5, or 1 cm^3) of each nanomaterial was placed in a glass test tube and agitated using a laboratory vortex shaker (Vortex Genie 2 Shaker, Cole-Parmer Instrument Co.). The number concentrations and size distributions (~10->10000 nm) of aerosol particles released during agitation with passing air through the test tube were continuously measured using an optical particle counter (OPC, Handheld 3016, Lighthouse Worldwide Solutions Inc.), an aerodynamic particle sizer (APS 3321, TSI Inc.), a condensation particle counter (CPC 3007, TSI, Inc.), and a scanning mobility particle sizer (SMPS) comprising a differential mobility analyzer (DMA 3081, TSI Inc.) and a condensation particle counter (CPC 3010 or 3772, TSI Inc.). The size distribution of nanometer scale particles was characterized using the SMPS (~13-760 nm), and that of larger particles was characterized using the OPC (~300->10 000 nm) and the APS (~700-20 000 nm).

Table 1. List of test nanomaterials.

Symbol	Material	Trade name, grade	Manufacturer	Properties
SWCNTs-1	Single-walled carbon nanotubes	HiPco, as-produced, 35wt% residuals as Fe	Carbon Nanotechnologies, Inc.	Diameter: ~1 nm, Specific surface area: 310 m^2/g
SWCNTs-2	Single-walled carbon nanotubes	HiPco, purified	Carbon Nanotechnologies, Inc.	Diameter: ~1 nm, Specific surface area: 930 m^2/g
SWCNTs-3	Single-walled carbon nanotubes	Super Growth	National Institute of Advanced Industrial Science and Technology (AIST)	Diameter: ~3 nm, Specific surface area: 890 m^2/g
MWCNTs	Multi-walled carbon nanotubes		Nikkiso Co., Ltd.	Diameter: ~30 nm, Specific surface area: 69 m^2/g
Fullerenes	Fullerene mixture	nanom mix ST-F	Frontier Carbon Corporation	Average particle diameter: ~1 µm, Specific surface area: 4.6 m^2/g
C_{60}	Fullerene C_{60}	nanom purple SU, purified	Frontier Carbon Corporation	Average particle diameter: ~1 µm, Specific surface area: 27 m^2/g
ZnO	Zinc oxide	FZO-50	Ishihara Sangyo Kaisha, Ltd.	Primary particle diameter: ~21 nm, Specific surface area: 32 m^2/g
TiO_2	Titanium dioxide	ST-01, anatase	Ishihara Sangyo Kaisha, Ltd.	Primary particle diameter: ~7 nm, Specific surface area: 300 m^2/g

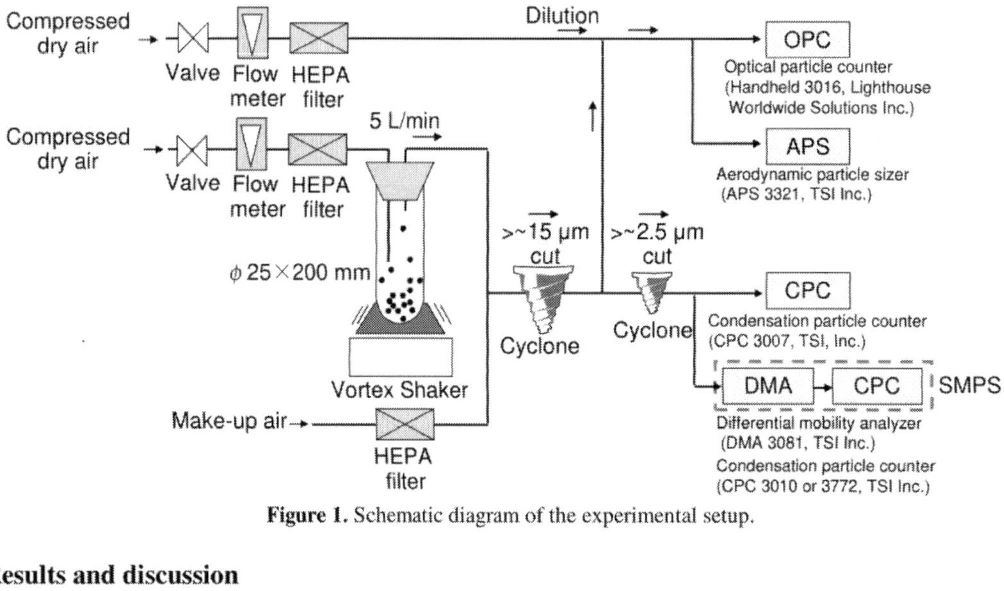

Figure 1. Schematic diagram of the experimental setup.

3. Results and discussion

The particle size distributions of aerosols generated from test nanomaterials are shown in Figure 2. Particle sizes are expressed by electrical mobility diameter, optical diameter, and aerodynamic diameter for SMPS, OPC, and APS, respectively. For the particle size distributions by SMPS, unrealistic distribution curves with a dip in the 200-300 nm diameter range were often observed when the multiple charge correction of the SMPS program was turned on, which appear to be due to an artifact by the correction. The modal diameter was greater than 100 nm for all tested nanomaterials, and for most of them some sub-100 nm particles were observed by SMPS.

For most of the test nanomaterials, aerosols were generated over a few hours of agitation and there were no obvious changes in the particle size distributions of aerosols, regardless of the level of agitation and the amount of the test nanomaterials placed in the test tube.

To evaluate differences in dustiness among the test nanomaterials, the number concentrations per the volume and mass of the test nanomaterials placed in the test tube are shown in Figures 3 and 4, respectively. The dustiness differed by two (or three) orders of magnitude among the test nanomaterials. It needs further consideration to determine which indicator is more appropriate.

The relationships between properties of nanomaterials and their dustiness will be investigated in the future.

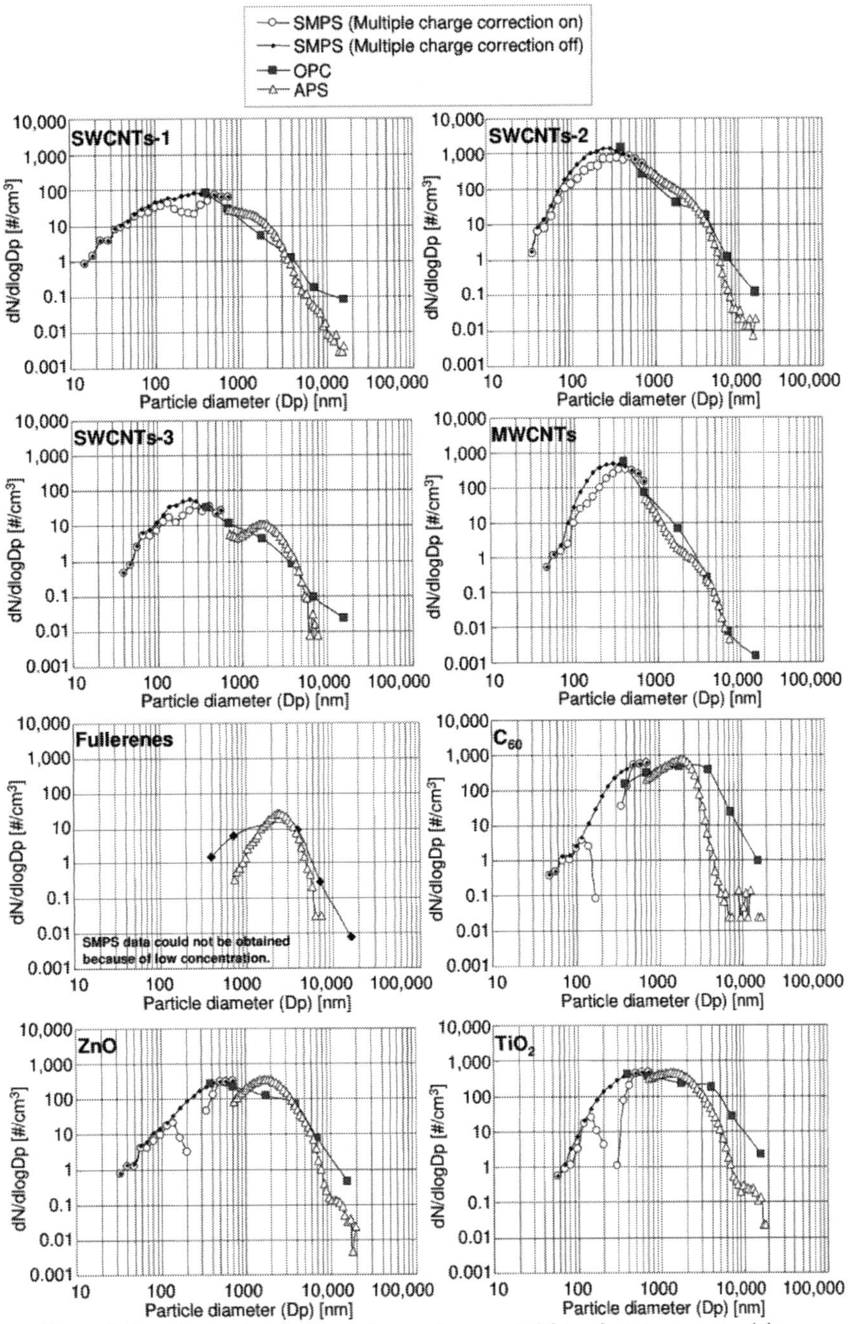

Figure 2. Particle size distributions of aerosols generated from the test nanomaterials.

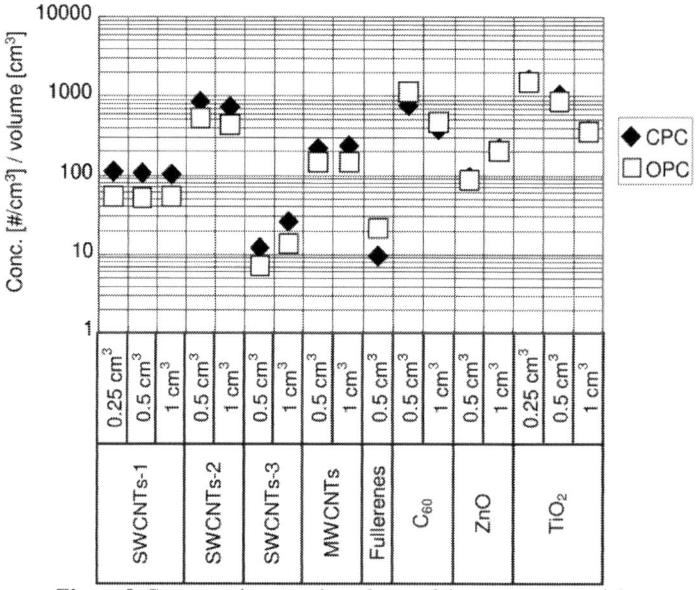

Figure 3. Concentrations per the volume of the test nanomaterials.

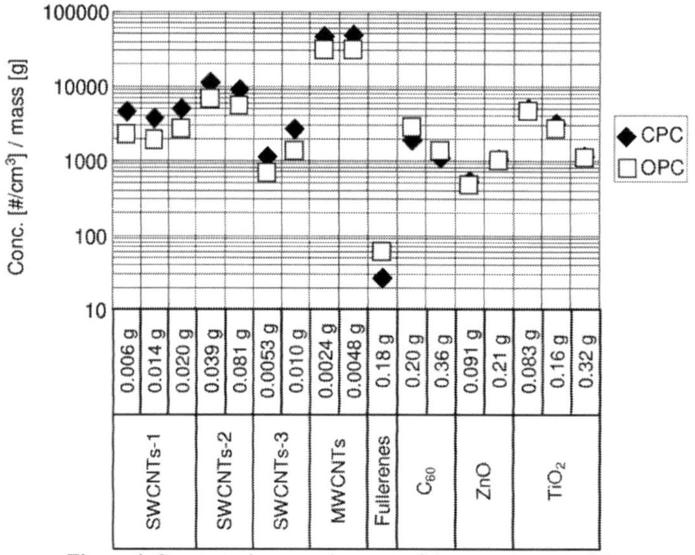

Figure 4. Concentrations per the mass of the test nanomaterials.

4. Conclusions

Aerosol particles generated from various nanomaterials during agitation were measured. The modal diameter was greater than 100 nm for all tested nanomaterials, and for most of them some sub-100 nm particles were observed. The dustiness differed by two (or three) orders of magnitude among the test nanomaterials.

References

[1] Lidén G 2006 *Annals of Occupational Hygiene* **50(5)** 437
[2] European Committee for Standardization, EN 15051, "Workplace atmospheres–Measurement of the dustiness of bulk materials–Requirements and reference test methods" (2006).
[3] Maynard A D, Baron P A, Foley M, Shvedova A A, Kisin E R and Castranova V 2004 *J Toxicol Environ Health A* **67(1)** 87

Nanosafe 2008: International Conference on Safe production and use of nanomaterials IOP Publishing
Journal of Physics: Conference Series **170** (2009) 012004 doi:10.1088/1742-6596/170/1/012004

Workplace exposure at nanomaterial production processes

Carsten Möhlmann[1], Johannes Welter[1], Martin Klenke[2] and Jürgen Sander[3]

[1]BGIA-Institute for Occupational Health and Safety, Alte Heerstraße 111, 53757 Sankt Augustin, Germany

[2]Nanogate Advanced Materials GmbH, Zum Schacht 3, 66287 Göttelborn, Germany

[3]Nanogate AG, Zum Schacht 3, 66287 Göttelborn, Germany

E-mail: carsten.moehlmann@dguv.de

Abstract. Typical nanomaterial production processes from daily practice had been performed in order to determine simultaneously the exposure to nanoparticles. They involve mixing of ZnO powder into a liquid, filling and emptying an oven with indium tin oxide (ITO), spraying a suspension of nanoparticles, flame spraying of silanes, and an outside location as comparison.

Parameters measured are the total number concentration for certain size ranges below 1 µm, the size distribution of the number concentration, the size distribution of the mass concentration, the size distribution of micrometer particles, and the mass concentrations of respirable and inhalable dust fractions according to EN 481. The particle number concentrations were measured with a scanning mobility particle sizer (SMPS from TSI type 3080 with long DMA 3081 and condensation particle counter CPC 3025A), a handheld condensation particle counter (TSI CPC 3007) and an aerosol particle sizer (TSI APS 3321). The SMPS and CPC cover a particle diameter range from a few nm up to a few µm, the APS a particle diameter range from 0.5 to 20 µm. Additionally the mass concentrations were measured at the person with personal air samplers and next to the operator with static dust samplers.

Results

The concentrations during manual mixing of ZnO powder with a liquid binder in small vessels of 2 liter were lower than 9 000 particles per cm³ in the entire measurement range of 14 to 673 nm (measurement distance approximately 0.5 m). The median diameter of the distribution was 75 nm, whereas the primary particle diameter of ZnO was approximately 30 nm. Even lower concentrations appeared while mechanically quenching the ZnO paste between rolls (Table 1, Figure 1). The mass concentrations next to the table where the mixing operations took place were 0.19 mg/m³ for the respirable and 0.24 mg/m³ for the inhalable aerosol fraction.

For post-treatment, indium tin oxide (ITO) powder was filled into an oven. At this operation the total number concentrations ranged between 9 000 and 10 500 particles per cm³ with maxima between 60 and 70 nm. The primary particle size of ITO after treatment was 40 to 50 nm.

Both for handling ZnO and ITO there seems to be no evidence of aerosolized primary nanoparticles. The total concentrations were similar to those measured outside the building, where traffic emissions were predominant (Table 1). Other aerosol sources like ovens located in the adjacent production hall can have a contribution to the measured size distributions.

In contrast the spraying process of a nanoparticle dispersion (organically modified SiO_2 in alcohol) onto a roll showed in a distance of 0.4 m to the roll total concentrations of up to 400 000 cm^{-3} with high concentrations from 30 nm upward. Flame spraying of silanes, which is often used in general coating applications, even showed very high concentrations of up to 18 000 000 cm^{-3} near to the burner and the operator's position.

Table 1. Aerosol data from SMPS measurements of different nanomaterial production processes.

Mean values	Median	Maximum	Total conc.	Total conc.
	nm	nm	cm^{-3}	cm^{-3}
Operation			14-673 nm	< 100 nm
outside	83	120	8 434	4 586
filling formation oven, ITO	73	70	10 050	6 581
emptying formation oven, ITO	48	62	9 198	6 470
mixing of ZnO powder and liquid	75	87	8 462	5 360
quenching of ZnO paste between rolls	78	96	5 458	3 379
flame spraying of silanes	54	84	5 071 333	3 280 582
spraying of suspension	99	94	376 558	190 299

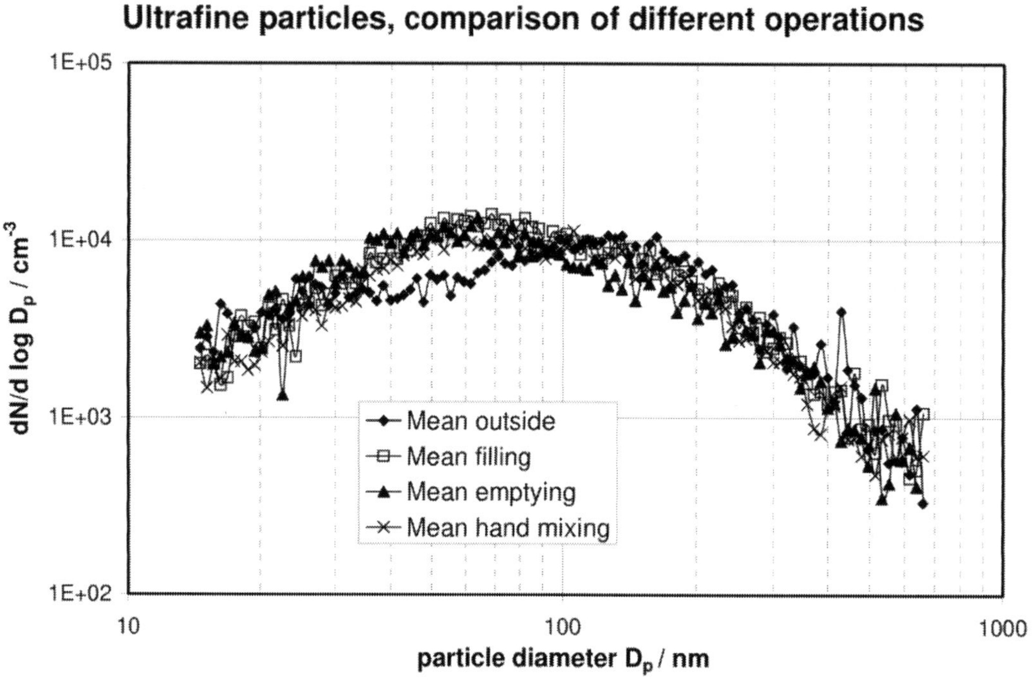

Figure 1. Differential number concentrations, measured with SMPS, at different nanomaterial production processes.

The long term simulation (approximately 13 min) of the pyrosilanisation gave total number concentrations of up to 413 000 particles per cm³ with a maximum particle size of 55 nm. Figure 2 presents the time averaged number concentration (differential scale, with error bars) and their calculated mass distribution (density 1 g/cm³ and spherical particles assumed).

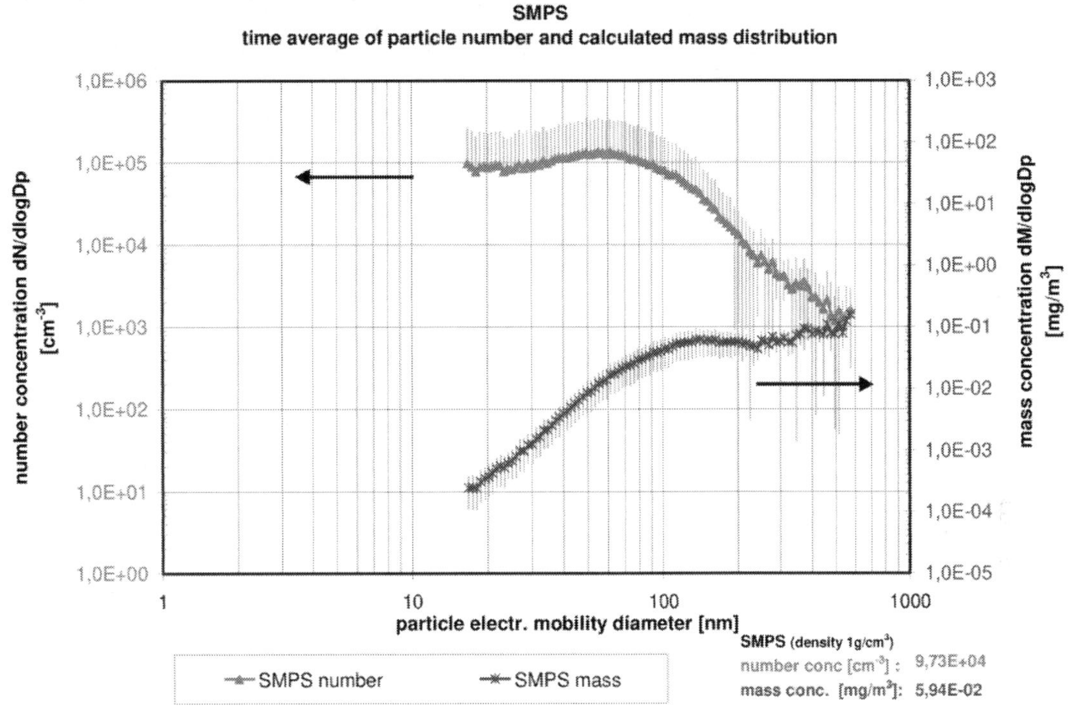

Figure 2. Particle size distribution during long term pyrosilanisation.

The particles were sampled in parallel on a TEM grid. Figure 3 presents a typical distribution, mainly with particle sizes below 100 nm as given in Figure 2. An elemental analysis was not possible so far.

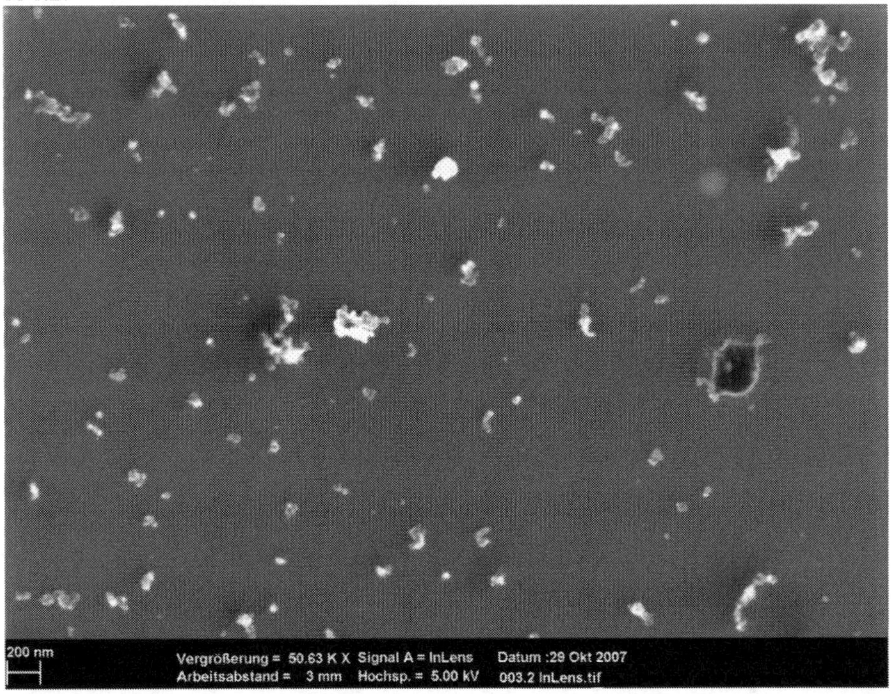

Figure 3. SEM analysis of pyrolized particles.

Sampling of the respirable and inhalable dust fraction on the left hand side from the process (approximately 1 m distance, direction of fresh air) and at the person revealed the mass concentrations given in Table 2. Especially the personal values are of interest. The sampling times comprise two entire coating procedures of two rolls (pyrosilanisation and sol-gel spraying, 2 x 5 min), an extended, simulated pyrosilanisation (13 min), and a simulated sol-gel spraying of 3 min.

Table 2. Mass concentrations during roll coating.

	Respirable dust / mg/m^3	Inhalable dust / mg/m^3	Sampling time
Static sampling (left side)	0.07	0.07	4.42 h
Personal sampling	1.37	2.63	1.5 h
Limit values	3	10	

The spatial distribution during sol-gel spraying was determined with a mobile condensation particle counter (TSI CPC 3007) next to the person and in its vicinity (Figure 4). During local exhaust ventilation horizontally behind the roll the total number concentrations stay low up to 12 000 particles per cm^3. Without exhaust ventilation the concentrations reach 225 000 particles per cm^3 just near to the spray gun. The aerosols were distributed in a cloud of approximately 2 m diameter and moved upward. The more distant locations (*e.g.* behind the person) stayed below 100 000 cm^{-3}.

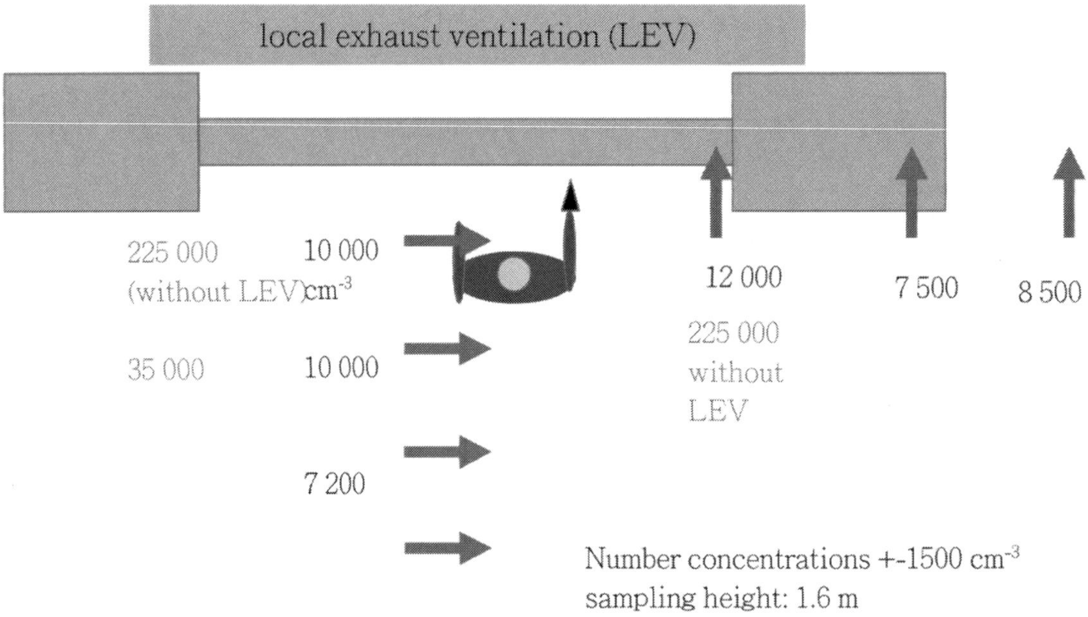

Figure 4. Spatial distribution of number concentration.

The SEM picture of the aerosols from sol-gel coating is presented in Figure 5. Mainly aggregates and/or agglomerates can be detected with primary particle sizes below approximately 2 µm. The aerosol was sampled from a cloud inside a container and was more dense than freely propagating.

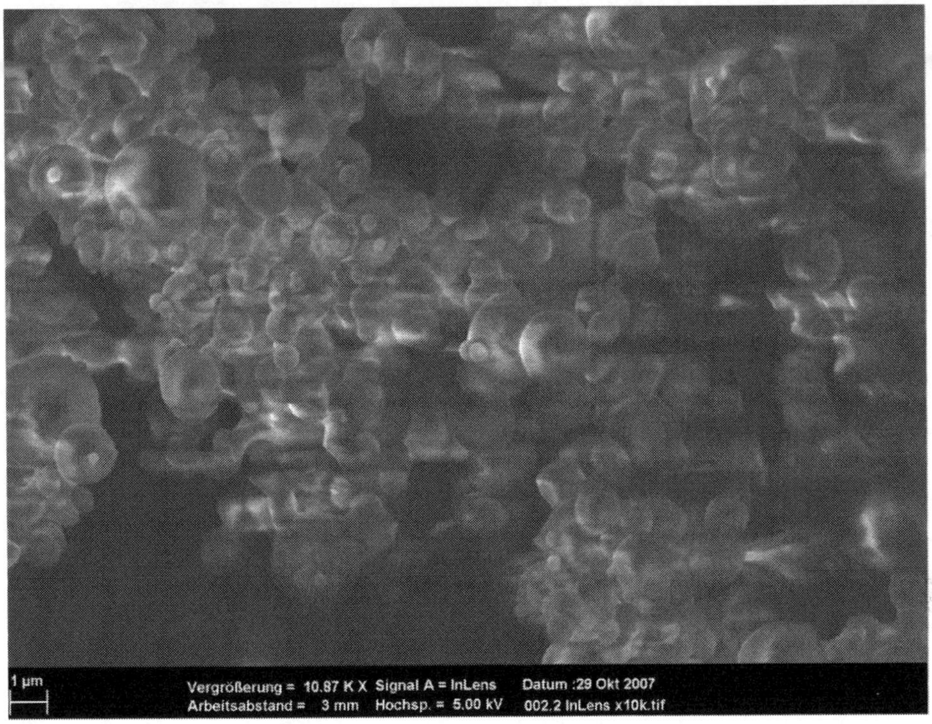

Figure 5. SEM picture of sol-gel spray aerosols

As strategy for further exposure measurements multiple point measurements for localising sources and background subtraction are recommended. This would demand also the application of TEM analysis for the particle's composition in order to determine the amount of environmental and process particles. The determination of surface area concentration is additionally of interest, as both the number and the surface area are seen to be major parameters of biological impact. Meanwhile new instruments (diffusion chargers) fulfil this demand and can be applied at workplaces. Special instrumentation for personal sampling of number and surface area concentration is to be developed and can be applied in a later stage of investigation.

Acknowledgements
This work was carried out under the European NANOSAFE 2 project with support from the European Commission (contract no. NMP2-CT-2005-515843).

On the performance of a calibrated nanoparticle generator

Ulrika Backman[1], Jussi Lyyränen[1], Unto Tapper[1], Ari Auvinen[1] and Jorma Jokiniemi[1,2]

[1]VTT Technical Research Centre of Finland, Fine Particles, P.O. Box 1000, FI-02044 VTT, Finland,

[2]University of Kuopio, Department of Environmental Science, Fine Particle and Aerosol Technology Laboratory, P.O. Box 1627, FI-70211 Kuopio, Finland.

E-mail: Ulrika.Backman@vtt.fi

Abstract. There is a need for nanoparticle generators with well characterised properties in many fields. For instance the calibration of measurement instruments can be done in place and the downtime for the instrument hence decreased. Also in nanoparticle toxicity experiments it is very important to have a well characterised particle source [1]. The aim of this study was to develop a calibrated nanoparticle generator with stable particle production. The number concentration should be regulated over many orders of magnitude and the particle size should also be adjustable. In this paper the design of the nanoparticle generator and the properties of the produced nanoparticles at one furnace temperature are presented.

The nanoparticle generator designed and developed is based on the formation of titania nanoparticles via thermal decomposition of titanium tetraisopropoxide (TTIP) at atmospheric pressure. This process was chosen as it is a well known and studied process [2-5]. The effect of the residence time and reaction temperature on the produced titania nanoparticles has among other parameters been studied. Also the decomposition reaction of TTIP has been studied and it has been reported that in addition to TiO_2 particles also the following gaseous components form: acetone (C_3H_6O), isopropanol (C_3H_8O), propene (C_3H_6) and water (H_2O) [6, 7].

Our setup, presented in figure 1, consists of a bubbler and a laminar flow reactor. The bubbler with the precursor, titanium tetraisopropoxide (TTIP, CAS# 546-68-9, 97%, Sigma-Aldrich Co.), is immersed in a water bath in order to keep it at a constant temperature. The temperature in these experiments was kept at 30°C. The precursor is vaporised by bubbling nitrogen gas through the liquid TTIP at a flow rate (MFC1 in Figure 1 and Table 1) between 1 l/min and 5 l/min. This gas flow was heated prior to the bubbler using heating tape to the same temperature as the water bath i.e. in this case 30°C.

The tubing from the bubbler to the furnace was heated with heating tape to 155°C in order to prevent the precursor from condensing before reaching the furnace. An additional nitrogen carrier gas flow (MFC2 in Figure 1 and Table 1) is fed from the bottom of the furnace at a flow rate between 2 l/min and 10 l/min, always being twice as much as the bubbler flow (MFC1). This flow (MFC2) is pre-heated to 155°C before entering the furnace.

The furnace consists of two heating zones, the preheating zone was set to 155°C and the reaction zone was set to 600°C, 700°C or 800°C. The preheating zone is 17 cm long and the reaction zone 63 cm long. The inner diameter of the quartz glass furnace tube is 18 mm. After the furnace the particle-laden gas is diluted with air (MFC3 in Figure 1 and Table 1) using a coaxial glass diluter. Before the sampling further dilution is done using a porous tube diluter (MFC4 in Figure 1 and Table 1) and two ejector diluters. The dilution gas is air in all three diluters. The flows to the ejector diluters were kept constant in all experiments at 50 Nl/min and 68 Nl/min, respectively. The total dilution ratio as compared to the bubbler flow is kept rather constant, between 2200 and 2600, in all experiments. All the flow rates are controlled with mass flow controllers (Brooks Instruments BV.), except the second ejector diluter flow which is controlled using a critical orifice. The different flow rates and the experimental parameters used in the experiments are presented in Table 1. In this paper only the results from experiments done at a furnace temperature of 700°C are presented.

The particle number concentration of the produced titania nanoparticles was measured with a Condensation Particle Counter (CPC, TSI Inc. model 3022). With this instrument particles above 7 nm in size at maximum concentrations of 10^7 particles/cm^3 can be measured. The instrument was operated at a flow rate of 0.3 l/min. The particle number size distribution was measured with a Scanning Mobility Particle Sizer (SMPS) consisting of a Differential Mobility Analyser (DMA, TSI Inc. model 3081) and a Condensation Particle Counter (CPC, TSI Inc. model 3775). With this setup particles between 6 nm and 225 nm can be detected, when the CPC was operated at a high flow rate, *i.e* 1.5 l/min and a sheath flow of 15 l/min is used in the DMA. The system was controlled with the Aerosol Instrument Manager (TSI) and in the data processing both multiple charge correction as well as diffusion correction was taken into account.

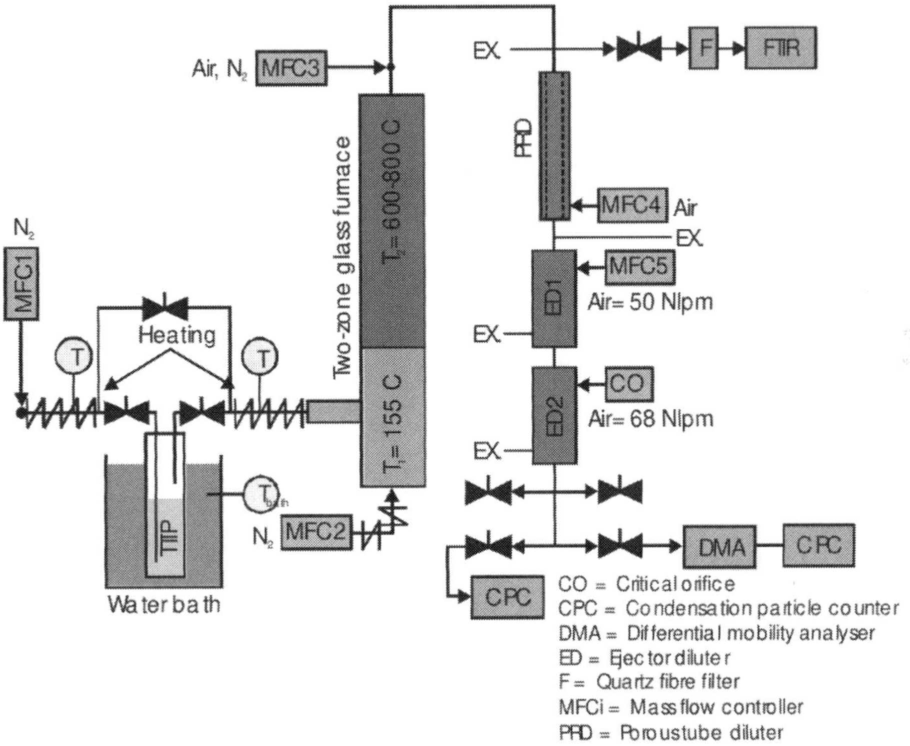

Figure 1. Schematic figure of the experimental setup used in this study.

The composition of the gas exiting the furnace was monitored using a Fourier Transform Infrared (FTIR) Spectrometer (Gasmet Dx-4000). Monitoring the gas phase compounds using FTIR gives us an opportunity to see if the reaction temperature is high enough and/or the residence time is long enough for the decomposition reaction to be complete. This is important because if unreacted TTIP exits the furnace, large, several hundred nanometer size particles are formed when the vapour phase TTIP condenses.

Table 1. The parameters of the experiments carried out for this study. For more detail on the flow rates MFCi, see Figure 1.

Experiment #	Furnace temperature [C]	Bubbler flow MFC1 [l/min]	Flow MFC2 [l/min]	Dilution flow MFC3 [l/min]	Porous tube diluter MFC4 [l/min]
1	700	1	2	9.98	5.988
2	700	2	4	19.96	11.976
3	700	3	6	29.94	17.964
4	700	4	8	39.92	23.952
5	700	5	10	49.9	29.94

The particle number concentration of two consecutive experiments is presented in Figure 2 for the five different flow rates at a furnace temperature of 700°C. The results are corrected for the dilution. We can see that the number concentration stays stable during one condition and from day to day. We can also see from the figure that the total particle number concentration first increases when increasing the flow rate through the bubbler, up to a flow rate of 3 l/min. After that the particle number concentration decreases when increasing the flow rate. This decrease in the particle number concentration indicates that the decomposition reaction of TTIP is not complete.

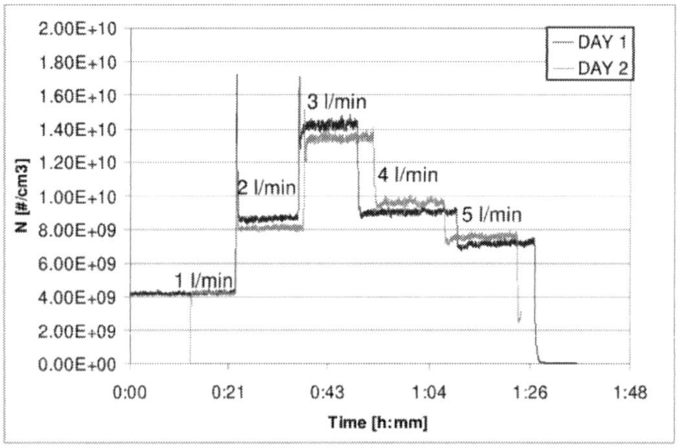

Figure 2. Total particle number concentration of titania particles produced at 700°C and different bubbler flow rates.

The particle number size distribution for the five different bubbler flow rates at a furnace temperature of 700°C is presented in Figure 3. From the figure we can see that smaller particles are produced when increasing the flow rate and thus decreasing the residence time, allowing less time for agglomeration and reaction. Also here the decrease in total number concentration at flow rates higher than 3 l/min is seen.

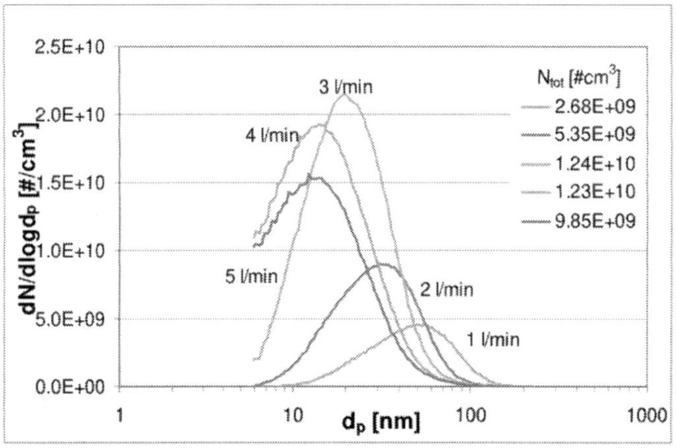

Figure 3. Number particle size distribution of titania particles produced at 700°C and varying bubbler flow rate.

The gas-phase compounds produced during the experiments at a furnace temperature at 700°C are presented in Figure 4. From the figure we can see that increasing the bubbler flow rate from 1 l/min to 2 l/min does not have an effect on the concentration of propene. However, a further increase in the flow rate decreases the propene concentration. This is due to an uncompleted decomposition of the precursor. The increase in isopropanol concentration at higher flow rates also indicates this. The level of acetone is rather low throughout the experiment.

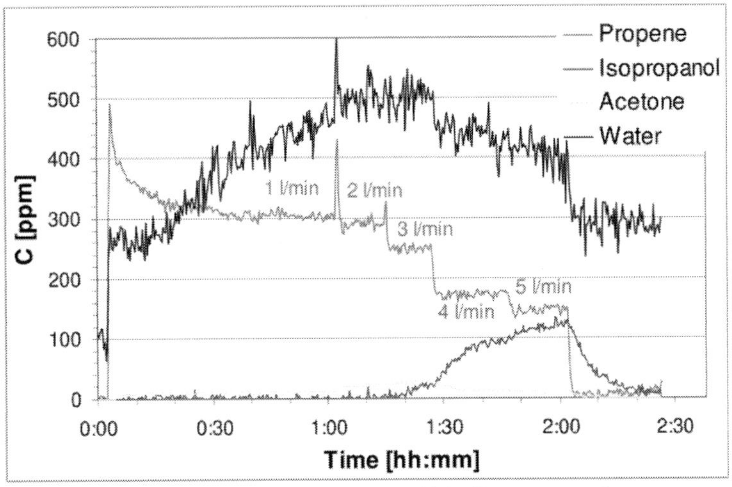

Figure 4. Variation in gas phase composition with different bubbler flow rates at a furnace temperature of 700°C.

Conclusions

An aerosol generator producing TiO_2 nanoparticles via thermal decomposition of titanium tetraisopropoxide was developed and characterised. The particle number concentration, the particle number size distribution and the gas phase compounds were measured at different bubbler flow rates. It was seen that when increasing the bubbler flow rate the residence time in the furnace decreased so that there was not enough time for the decomposition reaction to be completed. An increase in the flow rate and so decrease in the residence time also resulted in smaller agglomerates as there was less time for agglomeration.

References

[1] Oberdörster G, Maynard A, Donaldson K, Castranova V, Fitzpatrick J, Ausman K, Carter J, Karn B, Kreyling W, Lai D, Olin S, Monteiro-Riviere N, Warheit D and Yang H 2005 *Particle and Fibre Toxicology* **2**:8
[2] Komiyama H, Kanai T and Inoue H 1984 Chem. Let. 1283
[3] Okuyama K, Kousaka Y, Toghe N, Yamamoto S, Wu J J, Flagan RC and Seinfeld J H 1986 *AIChE Journal* **32(12)** 2010
[4] Kirkbir F and Komiyama H 1987 *The Canadian J Chem Eng* **65** 759
[5] Backman U, Tapper U and Jokiniemi J K 2004 *Synthetic Metals* **142** 169
[6] Fictorie C P, Evans J F and Gladfelter W L 1994 *J.Vac. Sci. Technol. A* **12(4)** 1108
[7] Wu Y-M, Bradley D C and Nix M 1993 *Appl. Surf. Sci* **64** 21

Experimental study of the response functions of direct-reading instruments measuring surface-area concentration of airborne nanostructured particles

Sébastien Bau[1], Olivier Witschger[1], François Gensdarmes[2] and Dominique Thomas[3]

[1]Institut National de Recherche et de Sécurité, INRS, Laboratoire de Métrologie des Aérosols, Rue du Morvan, CS 60027, 54519 Vandoeuvre Cedex, France.

2Institut de Radioprotection et de Sûreté Nucléaire, IRSN, Laboratoire de Physique et de Métrologie des Aérosols, BP 68, 91192 Gif-sur-Yvette Cedex, France.

3Laboratoire des Sciences du Génie Chimique, LSGC/CNRS, Nancy Université, BP 2041, 54001 Nancy Cedex, France.

E-mail: sebastien.bau@inrs.fr

Abstract. An increasing number of experimental and theoretical studies focus on airborne nanoparticles (NP) in relation with many aspects of risk assessment to move forward our understanding of the hazards, the actual exposures in the workplace, and the limits of engineering controls and personal protective equipment with regard to NP. As a consequence, generating airborne NP with controlled properties constitutes an important challenge.
In parallel, toxicological studies have been carried out, and most of them support the concept that surface-area could be a relevant metric for characterizing exposure to airborne NP [1]. To provide NP surface-area concentration measurements, some direct-reading instruments have been designed, based on attachment rate of unipolar ions to NP by diffusion. However, very few information is available concerning the performances of these instruments and the parameters that could affect their responses.
In this context, our work aims at characterizing the actual available instruments providing airborne NP surface-area concentration. The instruments (a- LQ1-DC, Matter Engineering; b- AeroTrak™ 9000, TSI; c- NSAM, TSI model 3550;) are thought to be relevant for further workplace exposure characterization and monitoring. To achieve our work, an experimental facility (named CAIMAN) was specially designed, built and characterized.

1. Materials and methods

1.1. Description of the experimental set-up

A versatile experimental facility, called CAIMAN (for 'Characterization of Instruments Measuring Aerosols of Nanoparticles') was designed and built to generate stable and reproducible airborne NP. Within CAIMAN, airborne NP are produced by means of a spark-discharge generator (GFG-1000,

PALAS). The production of airborne NP by spark discharge evaporation has been studied extensively by several authors in different situations [2-12].

This NP generation system allows a variation of both the particle concentration (*via* the argon and air flow rates, spark-discharge frequency) and their chemical nature by changing the material of the electrodes. In our paper, the results obtained for pure graphite electrodes are presented.

To cover a wide range of electrical charge level and morphology, a bipolar ion generator (EAN 581, TOPAS) and a high-temperature furnace (VECSTAR VTF 7) are included in CAIMAN. Finally, an ageing 2-liter volume is located at the end of the facility from which airborne NP can be sampled. The nanoparticle-free air introduced into the set-up comes from a purification unit (TSI model 3074B), and the excess aerosol is filtered by means of HEPA filters (CAMFIL, model 'filtre BAG').

Figure 1 shows a schematic of the CAIMAN experimental facility and lists the properties of the particles that can be varied within each element of the set-up.

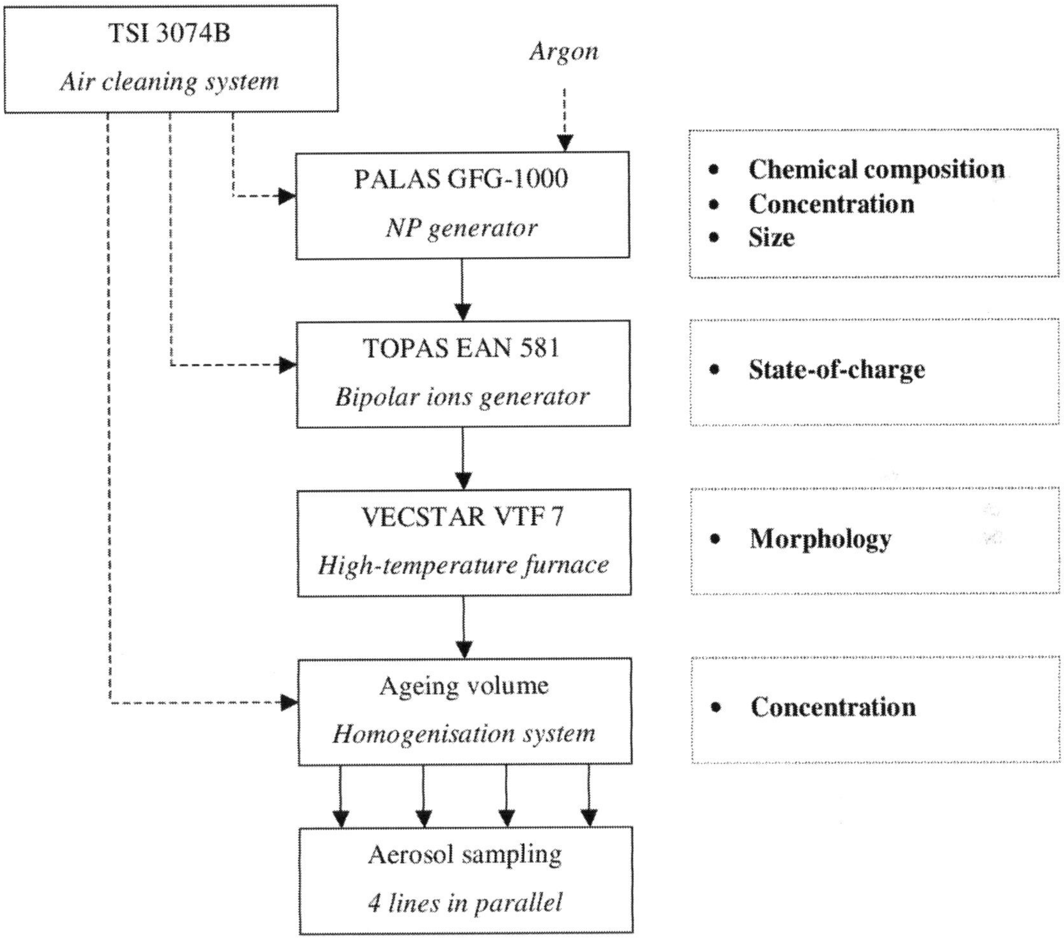

Figure 1. Schematic of the CAIMAN facility.

In this paper, the CAIMAN facility was only operated on carbon NP; neither the bipolar ions generator nor the high-temperature furnace were used.

1.2. Characterization of the generated airborne NP

The size distribution of the generated airborne NP was measured by a SMPS system (GRIMM SMPS+C, composed of a DMA 'Vienna Type' and a CNC model 5.403) operating at a flow rate of

0.3 L/min. This system allows the measurement of the number of particles according to their electrical mobility diameter. Figure 2 presents the count median diameters (CMD) for graphite electrodes as a function of the spark-discharge frequency obtained in our facility operated at constant Argon pressure (1 bar) and two different air pressures (none and 1 bar).

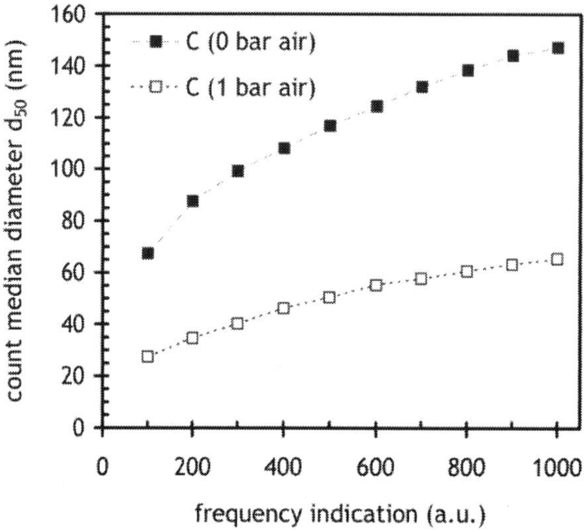

Figure 2. Evolution of the count median diameter of carbon particles produced within the CAIMAN facility (operating conditions: P_{Ar} = 1 bar, P_{air} = 0 and 1 bar)

It can be observed a significant evolution of the CMD as a function of the spark frequency and pressure conditions imposed. An increase of the discharge frequency leads to an increase of the CMD, mainly due to coagulation. This can be partially reduced by adding dilution air, as illustrated in Figure 2 (series corresponding to 1 bar of air), that divides the residence time by a factor of 7. Over all the operating conditions, a wide range of CMD extending from 25 nm to over 150 nm can be obtained. Furthermore, the NP size distributions are quasi lognormal, with geometric standard deviations (σ_g) ranging from 1.56 to 1.61 in absence of air, and from 1.62 to 1.68 when operating under 1 bar of air. Consequently, due to this slight polydispersion, carbon NP from about 20 to more than 500 nm can be produced within CAIMAN.

In terms of concentrations, real-time measurements were performed to quantify the concentration levels of airborne NP within the facility. Number concentrations in the range 10^4 to more than 10^7 #/cm^3 and mass concentrations from 0.6 to 6.4 mg/m^3 were measured respectively by means of a CNC (Condensation Nucleus Counter GRIMM model 5.403) and TEOM (Tapered Element Oscillating Microbalance, Rupprecht & Patashnick model 1400A).

Furthermore, the airborne NP sampled within the ageing volume were characterized in terms of primary particle size and morphology through the determination of the 2D-fractal dimension. To collect the particles, copper TEM-grids (200 mesh with a carbon film) were taped onto a 25 mm diameter PVC filter (0.08-µm pore size) and air flow was driven from the ageing volume at a flow rate of 1.5 L/min. The primary particle mean size is 16 ± 2 nm, while the 2D-fractal dimension was found to be 1.67 ± 0.03. The fractal dimension was determined by the Box Counting Method [13], which has been previously validated on simulated nanoparticles [14] and here associated to an optimised calculation procedure as proposed by Foroutan-Pour *et al.* [15].

1.3. Experimental measurement of instruments response functions

The response function, notated $\mathcal{R}$, is defined by:

$$\mathcal{R}(d_m) = \frac{\text{mean instrument indication}}{\text{mean number concentration}} = \frac{S(d_m)}{N(d_m)},$$

where S and N are respectively the NP surface-area and number concentrations.

Experimentally, a parallel measurement of particle number (N) and surface-area (S) mean concentrations is performed on a monodisperse aerosol of selected mobility size d_m (by using a DMA – Differential Mobility Analyzer GRIMM 'Vienna Type') over a stable period of 15 minutes, as shown in Figure 3.

For carbon electrodes, NP with electrical mobility diameters from 20 up to 520 nm have been selected, providing surface-area concentrations significantly over the limit of quantification of the different instruments.

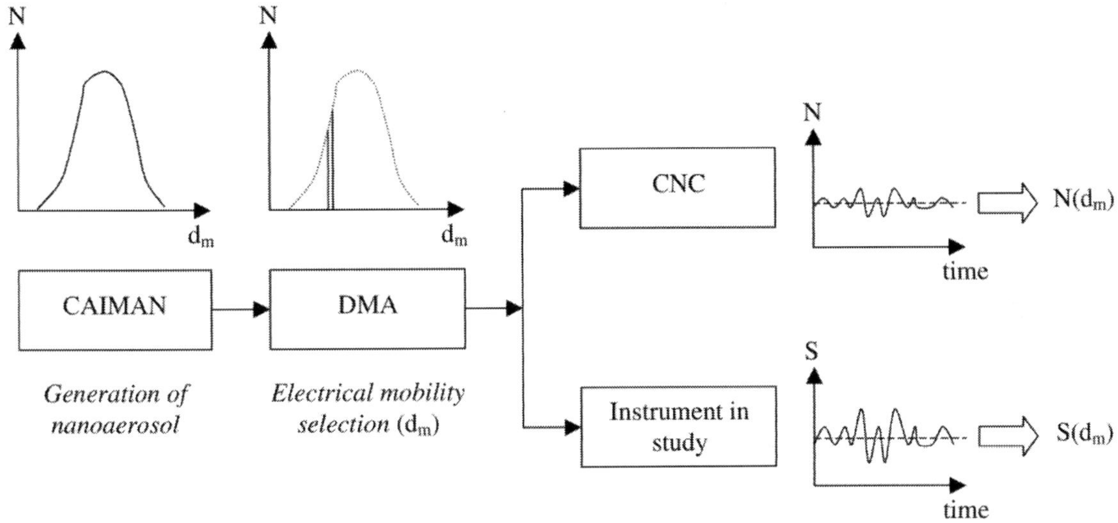

Figure 3. Experimental set-up for the measurement of the response functions of the instruments.

2. Experimental results

Experimental measurements were performed in monodisperse mode to determine the response functions of the selected instruments as a function of the mobility diameter, as presented in Figure 3.

The results were compared with the existing theoretical responses: lung-deposited surface-areas in alveolar or tracheobronchiolar region for instruments (b) and (c), and active surface-area for instrument (a). The lung-deposited surface-areas were calculated using the lung deposition model developed by ICRP [16] using physiological, worker activity related and aerosol parameters considered by the manufacturer of the NSAM and Aerotrak^{TM} 9000 [17], while active surface-area was obtained with the simplified expression proposed by Jung and Kittelson [18] with λ_{ion} = 15.2 nm.

Theoretical considerations are then required since raw data have to be corrected. Indeed, even if particles of unique electrical mobility are selected within the DMA, a polydisperse aerosol is obtained at its outlet due to multiple charging of NP with diameters above 50 nm. Consequently, the measured number and surface-area concentrations are due to the contributions of all particles exiting the DMA, *i.e.* NP of selected mobility carrying i (one and more) electrical charges. The response function experimentally measured is hence defined by:

$$\mathcal{R}_{measured} = \frac{\sum_{i=1}^{n} S_i}{\sum_{i=1}^{n} N_i} \approx \frac{S_1 + S_2}{N_1 + N_2},$$

where indexes 1 and 2 correspond to particles carrying respectively 1 and 2 electrical charges. It should be notified that in this study, particles carrying 3 electrical charges and more are neglected, since they represent a very little fraction of the aerosol.

The fraction of NP at the outlet of the DMA carrying 1 and 2 elementary charges was determined from the DMA transfer function obtained experimentally by Heim *et al.* [19]. This transfer function is a triangular function taking into account the diffusion of the NP within the DMA. Thus, for a given aerosol entering the DMA, the number and surface-area concentrations of particles carrying 1 and 2 electrical charges can be calculated. Then, the true response functions of the instruments were calculated by correcting the raw data as:

$$\mathcal{R}_{\text{corrected}} = \frac{S_1}{N_1} = \mathcal{R}_{\text{measured}} \cdot \frac{\dfrac{S_1}{S_1 + S_2}}{\dfrac{N_1}{N_1 + N_2}}.$$

To calculate the number concentrations ratio, the bipolar Fuchs extended model, as described by Wiedensohler [20], was coupled to the DMA transfer function. Notating $f_i(d_m)$ the fraction of particles carrying i elementary charges and $\eta(d_m)$ the DMA transfer function, then the number concentration of particles at the outlet of the DMA carrying i electrical charges is given by:

$$N_i(d_m) = N_0(d_m) \cdot f_i(d_m) \cdot \eta(d_m),$$

where $N_0(d_m)$ correspond to the NP size distribution at the inlet of the DMA. The overall number concentration of particles carrying i electrical charges hence corresponds to:

$$N_i = \sum_{d_m} N_i(d_m).$$

Concerning the fraction of particles carrying i elementary charges, the formalism coming from Wiedensohler [20] and reported in Flagan [21] was used:

$$\log(f_i) = \sum_{j=0}^{j=5} a_j(i) \cdot \left[\log\left(\frac{d_p}{d_0} \right) \right]^j,$$

where d_p is the particle diameter expressed in nm, d_0 = 1 nm is a reference diameter, and $a_j(p)$ are the coefficients of adjustment, reported in Flagan [21]. It should be noted that this relationship is valid for i = –1, 0, 1 in the size range 1 to 1000 nm, and for i = –2, 2 in the range 20 to 1000 nm.

Then, the ratio in terms of surface-area concentrations can be obtained assuming spherical particles, *i.e.*:

$$S_i = \sum_{d_m} S_i(d_m) = \sum_{d_m} N_i(d_m) \cdot \pi \cdot d_m^2.$$

Let us take for example an aerosol with count median diameter d_{50} = 100 nm and a geometric standard deviation of $\sigma_g = 1.5$. If the selected size within the DMA is d_m = 100 nm, then particles of 156 nm in diameter carrying 2 elementary charges will exit the DMA too. These doubly-charged NP represent 17 % of the aerosol number concentration and 29 % of its surface-area concentration, *i.e.:*

$$\frac{N_1}{N_1 + N_2} = 0.83,$$

and:

$$\frac{S_1}{S_1 + S_2} = 0.71.$$

Consequently, the measured response function will be corrected by a factor of 0.86 to take into account the contribution of doubly-charged NP.

To perform this multiple charging correction, the size distribution of the aerosol entering the DMA is required, notated previously $N_0(d_m)$. In our work, the spark-generated NP are not spherical individual particles, but were identified as agglomerates. Lall and Friedlander [22] showed that the electrical mobility diameter of an agglomerate is related to the number and size of the primary particles it is constituted of. Hence, SMPS-measured particle size distributions $N_{sph}(d_m)$ must be corrected by bipolar charging efficiencies of both spheres and agglomerates to obtain the agglomerates size distribution:

$$N_{agg}(d_m) = N_{sph}(d_m) \cdot \frac{\varepsilon_{sph}}{\varepsilon_{agg}},$$

where ε correspond to the charging efficiencies. To perform this correction, a primary particle size of 16 nm (see section 1.2.) was used in the calculations of NP concentrations. For example, for the same previous polydisperse aerosol of count median diameter d_{50} = 100 nm and geometric standard deviation of $\sigma_g = 1.5$, the ratio of total number concentrations is:

$$\frac{N_{agg}(d_m)}{N_{sph}(d_m)} = 1.02,$$

value obtained for primary particles of 16 nm.

Both corrections lead to a variation in the response functions (relative difference between raw and corrected data) reaching at maximum 15%.

Figure 4 presents the response functions of the three selected instruments for carbon airborne NP and both configurations (alveolar 'alv' or tracheobronchiolar 'tb') for instruments (b) and (c).

Figure 4 shows a good agreement between theoretical curves and corrected experimental responses obtained with carbon airborne NP for particle sizes in the range 20 – 520 nm. Indeed, the biases between experimental data and theory are in the range -45% to +30%.

More precisely, the biases observed for LQ1-DC (Figure 4a) are in the range –40% - +30%, for AeroTrak™ (Figure 4b) between –45% and +30%, and for NSAM (Figure 4c) between –10% and +30%. It should be noted that the three-quarters of the experimental data are in a range of biases to theory of ± 25%.

These new experimental data enrich the actual data base from the works of Shin *et al.* [23] and Asbach *et al.* [24] on the NSAM in the range 10 to 400 nm for both 'alv' and 'tb' configurations. In their studies, NP of sodium chloride (NaCl) were generated by means of an atomizer, and agglomerates of silver (Ag) were produced within a furnace. The NP were then mobility-selected to characterize the instrument.

Concerning the LQ1-DC, the only published data correspond to the size range 20 to 200 nm [25]. In their work, the authors generated silver NP of varying morphology (fractal-like to spherical, 2-D fractal dimensions from 1.58 to 1.94) by means of two furnaces in series.

To our knowledge, our experimental data on the AeroTrak™ 9000 (b) are the first ever published.

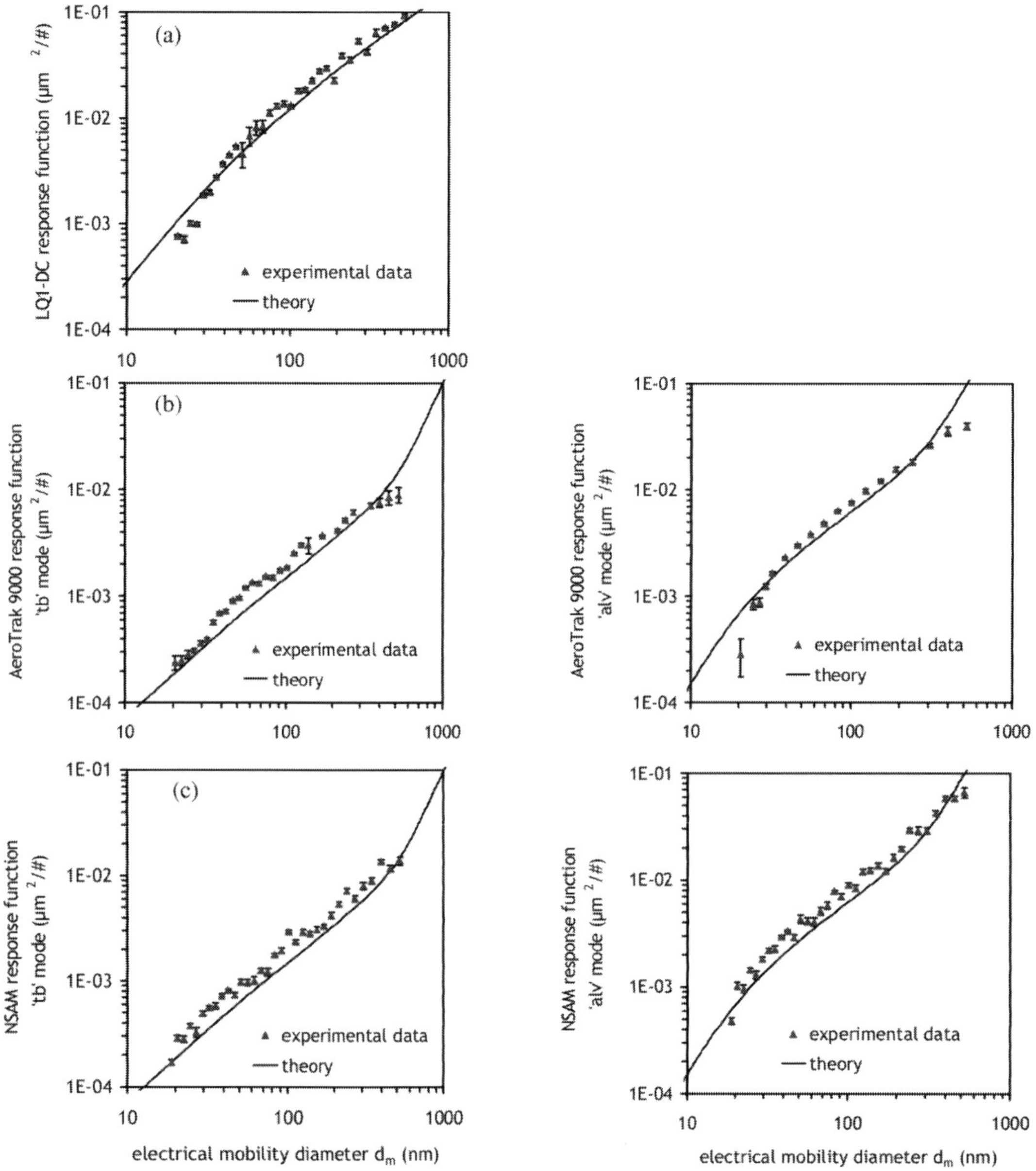

Figure 4. Response functions of (a) LQ1-DC, (b) AeroTrak™ 9000 and (c) NSAM measured with monodisperse carbon NP produced in the CAIMAN facility and mobility-selected

To conclude, this experimental study provides new results leading to a better understanding and characterization of three devices devoted to the real-time measurement of airborne NP surface-area concentration. For two of them (NSAM and LQ1-DC), our data enrich and enlarge the size range of the previously published data. Concerning the third instrument, these data constitute the first ever published.

References

[1] Maynard A D and Aitken R J 2007 *Nanotoxicology* **1(1)** 26
[2] Schwyn S, Garwin E and Schmidt-Ott A 1988 *Journal of Aerosol Science* **19(5)** 639
[3] Helsper C, Molter W, Loffler F, Wadenpohl C, Kaufmann S and Wenninger G 1993 *Atmospheric Environment* **27** 1271

[4] Roth C, Karg E and Heyder J 1998 *Journal of Aerosol Science* **29**(Suppl 1) S679
[5] Roth C, Ferron G A, Karg E, Lentner B, Schumann G, Takenaka S and Heyder J 2004 *Aerosol Science and Technology* **38** 228
[6] Brown J S, Kim C S, Reist P C, Zeman K L and Bennett W D 2000 *Aerosol Science and Technology* **32** 325
[7] Evans D E, Harrison R M and Ayres J G 2003 *Journal of Aerosol Science* **34** 1023
[8] Evans D E, Harrison R M and Ayres J G 2003 *Aerosol Science and Technology* **37** 975
[9] Horwath H and Gangl M 2003 *Journal of Aerosol Science* **34** 1581
[10] Borra J P 2006 *Journal of Physics D* **39** R19
[11] Byeon J H, Park J H and Hwang J 2008 *Journal of Aerosol Science* **39** 888
[12] Tabrizi N S, Ullman M, Vons V A, Lafont U, Schmidt-Ott A 2009 *Journal of Nanoparticle Research* **11**(2) 315
[13] Giorgilli A, Casata D, Sironi L and Galgani L 1986 *Physics Letters A* **115**(5) 202
[14] Oh C and Sorensen M 1997 *Journal of Colloid and Interface Science* **193** 17
[15] Foroutan-Pour K, Dutilleul P and Smith D L 1999 *Applied Mathematics and Computation* **105** 195
[16] ICRP 1994, Publication 66: Human respiratory tract model for radiological protection. Oxford: Pergamon
[17] TSI Application Note 2008. Measuring Nanoparticle Exposure. Application Note NSAM-001. 4/11/2008: 8p
[18] Jung H and Kittelson D B 2005 *Aerosol Science and Technology* **39** 902
[19] Heim M, Kasper G, Reischl G P and Gerhart C 2004 *Aerosol Science and Technology* **38**(Suppl 2) 3
[20] Wiedensohler A 1988 *Journal of Aerosol Science* **19** 387
[21] Flagan R C 2001 *Electrical Techniques*, in: Baron P A and Willeke K (Eds), *Aerosol Measurement: Principles, Techniques and Applications* (2nd Ed.). New York: Wiley Interscience, pp. 537-568
[22] Lall A A and Friedlander S K 2006 *Journal of Aerosol Science* **37**(3) 260
[23] Shin W G, Pui D Y H, Fissan H, Neumann S and Trampe A 2007 *Journal of Nanoparticle Research* **9**(1) 61
[24] Asbach C, Fissan H, Stahlmecke B, Kuhlbusch T A J and Pui D Y H 2008 *Journal of Nanoparticle Research* DOI 10.1007/s11051-008-9479-8
[25] Ku B K and Maynard A D 2005 *Journal of Aerosol Science* **36**(9) 1108

From nanoparticles to large aerosols: Ultrafast measurement methods for size and concentration

Lothar Keck, Jürgen Spielvogel and Hans Grimm

Grimm Aerosol Technik GmbH, 83404 Ainring, Dorfstrasse 9, Germany.

E-mail: lk@grimm-aerosol.com

Abstract. A major challenge in aerosol technology is the fast measurement of number size distributions with good accuracy and size resolution. The dedicated instruments are frequently based on particle charging and electric detection. Established fast systems, however, still feature a number of shortcomings. We have developed a new instrument that constitutes of a high flow Differential Mobility Analyser (high flow DMA) and a high sensitivity Faraday Cup Electrometer (FCE). The system enables variable flow rates of up to 150 lpm, and the scan time for size distribution can be shortened considerably due to the short residence time of the particles in the DMA. Three different electrodes can be employed in order to cover a large size range. First test results demonstrate that the scan time can be reduced to less than 1 s for small particles, and that the results from the fast scans feature no significant difference to the results from established slow method. The fields of application for the new instrument comprise the precise monitoring of fast processes with nanoparticles, including monitoring of engine exhaust in automotive research.

1. Introduction

Fine and ultrafine particles are commonly classified with a Differential Mobility Analyser (DMA). The classification (*i.e.* the selection of a well defined fraction from a broad size distribution) occurs in the electrostatic electric field in the annulus between inner and outer electrode of the DMA, where the particles in the sample air drift through the particle free sheath air. Only particles of a certain size reach a narrow slit at the bottom of the inner electrode and thus a downstream detection system. The size distribution of the particles can be measured by a stepwise or stepless change of the DMA voltage. If the DMA voltage is fixed, the DMA acts as particle generator for particles of well defined mobility diameter.

Such conventional DMAs are common and reliable instruments. They feature, however, several shortcomings which limit the field of application:

(1) The resolution of particle size, being controlled by the ratio of sheath air flow to sample air flow (at least when this ratio not too high), is limited due to the relatively low sheath air flow rate of typically 3 – 20 lpm. The resolution is typically around 10.

(2) The size resolution for very small particles ($D < \sim 20$ nm) is additionally deteriorated due to the effect of particle diffusion.

(3) The diffusion of particles causes as well high losses of these smallest particles and thus a low sensitivity of the instrument in that size range.

(4) The size range of the DMA, being controlled by the sheath air flow rate, features a lower limit of typically 5 nm.

(5) The measurement of size distributions is slow due to the long residence time of the particles in the DMA.

The nearly monodisperse particles downstream of a DMA are frequently detected with a Condensation Particle Counter (CPC). A typical CPC, however, detects only particles larger than ~ 5 nm and the response time is slow. Thus, a CPC is not the optimal detection system for a high flow DMA system for fast measurements of particles down to the 1 nm size range.

Hence, for the high flow DMA, a Faraday Cup Electrometer (FCE) is used to detect charged particles. Such a FCE features a higher detection limit for particle concentrations than a CPC, but the response time is below 100 ms and there is essentially no lower limit for the detected particle size. Figure 1 illustrates the principle of a FCE. Sample air is drawn through a filter inside the Faraday Cup and the Faraday Cup gains a net charge whenever charged particles enter it. The well isolated Faraday Cup is then discharged with a current passing through a 1 TΩ resistor. This current is converted to a voltage and amplified with a low-noise electrometer. The particles are permanently retained in the Faraday Cup on a filter of high loading capacity.

Minimizing the noise of the electrometer requires a sophisticated geometry with a space between Faraday Cup and outer casing. For the GRIMM FCE, this space is continuously flushed with a rinse air flow in order to achieve the shortest possible response time and to avoid any contamination of the surface of the insulators [1].

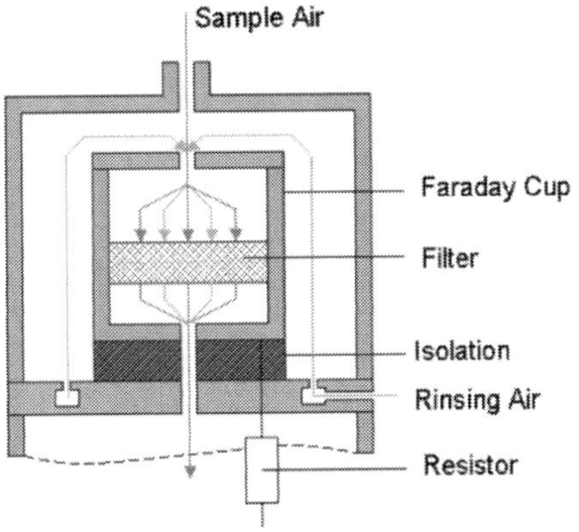

Figure 1. Schematic of the FCE.

2. High flow DMAs

Rosell-Llompart *et al.* [2] have demonstrated that the shortcomings listed above can be resolved by using a high sheath air flow rate for a short Vienna type DMA. Rosser [3, 4] presented a high flow DMA, which was as well based on the Vienna type design but which featured a considerably reduced exhaust pressure drop. A further improved design was presented by Fernandez de la Mora [5], where a new widely open sheath gas exhaust system produces essentially no pressure drop at all. The main challenge for high flow DMAs is keeping the flow between the electrodes laminar despite of the very high Reynolds numbers. For the later designs, laminar flow was accomplished with an accelerating flow in the working region. Another problem of the high flow DMA is that accurate measurements of the very high sheath air flow rates become very difficult.

The design of the new GRIMM high flow DMA is based on the commercial GRIMM Vienna-type DMAs. Head and bottom of the DMA were however adapted to the much higher sheath air flow rate. To establish a symmetric flow in the annulus between the electrodes, the head of the high flow DMA features a special slit as an inlet for the sample air.

The sheath air is generated by a membrane pump with eight pump heads. Therefore, critical nozzles can be employed to achieve a well defined flow rate of the sheath air. Using two parallel critical nozzles, flow rates of 75 lpm and 150 lpm can be selected. The high flow DMA is operated in the recirculation mode, *i.e.* the excess air from the end of the DMA is passed through a filter and reused as sheath air. To prevent a heating of the system by the pumps, two aerosol coolers are employed downstream of the pump. The complete pump module is separated from the DMA in order to have only a minimum of equipment on the table.

Table 1. The size range of the GRIMM high flow DMA.

DMA type	Active length of electrodes [mm]	Size range at 75 lpm sheath air [nm]	Size range at 150 lpm sheath air [nm]
S-DMA	15	0.4 – 19.6	0.3 – 13.8
M-DMA	88	1.1 - 50	0.8 - 34
L-DMA	350	2.1 - 107	1.5 - 72

Three electrodes of different lengths are available, termed S-DMA, M-DMA and L-DMA. The short electrode (S-DMA) features a length of 14 mm, the middle electrode (M-DMA) a length of 88 mm, and the long electrode (L-DMA) a length of 350 mm. The size range of the three DMAs in the high flow mode is shown in Table 1. As shown in Figure 2, the DMA is directly attached to the FCE in order to minimize the particle losses between outlet of the DMA and detection system.

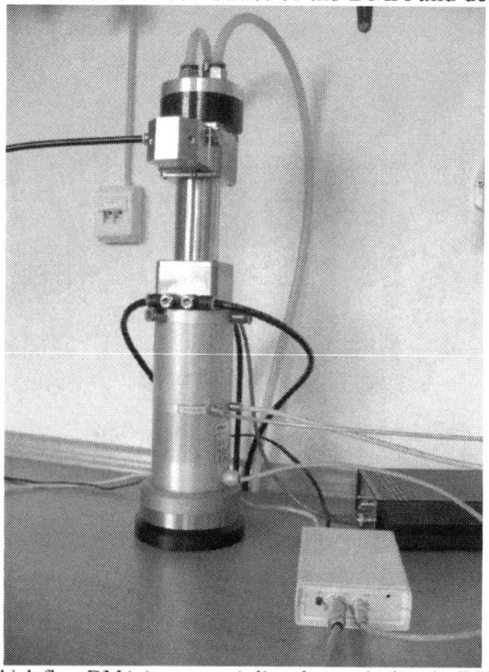

Figure 2. The high flow DMA (upper part) directly attached to the FCE (lower Part). The height of the arrangement is approximately 50 cm.

3. Application examples for high flow DMAs

3.1. Characterisation of aerosol standards with nanometer dimensions

Particles of standardized size are very useful in aerosol studies for instrument evaluation and calibration. For particle sizes larger than 100 nm, such standard aerosols are generated by atomizing suspensions of polystyrene latex particles. Below that size range, however, no precise size standards can be generated with this method mainly because the surfactants dissolved in the suspension remains on the surface of the particles. Much smaller size standards are achievable with electrospraying solutions of large organic salts. These organic salts end up in individual ions of the dissolved cation, or in ion clusters with one or more charges. Thus the size range of 1 – 3 nm can be covered. Ude *et al.* [6] have characterized such size standards using a high flow DMA and the results suggest that the particle mobility diameter can be determined with an accuracy of approximately 1%.

3.2. 2. Fast measurements of size distributions

We have conducted a number of experiments with the main aim of showing that our high flow DMA permits a considerable shortening of the sampling time without significant loss of resolution and sensitivity. To generate the test aerosol, we have adapted the methods described by Ude [6] and we have electrosprayed solutions of Tetra-butyl ammonium iodine (TBAI) and Tetra-pentyl ammonium iodine. Size distributions were measured both with the M-DMA and the S-DMA, sheath air flow rate was 150 lpm and the sample flow rate 5 lpm. No neutralizer was used for these measurements. The sampling time for a single channel was varied from 250 ms to 62.5 ms, the later value corresponds to the sampling frequency of the employed FCE, 1/16 Hz. Figure 3 shows that the shortening of the sampling time from 250 ms to 62.5 ms has no significant effects on the measured size distributions; this is only possible with the fast response time of the GRIMM FCE.

With the 62.5 ms sampling time, the size distributions shown below, representing 62 channels for the M-DMA and 44 channels for the S-DMA, are measured within 3.9 s and 2.8 s. It is obvious that a wider spacing of the measured particles diameters or a reduced size range would enable shorter scan times of well below 1 s. For these measurements the sampling time was limited by the FCE, however, a higher sampling frequency version of the FCE is under construction and with such an instrument the speed of the measurements would be limited only by the residence time of the particles in the DMA. The residence time is, for the sheath air flow rate of 150 lpm, 0.04 s for the S-DMA and 0.25 s for the M-DMA, thus the time for a scan could be decreases to well below 1 s.

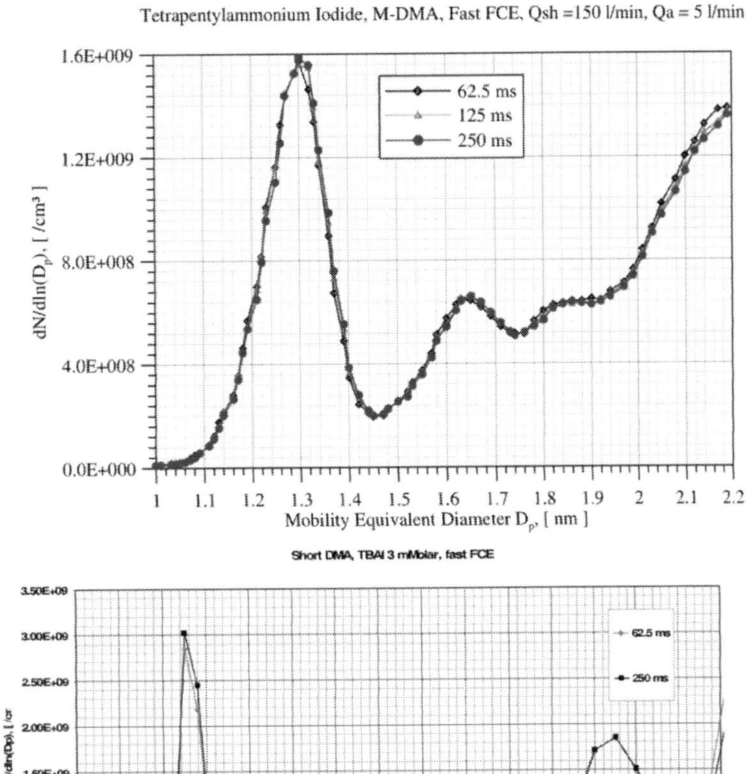

Figure 3. Size distributions of ions and clusters measured with the GRIMM high flow DMA. Upper panel: Scan measured with the M-DMA, lower panel: measurement with the S-DMA.

3.3. Generation of monodisperse soot particles for automotive applications

The EURO 5 regulations for Diesel emissions include the measurements of particle number concentrations [7]. The measurements are accomplished with a CPC which has to be calibrated. For calibration, the candidate CPC is operated, as illustrated in Figure 4, in parallel with a FCE that serves as a primary reference. Both counters - candidate CPC and reference FCE – must be supplied with identical concentrations of monodisperse soot particles from a Combustion Aerosol Standard (CAST). Number concentrations measured by the CPC are then evaluated against the concentrations measured by the reference FCE. These calibrations employ normally several instruments downstream of a single DMA, and therefore the sum of the sample flows may easily be as high as several litres per minute. The high total sample flow makes it difficult to achieve monodisperse aerosols of narrow size distribution with a conventional DMA; narrow size distributions can however be achieved if a high flow DMA is employed for the classification.

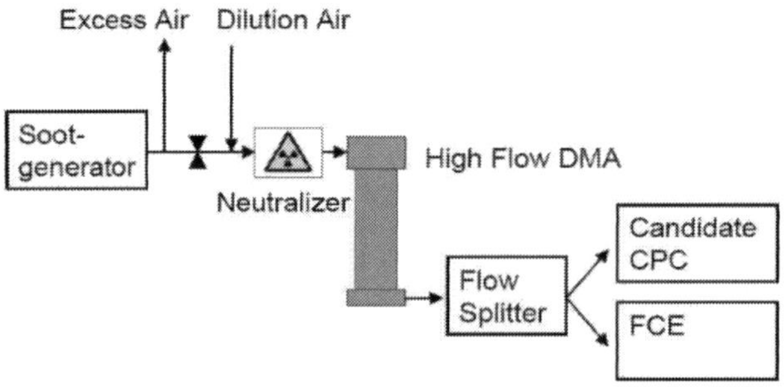

Figure 4. Setup for the calibration of the CPC using the FCE as a reference.

4. Conclusions and Outlook

Using a conventional Vienna-type DMA with a sheath air flow rate of 150 lpm expands the possibilities of the instrument considerably. It enables the generation of very narrow size distribution even when a relatively high sample flow rate is needed, it enables measurements of size distributions in the range well below 1 nm with good size resolution, and it permits fast measurements of the size distribution with the time for a single scan being as low as 1 s.

References

[1] Wimmer D, Attoui M, Steiner G W, Sipilä M and Reischl G P 2008 *European Aerosol Conference EAC 2008* Tessaloniki Greece
[2] Rosell-Llompart J, Loscertales G, Bingham D and Fernandez de la Mora J 1996 *Journal of Aerosol Science* **27** 695
[3] Rosser S, Ramiro E, Ramiro F, Laviada C and J. F. de la Mora 2003 *Journal of Aerosol Science* **34(2)** 917
[4] Rosser S and de la Mora J F 2005 *Aerosol Science and Technology* **39(12)** 1191
[5] Fernandez de la Mora J and Attoui M 2007 *European Aerosol Conference 2007* Salzburg
[6] Ude S and de la Mora J F 2005 *Journal of Aerosol Science* **36** 1224
[7] AEA Energy & Environment 2007 Particle Number Counter Calibration Procedure. Report to the Department for Transport.

Lung toxicity determination by *in vitro* exposure at the air liquid interface with an integrated online dose measurement

Sonja Mülhopt[1], S. Diabaté[2], T. Krebs[3], C. Weiss[2] and H.-R. Paur[1]

[1]Institute for Technical Chemistry, Thermal Waste Treatment Division, Forschungszentrum Karlsruhe GmbH, 76021, Karlsruhe, Germany

[2]Institute for Toxicology and Genetics, Forschungszentrum Karlsruhe GmbH, 76021, Karlsruhe, Germany

[3]VITROCELL Systems GmbH, 79183, Waldkirch, Germany.

E-mail: muelhopt@itc-tab.fzk.de

Abstract. Epidemiological studies show an association between the concentration of ultrafine particles in the atmosphere and the rate of mortality or morbidity due to respiratory and cardiovascular diseases. For the quantitative assessment of the toxicity of airborne nanoparticles the dose–response relationship is tested in *in vitro* test systems using bioassays of cell cultures as sensor.

For the air-liquid interface exposure of cell cultures towards aerosols the Karlsruhe exposure system was developed. The human lung cell cultures are exposed in VITROCELL® system modules with a constant flow of the conditioned aerosol. After exposure the cells are analyzed to measure the biological responses such as viability, inflammatory or oxidative stress.

For the determination of the dose response relationship the accurate knowledge of the deposited particle mass is essential. A new online method is developed in the Karlsruhe exposure system: the sensor of a quartz crystal microbalance is placed in an exposure chamber instead of the membrane insert and exposed to the aerosol in the same way as the cell cultures. The deposited mass per area unit is monitored as a function of exposure time showing a linear relationship for a constant aerosol flow with defined particle concentration.

A comparison of this new dose signal to a dosimetry method using fluorescein sodium particles shows a very good correlation between the sensor signal of the quartz crystal microbalance and the deposited mass on the membranes shown by spectroscopy. This system for the first time provides an online dose measurement for in vitro experiments with nanoparticles.

1. Background

Epidemiological studies show an association between the concentration of fine and ultrafine particles (PM10, PM2.5, PM0.1) in the atmosphere and the rate of mortality or morbidity due to respiratory and cardiovascular diseases [1]. The assessment of the risk of airborne nanoparticles in workplaces or other atmospheres containing ultrafine particles is therefore an urgent task. The causes of the toxicological effects of ultrafine and nanoparticles to the human organism are yet insufficiently known. Besides the chemical composition, the physical properties of the particles seem to be of particular importance for the effects. For the quantitative assessment of the toxicity of airborne nanoparticles the dose – response relationship is tested in *in vitro* test systems using bioassays of cell cultures as sensor.

2. Material and Methods

For the air-liquid interface exposure of cell cultures towards aerosols the Karlsruhe exposure system was developed [2, 3]. This system consists of an isokinetic sampling unit to collect the aerosol from the particle loaded atmosphere (Figure 1). The particles bigger than 1 μm are removed by passing a size selective inlet like a cyclone. In the next step the aerosol is humidified up to 85% by injecting water steam. The aerosol temperature is controlled to 37°C. From this conditioned aerosol, sample flows of 100 ml/min each are directed into VITROCELL® exposure modules [3, 4], which contain three Transwell® inserts with human lung cells on the microporous membrane (Figure 2). The aerosol flows perpendicular onto the surface of the cell culture and the particles deposit on it. After exposure experiments of 1 up to 8 hours the cell cultures were taken out of the system and tested for different biological responses. The responses of the cells were analyzed by measuring the viability (LDH, AlamarBlue) as well as the release of Interleukin-8 (IL-8) as a marker for pro-inflammatory changes [5].

Figure 1. Left: scheme of the Karlsruhe Exposure System with the (1) size selective inlet, (2) the water vapour dosage for humidification (3), the conditioning reactor for a constant aerosol with 37°C and 85% r.h., (4) the VITROCELL® exposure chambers containing the Transwell® membrane inserts and the sensor of the (5) quartz crystal microbalance.
Right: photograph of the fully automated and temperature controlled prototype during an outside measurement of environmental aerosols.

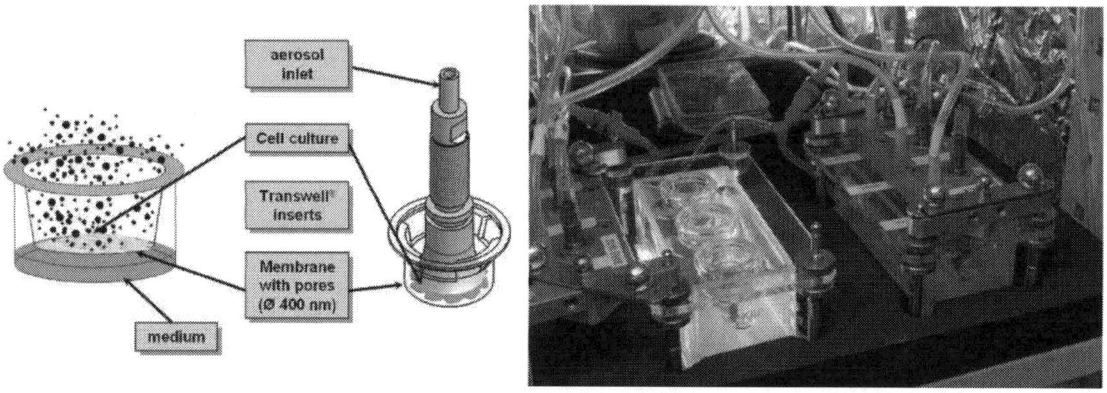

Figure 2. Left: scheme of the exposure of a cell culture towards aerosol at the air liquid interface consisting of the aerosol inlet, the cell culture insert and the medium container.
Right: opened and closed VITROCELL® exposure chambers with water temperature control and medium for supplying the cell cultures with nutrients.

One of the essential questions for the determination of the dose response relationship is the accurate determination of the dose [6]. Especially for nanoparticles with extremely low masses this is the challenging point. To measure accurately the dose of deposited particles on the exposed cell culture surface two alternatives were developed and compared:

1. **Fluorescein sodium dosimetry (FSD)**: The FSD is a highly sensitive method for a precise determination of the deposited mass exclusively on the membrane. Fluorescein sodium aerosol is generated as calibration aerosol by spraying a water solution of 3% of mass fluorescein sodium with a two phase nozzle in a heated air flow of 80°C. The water is dried out and the fluorescein sodium particles are sampled from the reactor to the Karlsruhe Exposure System and lead to the VITROCELL® exposure chambers containing Transwell® Clear membranes (Type 3450, Corning). After the exposure experiment the polyethylene membrane is cut out of the Transwell® insert, washed in 10 ml ultra pure water and 10 min treated by sonication. The fluorescein sodium mass in the solution is analyzed by a fluorescence spectrometer Aminco Bowman Series 2.
2. **Quartz crystal microbalance (QCM)**: Using a QCM for the online monitoring of particle dose the quartz is placed in a VITROCELL® exposure chamber instead of the Transwell® membrane insert and exposed to the aerosol in the same way as the cell cultures. In the Karlsruhe Exposure System in one of the 6 exposure chamber the membrane is replaced by the sensor crystal of a QCM. It monitors the deposited particle mass per area unit and time providing a control of the exposure quality.

3. Results

The deposited mass per area unit is monitored in µg/cm² as a function of exposure time in sec. It shows a linear increasing for a constant aerosol flow with defined particle concentration (Figure 3). For the comparison of the both methods, FSD and QCM, the calibration factor is determined as ratio of QCM versus FSD in µg/cm² (Table 1).

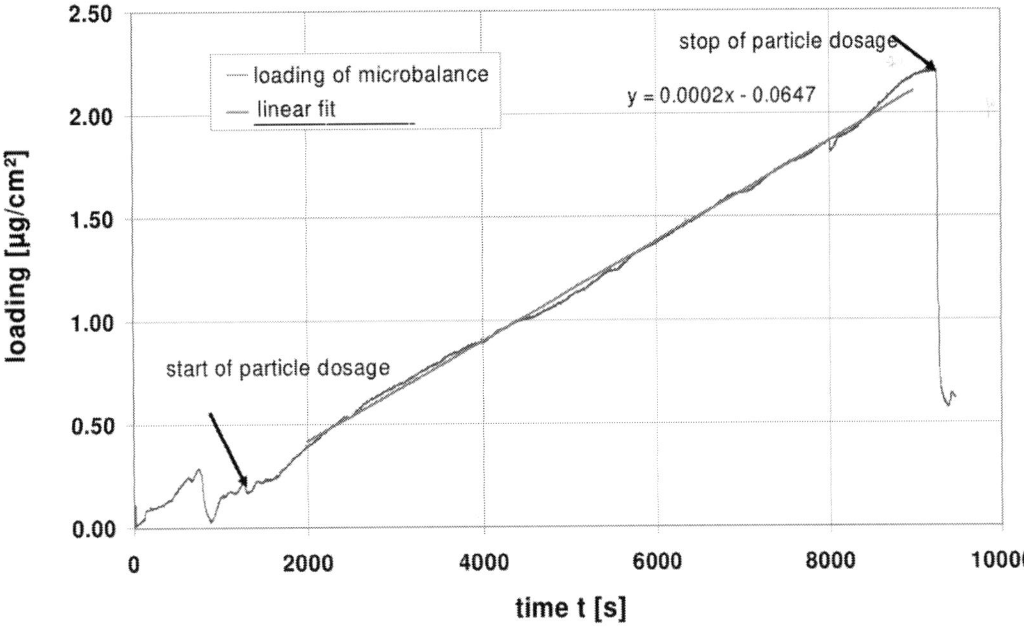

Figure 3. Loading of the quartz crystal microbalance in µg/(cm²*s) over time.

Table 1. Deposited masses determined by QCM and FSD methods in µg/cm² and calibration factor.

Deposited mass by spectroscopy [µg/cm²] Average of 5 values	1,50±18%	1,67±13%
Deposited mass by microbalance [µg/cm²] determined over 2 hours	1,74	2,37
Calibration factor $\frac{m(\text{QCM})}{m(\text{FSD})}$	1,16	1,42

4. Summary

The Karlsruhe exposure system and the lung specific bioassay are an appropriate tool to screen the biological effects of different industrial particles directly in the particle loaded atmosphere. The new online measurement of the deposited mass in dependence of time provides a direct control of the exposure parameters like aerosol concentration.

Acknowledgements
We thank A. Grotz for the calibration experiments.
This work was supported by the BMBF as part of the NanoCare project and by the DFG in NANO-SYNCC, part of DFG SPP1313.

References

[1] Peters A, Wichmann H E, Tuch T, Heinrich J and Heyder J 1997 *American Journal of Respiratory and Crititical Care Medicine* **155** 1376
Peters A, Wichmann H E, Tuch T, Heinrich J and Heyder J 1997 *The Annals of Occupational Hygiene* **41(1)** 19
[2] Mülhopt S, Paur H R, Diabaté S and Krug H F 2007 *Advanced Environmental Monitoring* Kim Y J and Platt U (eds.) Springer Netherlands 402
[3] Paur H R, Mülhopt S, Diabaté S, Weiss C 2008 *Journal für Verbraucherschutz und Lebensmittelsicherheit* **3(3)** 319
[4] Aufderheide M and Mohr U 1999 *Experimental and Toxicologic Pathology* **51** 489
[5] Aufderheide M 2005 *Experimental and Toxicologic Pathology* **57(1)** 213
[6] Diabaté S, Mülhopt S, Paur H-R and Krug H F 2008 *Alternatives To Laboratory Animals* 36 285
[7] Teeguarden J G, Hinderliter P M, Orr G, Brian D, Thrall B D and Pounds J G 2007 *Toxicological Sciences* **95(2)** 300

An overview of the characterization of occupational exposure to nanoaerosols in workplaces

Paola Castellano[1], Riccardo Ferrante[1], Roberta Curini[1] and Silvia Canepari[2]

[1]National Institute for Occupational Safety and Prevention, ISPESL, Via Fontana Candida 1, 00040 Monte Porzio Catone (Rome), Italy

[2]Chemistry Department, University of Rome "La Sapienza", P.le Aldo Moro 5, 00184 Rome, Italy.

E-mail: paola.castellano@ispesl.it

Abstract. Currently, there is a lack of standardized sampling and metric methods that can be applied to measure the level of exposure to nanosized aerosols. Therefore, any attempt to characterize exposure to nanoparticles (NP) in a workplace must involve a multifaceted approach characterized by different sampling and analytical techniques to measure all relevant characteristics of NP exposure. Furthermore, as NP aerosols are always complex mixtures of multiple origins, sampling and analytical methods need to be improved to selectively evaluate the apportionment from specific sources to the final nanomaterials.
An open question at the world's level is how to relate specific toxic effects of NP with one or more among several different parameters (such as particle size, mass, composition, surface area, number concentration, aggregation or agglomeration state, water solubility and surface chemistry).
As the evaluation of occupational exposure to NP in workplaces needs dimensional and chemical characterization, the main problem is the choice of the sampling and dimensional separation techniques. Therefore a convenient approach to allow a satisfactory risk assessment could be the contemporary use of different sampling and measuring techniques for particles with known toxicity in selected workplaces.
Despite the lack of specific NP exposure limit values, exposure metrics, appropriate to nanoaerosols, are discussed in the Technical Report *ISO/TR 27628:2007* with the aim to enable occupational hygienists to characterize and monitor nanoaerosols in workplaces. Moreover, NIOSH has developed the Document *Approaches to Safe Nanotechnology* (intended to be an information exchange with NIOSH) in order to address current and future research needs to understanding the potential risks that nanotechnology may have to workers.

1. Introduction

Does exposure to nanoparticles (NP) pose a health risk to workers?

Many searches have demonstrated various specific health risks due to the absorption of NP, but data are still limited and numerous uncertainties remain. A number of studies are already available from the literature [1] on the toxicological properties of NP and on the risk associated to their exposure, but yet a classification of NP based on their relative level of risk for living species and the environment is far from being reached.

Different hazardous and toxic effects have been documented particularly in animals and have evidenced that some kinds of NP can pass through the various protective barriers of living organisms: the inhaled NP can end up in the bloodstream after passing through all the respiratory or gastrointestinal protective mechanisms and then are distributed in various organs and accumulated at specific sites. Few studies have also evidenced that NP can travel along the olfactory nerves and penetrate directly into the brain and in the various protective barriers of living organisms, inducing morphological changes in cells. Citotoxic, nephrotoxic, genotoxic effects and disturbs on the reproduction have been identified from studies *in vitro* and NP of different composition have been reported to cause granulomas, fibrosis and tumoral reactions in lungs.

However, in general the toxicological data specific to NP are insufficient due to the small number of studies, the short period of exposure, the different composition of NP tested and, in particular, the often unusual exposure routes in workplaces. Therefore additional researches (on absorption, translocation to other tissues or organs, biopersistence, carcinogenicity etc.) must be performed to assess the risk associated with inhalation and dermal exposure of workers to NP.

Aerosol exposure has been generally characterized by the mass concentration of airborne material usually associated with specific size ranges corresponding to different deposition regions in the respiratory system. So, in recognition of the potential importance of particle size, the term "*ultrafine aerosol*" refers to particles smaller than 100 nm in diameter and it is widely used to individuate incidental aerosols where there are potential particle size-dependent health effects.

As NP are developed in research laboratories and produced in industrial plants, the terms "*engineered NP*" and "*engineered nanoaerosols*" are used loosely to describe particles and aerosols associated with engineered nanometre-structured purposely generated particles with nanometre diameters or nanoscale structures [2]. For clarity in this work the term "NP" will be used to describe all aerosol particles with diameter smaller than approximately 100 nm with a potential inhalation health hazard.

Another special feature of NP is that their toxicity is probably linked also to the surface beside the product's mass: NP are so tiny that small quantities could have major toxic effects due to their large surface. But, in addition, we have also to consider all the NP's chemical-physical characteristics, such as the morphologic structure, which can be contribute to cause potential health effects.

Currently, there is a lack of standardized sampling and metric methods that can be applied to measure the level of exposure to nanosized aerosols and there are no specific data related to the measurements of NP concerning occupational exposure levels in workplaces. For this reason it is necessary to establish a structured approach dealing with hazard NP identification, exposure characterization, risk assessment and management with the objective of a standardisation and implementation of a prevention program to follow up and to refine through iterative approach as new data become available in this kind of studies. In particular must be considered which steps have to be done to protect the health of the identified workers exposed to NP, beginning from the characterization of nanoaerosols and from the consequent appropriate measurements of them in workplaces, in order to provide best safety and health practices to protect workers and the environment.

2. Approach to evaluation of the professional exposure to NP and nanoaerosols

The evaluation of occupational exposure to NP in workplaces needs dimensional and chemical characterization. Therefore, any attempt to characterize exposure to NP in a workplace must involve a multifaceted approach, including different sampling and analytical techniques for measuring all relevant characteristics of nanoparticle exposure. Furthermore, as NP aerosols are always complex mixtures of multiple origins, sampling and analytical methods need to be improved to selectively evaluate the portion from specific sources to the final nanomaterials.

An open question at the world's level is how to relate specific toxic effects of NP with one or more among several different parameters, such as particle size and mass, composition, surface area, number concentration, aggregation or agglomeration state, water solubility and surface chemistry. In fact, for NP, a major difference from the usual situations exists, since their potential toxicity can not be exclusively linked only to their mass. The evaluation of occupational exposure to NP in workplaces needs dimensional and chemical characterization. The main problem is linked to choice of the sampling and dimensional separation techniques. The EU Scientific Committee on Emerging and Newly Identified Health Risks [3] (which was requested to answer the question about the appropriateness of existing methodologies to assess the potential risks associated with engineered and adventitious products of nanotechnologies) has produced in 2006 and 2007 two documents suggesting that there is insufficient knowledge and data concerning nanoparticle characterisation, detection and measurement, to allow a satisfactory risk assessment. A convenient approach to improve the knowledge on NP, including all health, physical and chemical aspects, could be the contemporary use of different sampling and measuring techniques for particles with known toxicity in selected workplaces.

Exposure metrics appropriate to nanoaerosols are discussed in ISO/TR 27628:2007 [2], a Technical Report with the principal aim to provide the necessary background information and sampling guidelines to enable occupational hygienists to characterize and monitor nanoaerosol exposure in the workplaces before the specific exposure limits and standards are developed and implemented. Furthermore, to ensure worker protection as nanotechnology develops, the Document Approaches to Safe Nanotechnology [4] represents an information exchange with NIOSH in order to raise potential safety and health concerns from exposure to nanomaterials in the workplaces.

3. Characterization of the professional exposure to NP and nanoaerosols: environmental and analysis methodologies in workplaces

In this work we want to underline how the efforts in this field must be focused in increasing the present knowledge of crucial aspects of NP-related health risks and in providing valuable new tools for evaluating the professional exposure. In particular these objectives will be reached by:

1. developing new sampling procedures, selective for specific classes of particles (hydrophilic or hydrophobic, soluble or insoluble, agglomerated or aggregated);
2. studying aggregation and agglomeration dynamics and their relations with chemical composition and surface chemistry of NP;
3. applying suitable method of statistical analysis for source apportionment from bulk analytical data;
4. developing a standardized protocol for the evaluation of professional exposure;
5. individuating best practices for working with NP following a precautionary approach in order to minimize possible adverse outcomes, until the studies establish threshold limit values for NP or aggregate/agglomerate of NP that will not pose undue risk.

The effects on health of NP are related to their chemical and physical characteristics, whose evaluation requires use of different advanced sampling and analytical tools.

An advanced particle characterization by an integrated approach is needed to understand the chemical-physical nature of particles, which involves chemical, morphological, dimensional and surface chemistry information. Surface extension and reactivity have been demonstrated to be the primary characteristic in determining NP interaction with cell barriers. Conventional surface analysis techniques, like BET (Brunauer, Emmett and Teller method) or epiphaniometer, give information not sufficiently detailed: advanced non-conventional spectroscopic techniques are needed for a full description of surface reactivity. This can help in understanding the strong inter-correlation among the different characteristics which address the interaction of NP among themselves, by accumulation, agglomeration and aggregation dynamics, and with the environment. The comprehension of accumulation, agglomeration and aggregation processes is the main objective, as they have a fundamental role in metrology. By now, there is not information about the possible differences in toxicity between aggregates/agglomerates and single NP, but when NP aggregates to ultrafine particles, the change of the dimensional distribution causes important artefacts in the measurements In fact, all the measurements and sampling devices for NP are based on a dimensional fractionation which is sensible to the overall dimension.

Sufficiently detailed characterization is available only for exposure from production and manipulation of engineered nanomaterials, but NP are ubiquitous in many workplaces as they are produced from particular emission's sources (such as hot processes, combustion, office machinery, cleaning fluids, mechanical processes, flame-based powder generation, material handling etc.) or can come by infiltration of outside air. In these cases, the identification and evaluation of the single sources is of primary importance, as differently produced particles are characterized by different toxicity. The available measurements systems are not able to distinguish the contribution from different sources: furthermore source apportionment may be obtained by the application of proper statistical analysis of bulk chemical analysis data, but the comparability of the results requires a harmonized approach.

The development of appropriate sampling and measurement procedures, selective to particular classes of particles (hydrophilic and hydrophobic, water-soluble or insoluble), would be then a relevant improvement in the risk evaluation.

For this reason it is necessary to develop a multifaceted protocol for the NP measurements and characterization in workplaces, in order to improve the evaluation of the professional exposure, by means of a full characterization of particles from a selected workplace, used to set up the methodological approach [5, 6]. The comprehension of relationship among morphological, chemical and dimensional characteristics and surface reactivity and the optimization of a measurement procedure able to distinguish among some chemical classes of particles will contribute, beside a proper statistical source apportionment tool based on bulk chemical analysis, to the identification of the main parameters which will need to be evaluated in routine NP monitoring activities.

4. Analytical procedure

The study design for developing the analytical procedure must start, after the selection of properly representative workplaces for the detailed study of particle characteristics, from parallel samplings with devices based on different physical principles and with different time resolution for the evaluation of the specific responses towards the same particle mixture. The following step must consist in the collection of dimensionally segregated samples (properly dimensioned considering the analytical techniques performances) and their preparation for bulk, morphological and surface analysis. In particular, for the acquisition of data based on particle dimensional distribution, also considering its time evolution must be used multistage cascade impactors and optical counters and, in order to evaluate NP number concentration, application of condensation particle counters (CPC) must be considered.

For the bulk analysis, used to evaluate organic and inorganic components expected in selected workplaces, different advanced hyphenated analytical techniques such as High Performance Liquid Chromatography (HPLC) interfaced with various detectors as Mass Spectrometer (HPLC-MS) and Diode Array Detector (HPLC-DAD), Gascromatography - Mass Spectrometer (GC-MS) and Inductively Coupled Plasma – Mass Spectrometer (ICP-MS), are available. In addition electron microscopy analysis provides a wide range of qualitative and quantitative analysis methods that are suited for a morphological characterization of individual collected nanoaerosol: characteristic X-ray generated through atomic electron excitation and decay are routinely used in Transmission Electron Microscopy (TEM) and Scanning Electron Microscopy (SEM) analysis to detect elemental components through Energy Dispersive X-ray (EDX). The surface chemical composition and electronic structure are usually studied by application of conventional and advanced spectro/nanoscopic tools as PhotoElectron Emission Microscopy (PEEM), X-ray-photoemission-spectroscopy (XPS) and Ultraviolet Photoelectron Spectroscopy (UPS).

Finally, the evaluation of all the obtained results, performed with particular attention to the inter-relations among measured physical-chemical properties, supported by the source apportionment approach, could contribute in the identification of ultrafine particles emission's sources in workplaces and could help in the individuation of the more representative measurements which are needed for the exposure assessment.

Conclusions

The standardization of methodological approach in collecting and analyzing the composition regarding NP represent a useful tool for the evaluation of professional exposure. An improvement of the guidelines by now available (i.e. NIOSH - Approaches to safe nanotechnologies [4]) can be obtained only through an advanced and full characterization of particles, which allows the knowledge of the fundamental basis of the interaction with cells and environment. This approach has been attempted in the last years only for the exposure to some engineered nanomaterials, but the knowledge is poor about non-intentionally generated NP, despite of the number of exposed workers.

The approach protocol we have presented in this work, suitable in specific workplaces in which different non-intentional sources are expected, can be considered as a preliminary basis for settling specific Italian guidelines concerning evaluation of professional exposure to NP in workplaces.

Furthermore, the identification of some of the main parameters which must be evaluated in routine NP monitoring activities, will furnish innovative opportunities for the development of new efficient Personal Protective Equipment (PPE) (gloves, respiratory protections, clothing etc.) and engineering controls (ventilation systems, fume hoods, etc.) and will allow to optimize all the procedural controls by means of appropriate working practices in workplaces characterized by NP exposures.

In this context it is also important to underline that for establishing threshold limit values for NP, exposure should be considered whether the biological impact of discrete NP depositing within the respiratory system is distinct from, or similar to, the impact of large agglomerates or aggregates of NP containing the same volume of material.

References

[1] IRSST "Health effects of Nanoparticles" R -469 (August 2006)

[2] Norma ISO/TR 27628/2007 "Workplace atmospheres – Ultrafine, nanoparticle and nano-structured aerosols – Inhalation exposure characterization and assessment"

[3] European Commission Scientific Committee on Emerging and Newly Identified Health Risks SCENIHR "Opinion on the appropriateness of existing methodologies to assess the potential risks associated with engineered and adventitious products of nanotechnologies". (2006).

[4] NIOSH "An information Exchange with NIOSH". Approaches to Safe Nanotechnology" (July 2006)

[5] Castellano P, Ferrante R, Canepari S and Astolfi M L, Chemical and dimensional characterization of nanoparticles in the evaluation of professional exposure. EuroNanOSH 2007 "European NanOSH Conference – Nanotechnologies: A Critical Area in Occupational Safety and Health", pag. 61 (Helsinki, 3 – 5 dicembre 2007)

[6] AIDII 14° Convegno di Igiene Industriale "Le Giornate di Corvara" (Corvara – Bolzano, 1 – 4 Aprile 2008), "Strategia di valutazione dell'esposizione professionale a nanoparticelle: approccio metrologico multiparametrico" (Paola Castellano, Anna Rita Proietto, Riccardo Ferrante, Silvia Canepari, Silvia Moretti, Ivano Ammoscato, Roberta Curini), pag. 18 -24

Specific response of SMPS particle counter to CNT

Luana Golanski[1], Stéphanie Bernard[1], Arnaud Guiot[1], Olivier Poncelet[1], Hélène Le Poche[2], François Tardif[1]

[1]CEA-Grenoble, Liten, Laboratory of Tracer Technologies, 17 Avenue des Martyrs, 38054, Cedex 9, Grenoble, France

[2]CEA-Grenoble, Liten, Laboratory of Hybrid Components, 17 Avenue des Martyrs, 38054, Cedex 9, Grenoble, France

E-mail: luana.golanski@cea.fr

Abstract. The aim of this work is to evaluate the ability of conventional particle detection like SMPS tools calibrated for spherical particles to measure CNTs which present both a nanometre and a micrometre size. Using such conventional measurement tools for the detection of CNTs would be of particular interest. The SMPS gives numbers in terms of concentration and electrical mobility diameter. A theoretical approach is first explored in order to predict the spherical equivalent diameter for CNTs as seen by a SMPS. Experimental measurements were performed in order to validate the theoretical estimations.

1. Introduction

The difficulty in characterizing CNTs from a SMPS (Scanning Mobility Particle Sizer) response is that both the diameter and the CNT length must be determined. SMPS is composed of two parts: the Differential Mobility Analyzer (DMA) + and the Condensation Particle Counter (CPC). DMA classifies particles and CPC counts the particles with a given size. The SMPS response is in direct relation with the CNTs (Carbon Nanotubes) transport thought the DMA. The latter is influenced by CNTs peculiar shape. Only few studies have been carried out about the effect of asphericity on the equivalent spherical mobility diameter. Some theories are developed for the transport of aggregates or aspherical particles in the DMA [1]. Zelenyk *et al.* [1] demonstrates that in intense electric fields, high aspect ratio particles tend to orient parallel to the electric field. At lower electric fields, particles are in random orientation [1]. Some approaches of the response of SMPS to carbon nanotubes have been also identified. CNTs of length 52 nm and diameter 6.5 nm were measured during the process at a spherical equivalent diameter of 23 nm [2]. Unrau *et al.* [3] produced Single Wall CNTs in diffusion flames and used a SMPS for online measurements of the synthesized particles. They were able to evaluate the effectiveness of the synthesis by observing a bimodal size distribution corresponding to catalyst particles and nanotubes. SWCNT of diameter 2 nm and length 1 μm were observed at an equivalent sphere diameter of 48 nm. Some studies about measurement of aspherical particles by SMPS show that this relation is confirmed experimentally for particles with small length and diameter [4, 5]. Kim *et al.* [6] calculated the behaviour of singly charged cylinders undergoing Brownian motion in an electric field considering their polarization to explain their travelling trough the DMA. Cylinders with small aspects ratios tend to rotate freely; they are undergoing the Brownian motion. When the aspect ratio increases and over a critical value of β, the cylinders tend to align with the electric field, β is the ratio of the cylinder length to the cylinder diameter.

The aim of this work is to evaluate the ability of conventional particle detection like SMPS tools calibrated for spherical particles to measure CNTs which present both a nanometre and a micrometre size. A theoretical approach is explored in order to predict the spherical equivalent diameter given by the SMPS for CNTs with different aspect ratio. Experimental measurements are performed in order to validate the theoretical approach. The detection limits of the SMPS to CNTs are also given. The detection limits of the apparatus in terms of CNTs dimensions are specific to the material used for SMPS (geometry, air flow rate).

2. Theoretical approach

A simplified version of calculations given by Kim *et al.* [6] is proposed in this paper. Prediction of the CNTs length can be carried out by two explicit expressions of the relationship between the nanotube length and the sphere equivalent diameter. Cylinders with $\beta<\beta_1$ and high aspect ratios have a random orientation when travelling through the DMA. β is the ratio of the cylinder length to the cylinder diameter. The effective length of the particle can be calculated by:

$$L = \frac{6Kd_{sph}}{d_{cyl}C(d_{sph})}$$

with d_{sph} the sphere equivalent diameter measured by the SMPS, d_{cyl} the diameter of the cylindrical particle and $C(d_{sph})$ the Cunningham slip correction in air equal to :

$$C(d_{sph}) = 1 + 2.492\frac{a}{d_{sph}} + 0.84\exp(-0.43d_{sph}/a)$$

with a the mean free path (dependant on the temperature and pressure) and K ~ 0.512.

Cylinders with $\beta>\beta_2$ and high aspect ratios are aligned with the electric field.

$$L = \frac{6ad_{sph}}{0.9d_{cyl}C(d_{sph})}$$

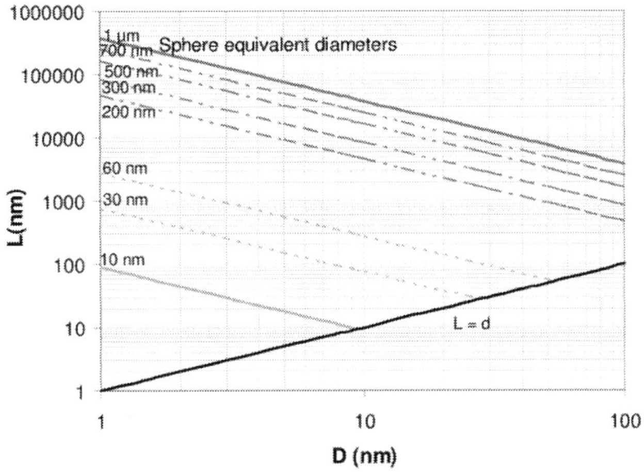

Figure 1. Estimation of equivalent sphere diameter for several nanotube lengths and diameters.

The estimation of the equivalent spherical diameter d_{sph} was performed for several nanotube lengths and diameters (Figure 1). The curves are plotted for aerosol in air, for a maximum d_{sph} of 1 µm corresponding to the 'long' DMA. In black is represented the limit of prolate nanotubes (L = d). In light respectively dark gray are represented the detection limits of the SMPS (d_{sph} = 10 nm and 1 µm). The transition from random orientation to aligned particles ($\beta_1<\beta$ and $\beta<\beta_2$) depends on the geometry of the SMPS used. For high sphere equivalent diameters (dark gray curves 200 nm – 1 µm) nanotubes are aligned with the electric field. For small sphere equivalent diameters (light gray curves 10 - 60 nm) nanotubes are randomly oriented.

3. Experimental procedure

3.1. Aerosol generation from a liquid dispersion

SMPS measurements were carried out on CNTs dispersed in air. Nanotubes were generated from CNTs dispersion in solutions and generated by atomisation. They present a diameter of 1.4 nm and a length of several microns. An atomizer model TSI 3076 is used to generate a spray of droplets containing CNTs. The aerosol was dried in a high flow rate of compressed air before entering the measuring tools. First, stable dispersions were obtained using surfactant (as shown in Figure 2). A surfactant was used to enhance electrostatic repulsion leading to a better dispersion stability. Two surfactants were tested: CTAB and Triton X-100. An ultrasonic probe of 200 W was used to disperse and break the CNTs.

Figure 2. Stable dispersions of CNT's.

3.2 Aerosol generation from CNTs deposited on a wafer – Dry way

The CNT aerosol was also generated by a dry process. The CNTs were grown by PECVD on a Si wafer as shown on the picture taken by SEM (Figure 3). They have a "bamboo-like structure". The fabrication process allows to vary the CNTs diameter and length. The dimensions of the CNTs are given in Table 1.

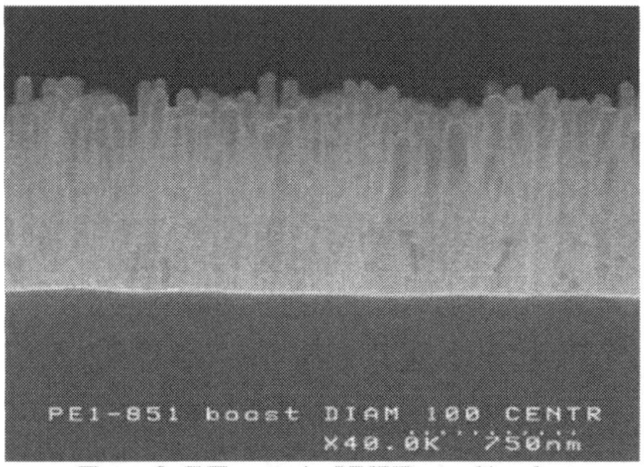

Figure 3. CNT grown by PECVD on a Si wafer.

Table 1. Characteristics of the tested CNTs.

Sample	Length (μm)	Diameter(nm)
1	0.3	90
2	0.9	90
3	1.2	100

In order to generate the aerosol, a rotating brush placed on a tool and turning at a speed of 10 000 tr/min was used to smoothly scratch the wafer in a glove box (Figure 4). The experiment was carried out in a glove box in order to get a low background level of nanoparticles: a HEPA filter maintained a clean air in the box.

The SMPS aspiration was directly positioned over the wafer to collect most of the released CNTs.

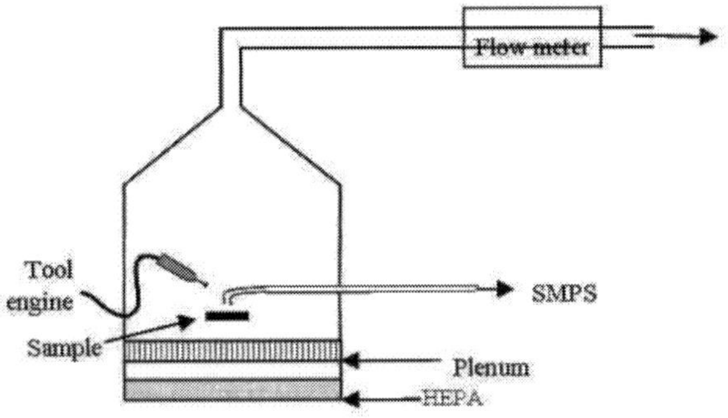

Figure 4. Schematic of the dry way aerolisation method.

3.3 Description of the detection instrument

The SMPS determines the particles size distribution according to their electrical mobility. The DMA is composed of a cylinder with a negatively charged rod at the centre. The particles move in a laminar flow toward the central rod if positively charged at a rate determined by their electrical mobility. The output flow contains a monodispersed aerosol which goes in a CPC. A DMA (model 5.5-900, Grimm) and a Condensation Particle Counter (CPC model 5.403, Grimm) is used. The concentration at one size is determined by counting the particle after their artificial grow in a buthanol cloud. A ^{85}Kr radioactive source is used in order to neutralize the nanoaerosol before introducing into the DMA. A long DMA measures particles between 10 nm and 1 μm. The characteristics are listed here after: sheath air flow = 3 L/min, dimensions of the electrode: length = 35 cm, internal diameter 10 mm, external diameter 20 mm.

4. Results and discussion

4.1. Detection of CNTs aerosolized from stable dispersions

Stable dispersions containing NTC were aerosolized in the air. First measurements of CTAB solutions with and without CNTs were carried out with the SMPS (Figure 5). No significant differences can be noted between the 2 curves. The surfactant seems to create an important parasitic signal: we can hardly draw conclusion about the detection zone of the CNTs. Results shows that the use of surfactant to stabilize the dispersion, can induce a very large amount of background noise, masking the CNT contribution. We did not reach to eliminate this background level even with low surfactant concentration. This is why a dry procedure was tested as well.

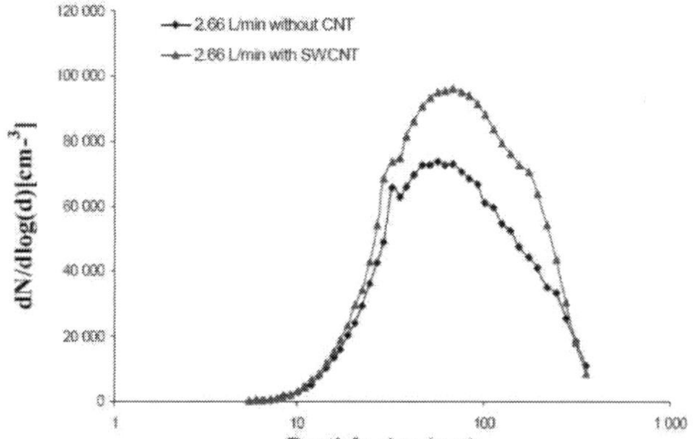

Figures 5. SMPS detection for solutions with and without CNTs.

4.2. Detection of CNTs aerosolized in air

Tests were performed on CNTs with different aspect ratio in order to validate the theoretical approach. First measurements were carried out on Sample 1, Sample 2, Sample 3 (Table 1). The background, including the rotation of the brush on the wafer, is less than 20 part/cc which is negligible. An example is given in the Figure 6: CNTs were detected with an equivalent sphere diameter centred around 200 nm. A second peak is also centred around 100 nm as shown on the second measure. This is supposed to be due to CNTs broken during the scratching or to the bad growth of some CNTs on the wafer.

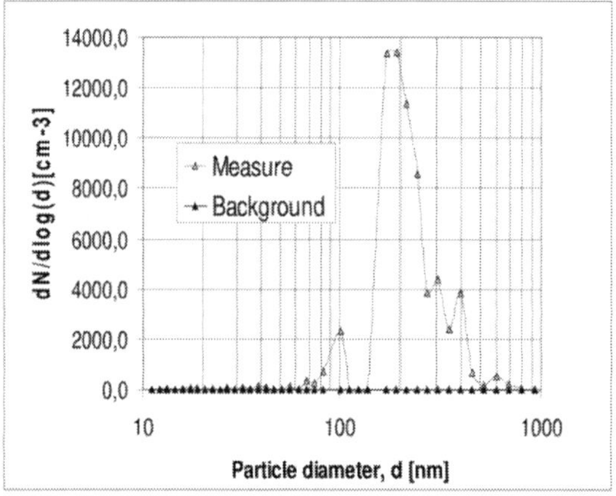

Figure 6. SMPS response to CNTs of length 300 nm and diameter 90 nm.

Figure 7 shows the estimation of the spherical equivalent diameter for the SMPS for CNTs. The detection zones for the aerosolized CNTs are represented also on the Figure 7. In order to define the detection zones: the width at half weight for the Gaussian peak detected by the SMPS was taken. The length variation induced by the fabrication process, is estimated from the SEM image taken on the wafer. The first parts of the curves are dotted because they do not agree with the assumption of high aspect ratio. The transition from the regime ''random orientation'' to ''aligned particles'' has been estimated. Experimental results fit quite well with the theoretical approach. Generally speaking, the measured equivalent sphere diameters were situated between the 'randomly oriented' curve and the 'aligned' curve. For CNTs (L = 300 nm, d = 90 nm), the prediction slightly underestimates the equivalent sphere diameter. This deviation may be due to the uncertainty of the transition zone from the 'randomly oriented' curve and the 'aligned' curve used. For longer CNTs (L = 900 nm, d = 90 nm, L = 1200 nm, d = 100 nm), the prediction is slightly overestimating the equivalent sphere diameter.

According to Kim *et al.*, this may be caused by the bent structure of the CNT which modifies the drag force.

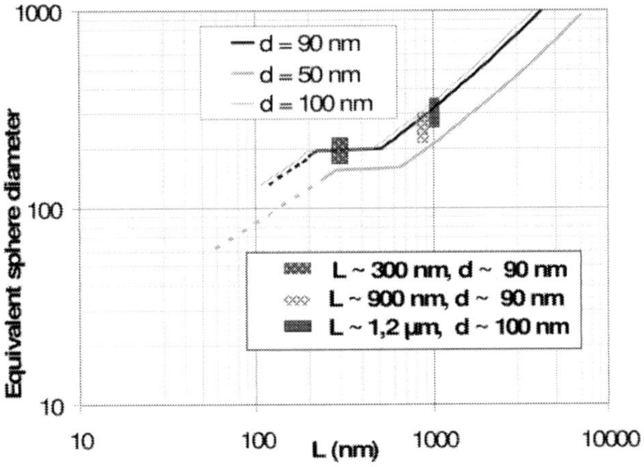

Figure 7. Estimation of the equivalent spherical diameter for Grimm SMPS and experimental results.

5. Conclusions

Detection of aerosolized CNTs from dispersed liquid appears masked by the surfactant. CNTs generation by a dry way allows to detect CNTs by the SMPS. The SMPS is able to detect, count and measure CNTs, however the equivalent sphere diameter response of the apparatus must be considered very carefully. Considering unitary and straight CNTs, a calculation of equivalent-spherical diameter (NTC length and diameter) is proposed for a standard SMPS (10 nm -1 µm). First experimental results fit quite well with this theoretical approach.

Acknowledgments
These results have been obtained in the frame of FP6 European program Nanosafe 2.

References

[1] Zelenyk A and Imre D 2007 *Aerosol Sci. Technol.* **41** 112
[2] Moisala A, Nasibulin A G, Shandakov S D, Jiang H and Kauppinen E I 2005 *Carbon* **43** 2066
[3] Unrau C J, Axelbaum R L, Biswas P and Fraundorf P 2007 *Proceed. Combus. Inst.* **31** 1865
[4] Kim S H and Zachariah M R 2005 *Nanotechnol.* **16** 2149
[5] Ku B B and Maynard A D 2005 *Aerosol Sci.* **36** 1108
[6] Kim S H, Mulholland G W and Zachariah M R 2007.*Aerosol Sci. Technol.* **38** 823

A size selective nanoparticle collection device based on diffusion and thermophoresis

Jussi Lyyränen[1], **Ulrika Backman**[1], **Unto Tapper**[1], **Ari Auvinen**[1] **and Jorma Jokiniemi**[1,2]

[1]VTT Technical Research Centre of Finland, Fine Particles, P.O. Box 1000, FI-02044 VTT, Finland

[2]University of Kuopio, Department of Environmental Sciences, Fine Particle and Aerosol Technology Laboratory, P.O. Box 1627, FI-70211 Kuopio, Finland

E-mail: Jussi.Lyyranen@vtt.fi

Abstract. There is a growing industry fabricating products that are based on nanoparticles (particle diameter $d_p \leq 100$ nm). The production of these particles requires detection, classification and characterisation of even smaller particles because of, *e.g.* preventing unwanted particle emissions from the processes and health issues. Monitoring of the processes is needed on one hand for product quality determinations, on the other hand to ensure safe and particle-free working conditions. Thus simple, fast and reliable measurement devices are needed for particle characterisation.

1. Introduction

The ideal "particle monitor" would be based on on-line detection, and would measure particle number (*e.g.* CPC) (and mass) size distribution (*e.g.* DMA connected to a CPC and controlled by SMPS software) and would also analyse the morphology and chemical composition (elemental, for a start) of the measured particles. This is important because it is known [1, 2] that there is no single particle property parameter that would describe, *e.g.* the health effects or particle exposure adequately. In addition, if the chemical composition of the particles could be determined at the same time as the physical particle parameters are measured this would indicate the source of the measured particles, *e.g.* do the particles originate from the process itself or is the source somewhere else. Another requirement for the "particle monitor" is that it should be small in size to make the transportation easy in industrial environments. However, research is still needed to achieve these features, although some promising techniques are already available, *e.g.* aerosol mass spectrometry (AMS) [3] and laser-induced breakdown spectroscopy (LIBS) [4]. The problem with these techniques is, however, that they are currently quite bulky and/or not so easy to operate.

In this study the target was to develop a particle sampler that would be simple to use, durable, preferably small in size, and should be suitable for use in industrial environments. Because there already are a number of particle measurement devices based on particle number and mass concentration and size distribution measurements, there is no need to develop a device based on similar techniques. Therefore, a different approach by sampling particles on a grid (d_p=3 mm) used for electron microscopy (EM) samples was developed. The grid is placed on top of a tube-like resistor heater. The collection of particles is based on diffusion of the nanoparticles and thermophoretic repulsion (antithermophoresis; temperature difference between the aerosol flow and the grid) of the large particles (d_p>200 nm). Samplers based on particle collection on electron microscopy grids have been developed before but these samplers are based on different collection techniques, *e.g.* electrostatic precipitation with a point-to-plate ESP [5, 6], and do not classify particles in any way.

2. Experimental methods

2.1. The heated electron microscopy (EM) grid sampler

The collection of the particles by the heated EM grid sampler is based on diffusion of the nanoparticles and thermophoretic repulsion [7, 8]; temperature difference between the aerosol flow and the grid) of the large particles (d_p>200 nm). The size of the heated EM grid sampler is a compact, pocket-size model approximately 80 mm by length and 20 mm by diameter with ½-inch Swagelok fittings. However, these dimensions do not include the resistor heater controller needed for accurate temperature control of the sampler. The temperature of the heated EM grid sampler is adjusted via a heating controller equipped with a CAL 3300 PID regulator. The heating of the EM grid is fast, typically on the order of minutes or less.

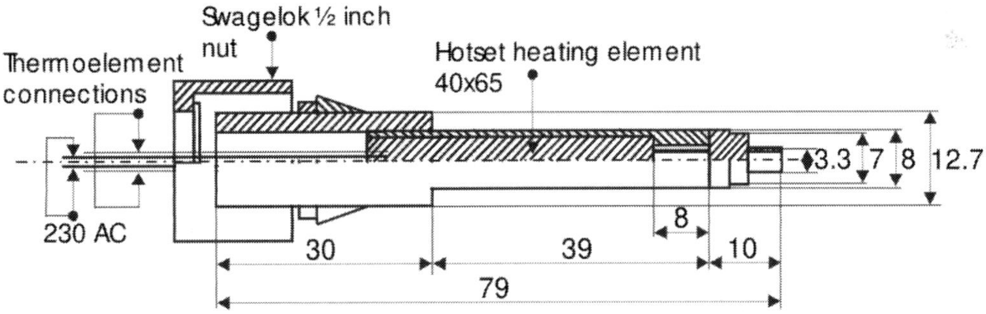

Figure 1. Schematics of the compact heated EM grid sampler.

The heating resistor element (P=100 W, $l_{element}$=40 mm, $d_{element}$=6.5 mm) in the heated EM grid sampler is a commercial, cylindrical shaped HHP-type made by Hotset GmbH. The shell of the resistor element is made of stainless steel. The resistor element is inserted inside a brass tube, whose outside diameter is $d_{tube1,\ out}$=8 mm and length l_{tube1}=40 mm. This 8 mm tube is inserted inside another tube of length l_{tube2}=30 mm and outside diameter of $d_{tube2,\ out}$=12.5 mm. This tube section has a ½-inch Swagelok tube fitting. The temperature of the heating resistor is controlled by a K-type thermoelement inserted inside the heating element. The head of the heated EM grid sampler (l_{head}=10 mm, d_{head}=3.3-7 mm), where the grid is mounted, is screwed into the brass tube ($d_{tube1,out}$=8 mm, l_{tube1}=40 mm) surrounding the heating resistor element making the thermal contact between the resistor element and the head (Figure 1).

a)  b) 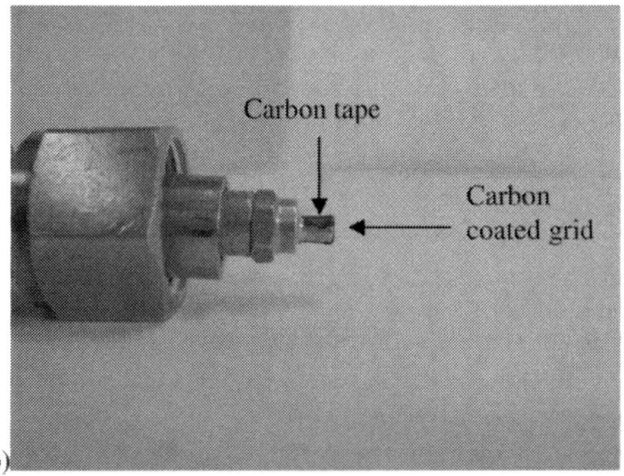

Figure 2. The heated EM grid sampler. a) The collecting head of the sampler and the brass hat used for mounting the grid. b) The collecting head of the sampler with carbon tape mounting and the grid.

Mounting of the EM grid on top of the heated sampler may be done in two ways: either with the help of a brass hat (Figure 2a) or with carbon tape (Figure 2b). In the case of the carbon tape the EM grid is glued directly on top of the collecting head of the EM grid sampler with the tape (Figure 2b). By using carbon tape instead of the brass hat the passing flow may sweep directly the surface of the grid. Because the dimensions controlling the balance between diffusion and anti-thermophoresis may be small, this may have an effect on the collection efficiency. In the case of the brass hat the grid is located in a pit of approx. 1 mm deep (the thickness of the brass hat; Figure 2a).

Sample collection with the heated EM grid sampler is simple. The grid is mounted on the EM grid sampler, and the sampler is mounted on the collection location, in this case a 1/2-inch T-fitting on the sampling tubing. Parallel or perpendicular mounting of the EM grid sampler in relation to the aerosol flow may be used. The heater controller is set to the desired temperature to achieve temperature difference between the heated EM grid sampler and the passing flow. After reaching the preset temperature the EM grid sampler is ready for sampling. The warm-up usually takes on the order of minutes. The sampling is initiated by opening the on-off valve upstream the sampler and switching on the pump.

2.2. Aspiration sampler for comparison measurements

To compare the performance of the heated EM grid sampler to another individual particle sampler an aspiration EM grid sampler was built and designed. The sampler consists of a commercial preimpactor (orifice diameter 0.0457 mm) normally inserted in front of a TSI electrostatic classifier model 3080, an EM grid mounting head soldered with silver into 6 mm tubing and Swagelok fittings (Figure 3). By regulating the flow through the aspiration sampler to approximately 0.3 Nlpm with a critical orifice (CO) the aerodynamic cut size of the preimpactor is approximately at $d_{50,ae}$=600 nm.

In the aspiration sampler the sample flow enters the preimpactor, where the particles whose aerodynamic size is approximately $d_{p,ae}$>600 nm are deposited on the impactor collection plate because of their inertia. The smaller particles are able follow the flow and are collected on the EM grid after the preimpactor. The grid is installed on the mounting head with the help of a hollow screw, and a copper seal is mounted between the screw and the EM grid.

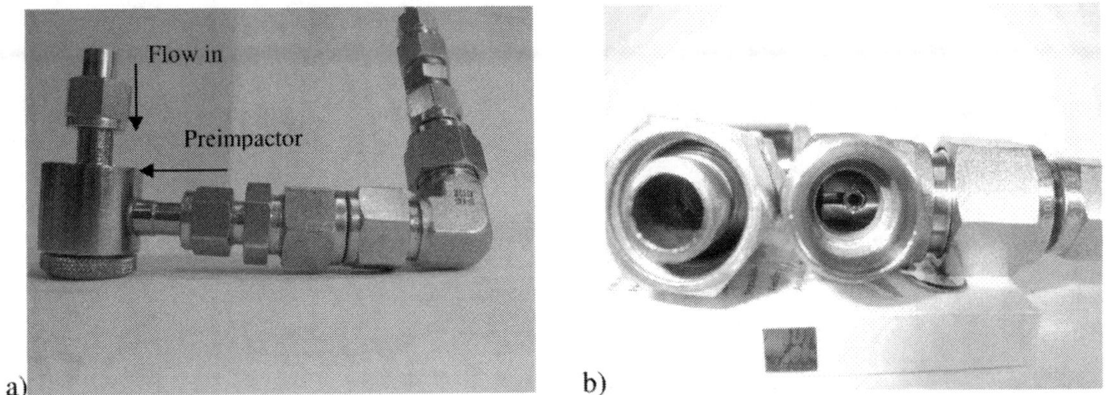

Figure 3. The aspiration sampler equipped with a preimpactor normally inserted in front of a TSI electrostatic classifier model 3080. a) The sampler with fittings. b) The sampler installed inside a ½-inch Swagelok 90 degree elbow, and the head of the aspiration sampler, where the grid is mounted (under the hollow screw).

2.3. Measurement set-up

The experimental measurements were carried out using TiO_2 particles produced by thermal decomposition of TTIP (titanium tetraisopropoxide). The nanoparticle generator used in the experiments is described in detail elsewhere [9, 10].

The test matrix to study the performance of the revised heated EM grid sampler consisted of two different mounting set-ups of the EM grid sampler in relation to the passing flow (perpendicular and parallel), two types of grid mounting (carbon tape and brass hat), different collection times (from 1 min to 6 min) and temperature differences (dT=0, 10, 30, 70°C) between the EM grid sampler and the aerosol flow. In addition, the performance of the heated EM grid sampler was compared to an aspiration sampler (Figure 3) equipped with a preimpactor. The comparison measurements were done by measuring simultaneously with both samplers after the two-stage dilution of the measurement set-up (Figure 4). The temperature difference between the EM grid sampler and the aerosol flow was set to dT=10°C that ensured sound collection efficiency. During all the measurements the total particle number concentration was measured with a TSI CPC model 3022. Typical collection times with the aspiration sampler with the preimpactor were 5 and 10 s, and for the heated EM grid sampler 30 and 45 min at the total particle number concentration of $N_{tot} \approx 6.0\text{-}6.5 \cdot 10^9$ #/cm^3 with dilution correction (DR=1001) and $N_{tot} \approx 6.0\text{-}6.5 \cdot 10^6$ #/cm^3 after two-stage dilution, i.e. at the measurement location.

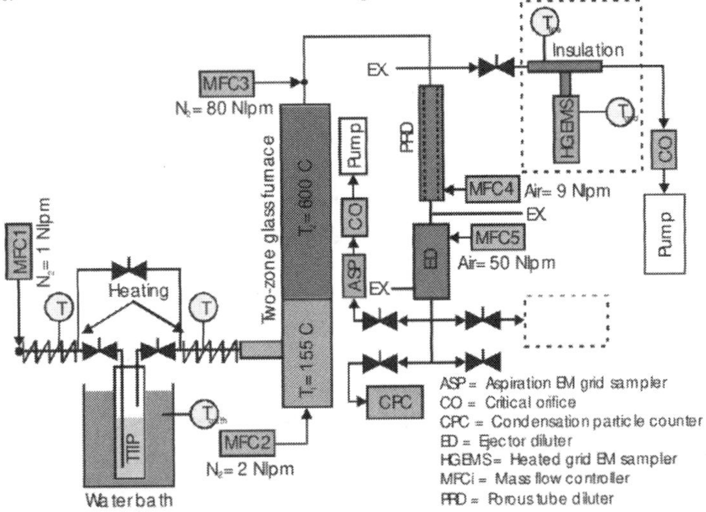

Figure 4. Schematics of the measurement set-up. The empty rectangle drawn with a broken line indicates the location of the heated EM grid sampler during the comparison measurements with the aspiration sampler.

3. Results

3.1. Modelling

Deposition velocities as a function of particle size were calculated for the heated EM grid sampler placed perpendicular to the passing aerosol flow. The model calculations were based on a previous study [11]. The results of the calculations for passing flow velocities of 1 (Figure 5a) and 10 m/s (Figure 5b) indicated that the larger the temperature difference between the collecting grid and the passing flow , the stronger is the dependence of the collection efficiency on the particle size. However, reasonable nanoparticle collection efficiency and size classification (up to approximately 100 nm) is achieved already at temperature differences of dT=10°C.

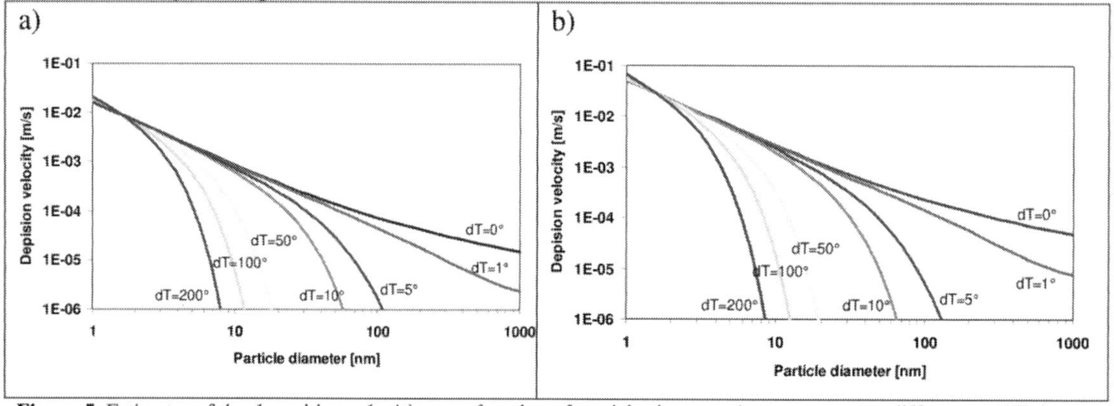

Figure 5. Estimates of the deposition velocities as a function of particle size at various temperature differences (dT) for the perpendicular flow towards the heated EM grid sampler. a) flow velocity 1 m/s b) flow velocity 10 m/s.

3.2. Experimental

3.2.1. Heated EM grid sampler

The modelling calculations indicated that as low temperature difference as dT=10°C between the passing flow and the heated electron microscopy (EM) grid sampler causes a significant, more than three orders of magnitude, difference in the deposition velocity compared to the unheated grid case, dT=0°C. However, the modelling results apply for the case where the flow was perpendicular to the heated EM grid sampler. In the following experimental measurements the heated EM grid sampler was also placed parallel to the passing flow to find out if there are any differences in collection efficiency and collected particles. Currently the flow velocity past (or towards in case of a perpendicular flow) the heated EM grid sampler is v=8.0 m/s.

The first experiments were carried out at temperature differences of dT=0, 10, 30, 70°C for perpendicular flow towards the heated EM grid sampler with a brass hat and a carbon tape mounting using collection times of 1 and 3 min. Based on the experimental and previous modelling results, the temperature differences in the following experiments were set to dT=0, 10, 30°C. The position of the heated EM grid sampler in relation to the aerosol flow was also changed from perpendicular to parallel based on the experiments with hat and tape at dT=10, 30°C. At these temperature differences with perpendicular flow with hat mounting mainly large, almost spherical particles of d_p=300-700 nm in diameter are observed. In the case of carbon tape mounting similar spherical particles are also seen together with chain-like aggregates. It seems that especially in the case of the hat mounting the collection of the particles is dominated by large particle impaction, and the temperature difference hinders the collection of the smaller particles (d_p<200 nm). The structure of these spherical particles (d_p=300-700 nm) is presented in Figure 6 for the case of perpendicular aerosol flow towards the collecting EM grid. Particle image velocimetry (PIV) was also carried out to found out the flow field around the heated EM grid sampler in the case of the perpendicular flow (Figure 7). Twisting of the flow vectors/flow field is observed at the edges of the sampler thus indicating possible particle impaction points.

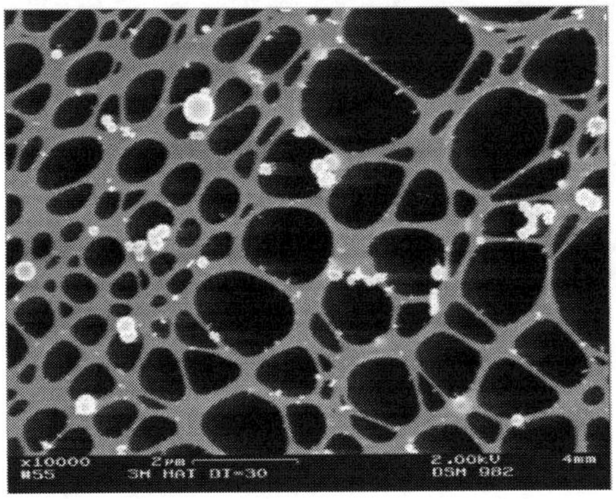

Figure 6. SEM image of TiO2 particles collected with the heated EM grid sampler, optimal collection time. Grid mounted on the heated resistor with a brass hat, aerosol flow perpendicular to the surface of the grid, sampling time 3 min, temperature difference dT=30°C.

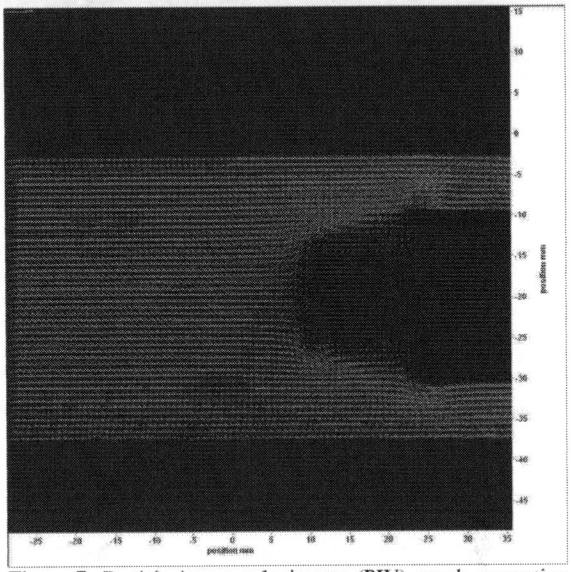

Figure 7. Particle image velocimetry (PIV) graph presenting the flow field (average velocity vectors) around the EM grid sampler mounted perpendicular to the flow.

For the carbon tape mounting this is not so evident, because the mounting of the carbon tape was not successful (one spot mount) thus making the thermal conduction from the heater resistor to the grid worse than in the case of the hat. This was fixed later on by applying two spot, opposite to each other carbon tape mounting (Figure 2). No significant difference in the size of the particles was found between dT=10°C and dT=30°C cases. Therefore, the further experiments with the heated EM grid sampler either with brass hat or carbon tape mounting were carried out at the temperature difference of dT=0 (reference), 10°C with parallel aerosol flow sweeping the surface of the grid.

The results of the collected particles for the heated EM grid sampler with carbon tape mounting and parallel flow over the grid for dT=0, 10°C (Figures 8 and 9; sampler placed before two-stage dilution, see Figure 4) indicate the effect of the temperature difference between the grid and the aerosol flow. For the dT=0°C case, collection time 3 min (Figure 8a; no thermophoretic repulsion), the total particle number concentration during measurement after two-stage dilution is $6.0 \cdot 10^9$ #/cm^3 ($6.0 \cdot 10^6$ #/cm^3 at the measurement location; Figure 11a), and the chain-like aggregates are larger than in the case of dT=10°C, total particle number concentration $6.3 \cdot 10^9$ #/cm^3 ($6.3 \cdot 10^6$ #/cm^3 at the measurement location), collection time 2 min (Figure 8b). In addition, in dT=0°C case there are large, spherical particles of 300-700 nm in diameter that are absent in the dT=10°C case. Longer collection times (Figure 9), in this case 6 min for dT=0, 10°C, lead to particle overlapping on the grid (on-top deposition).

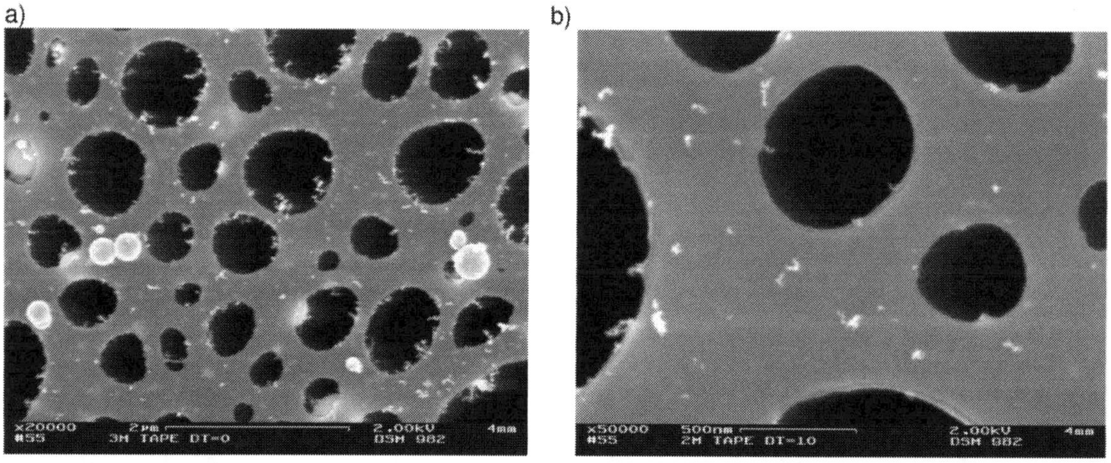

Figure 8. SEM images of TiO_2 particles collected with the heated EM grid sampler, optimal collection time. Grid mounted on the heated resistor with carbon tape, aerosol flow sweeping parallel to the surface of the grid. a) Sampling time 3 min, temperature difference dT=0°C, *i.e.* no thermophoretic repulsion, b) Sampling time 2 min, temperature difference dT=10°C.

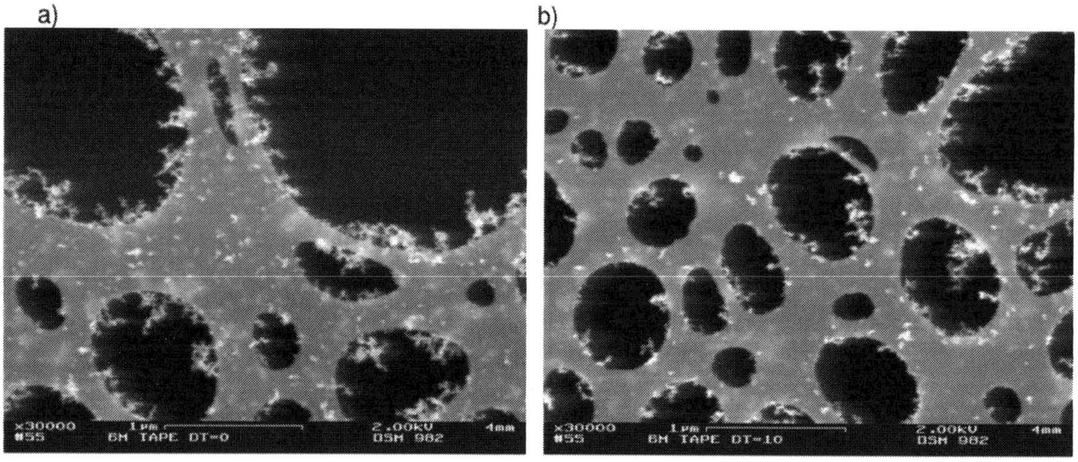

Figure 9. SEM images of TiO_2 particles collected with the heated EM grid sampler, too long collection time. Grid mounted on the heated resistor with carbon tape, aerosol flow sweeping parallel to the surface of the grid. a) Sampling time 6 min, temperature difference dT=0°C, *i.e.* no thermophoretic repulsion, b) Sampling time 6 min, temperature difference dT=10°C.

At identical experimental conditions as above but using the brass hat for mounting the grid almost similar effects on the collected particles are observed. For the dT=0°C case, collection time 2 min (Figure 10a), the total particle number concentration is $5.4 \cdot 10^9$ #/cm^3 ($5.4 \cdot 10^6$ #/cm^3 at the measurement location; Figure 11b), and similar chain-like aggregates as in the case of the carbon tape above are seen. However, the large, spherical particles of 300-700 nm in diameter are absent in the brass hat case. For dT=10°C case for brass hat mounting (Figure 10b), collection time also 2 min and total particle number concentration $5.5 \cdot 10^9$ #/cm^3 ($5.5 \cdot 10^6$ #/cm^3 at the measurement location; Figure 11b), the collected particles are almost identical in size compared to the carbon tape mounting case (Figure 8b), smaller than 200 nm (experimentally determined from the SEM micrographs) in diameter. Also the collected number of the particles seemed to be almost the same for the carbon and brass hat mounting. Thus it seems that under these collection conditions (dT=10°C, $N_{tot} \approx 5.5 \cdot 10^6$ #/cm^3 after two-stage dilution) the two different grid mounting types do not cause any significant difference on the collected particles.

a) b)

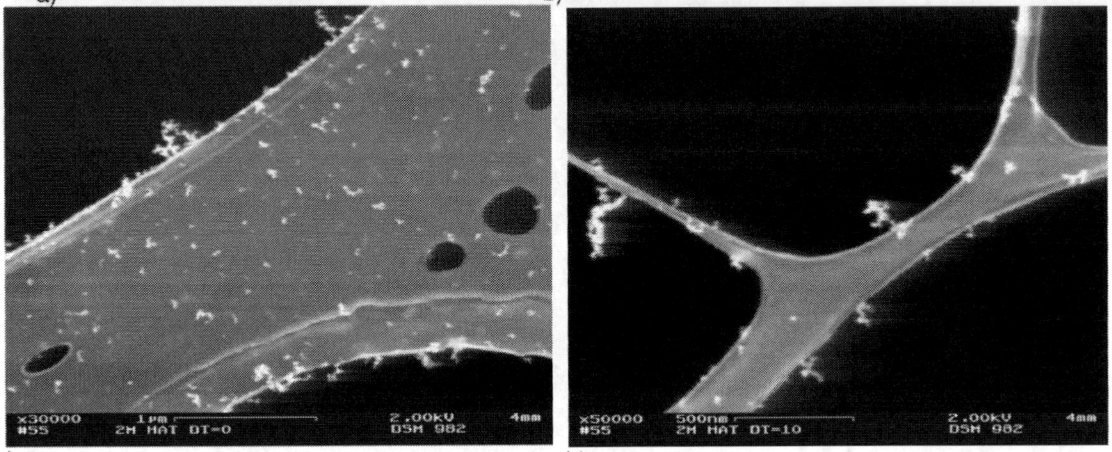

a) b)

Figure 10. SEM images of TiO_2 particles collected with the heated EM grid sampler, optimal collection time. Grid mounted on the heated resistor with a brass hat, aerosol flow sweeping parallel to the surface of the grid (compare to Figure 8).
a) Sampling time 2 min, temperature difference dT=0°C, i.e. no thermophoretic repulsion,
b) Sampling time 2 min, temperature difference dT=10°C.

a) b)

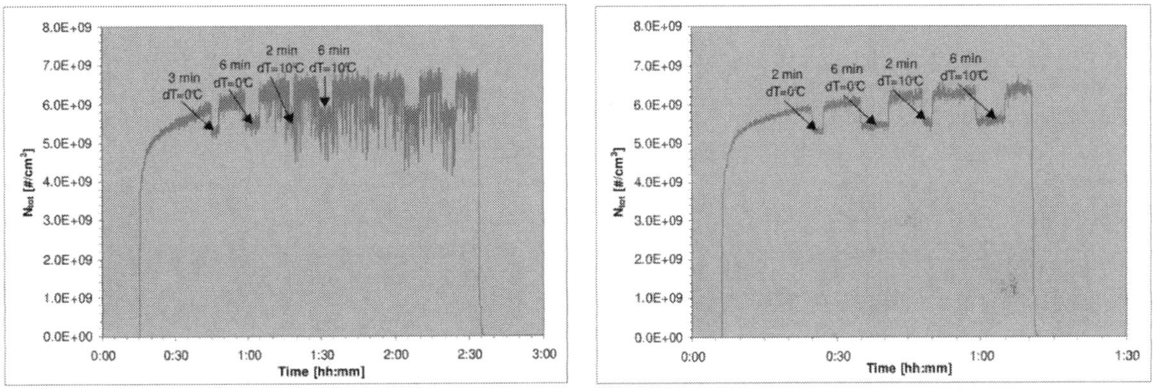

Figure 11. Total TiO_2 particle number concentration measured after the dilution, dilution ratio DR=1001 (Figure 4), aerosol flow sweeping parallel to surface of the grid, collection by heated EM grid sampler.
a) Grid mounted with carbon tape, (SEM Figures 6, 7). b) Grid mounted with a brass hat (SEM Figure 8).

3.2.2. Comparison between heated EM grid and aspiration sampler

To compare the performance of the heated EM grid sampler to other individual particle samplers an aspiration sampler (Figure 3) equipped with a preimpactor (orifice diameter 0.0457 mm) normally inserted in front of a TSI electrostatic classifier model 3080 was applied. Both samplers were operated simultaneously at identical locations of the sampling system (Figure 4). Because of the very different collection principles of these samplers, the collection times are very unequal. For the heated EM grid sampler at dT=10°C and collection time of 45 min after two-stage dilution (Figure 12) the collected particles are almost identical to the dT=10°C case with carbon tape mount before the two-stage dilution (Figure 8b). However, overlapping of the particles (longer chain-like aggregates than they actually are) in the case of the 45 min collection has occurred. Thus a shorter collection time should have been applied. The total particle number concentration during the measurement is approximately $6.9 \cdot 10^9$ #/cm^3 ($6.9 \cdot 10^6$ #/cm^3 at the measurement location; Figure 14).

Journal of Physics: Conference Series **170** (2009) 012011 doi:10.1088/1742-6596/170/1/012011

a) b)

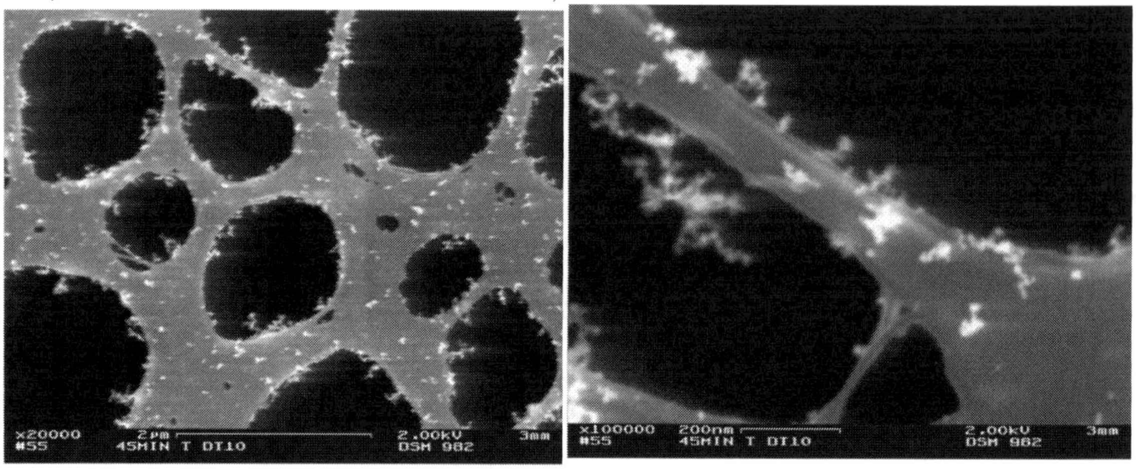

Figure 12. SEM images of TiO_2 particles collected with the heated EM grid sampler, too long collection time causing partial particle overlapping. Grid mounted on the heated resistor with carbon tape, aerosol flow sweeping parallel to the surface of the grid, sampling time 45 min after two-stage dilution (Figure 4), temperature difference dT=10°C. a) Overall view of the particles where overlapping is observed b) Detailed image of the individual particles, whose sizes are typically under d_p=200 nm when no mutual overlapping is present.

For the aspiration EM grid sampler the collection times are usually very short. No overlapping of the collected particles (particles collected on top of each other) is observed (Figure 13a), and individual particle structure is still visible (Figure 13b). The particles are generally smaller than 200 nm in diameter. At the total particle number concentration of approximately $6.6 \cdot 10^9$ #/cm^3 ($6.6 \cdot 10^6$ #/cm^3 at the measurement location; Figure 14) after two-stage dilution the optimal collection time is on the order of 5s.

a) b)

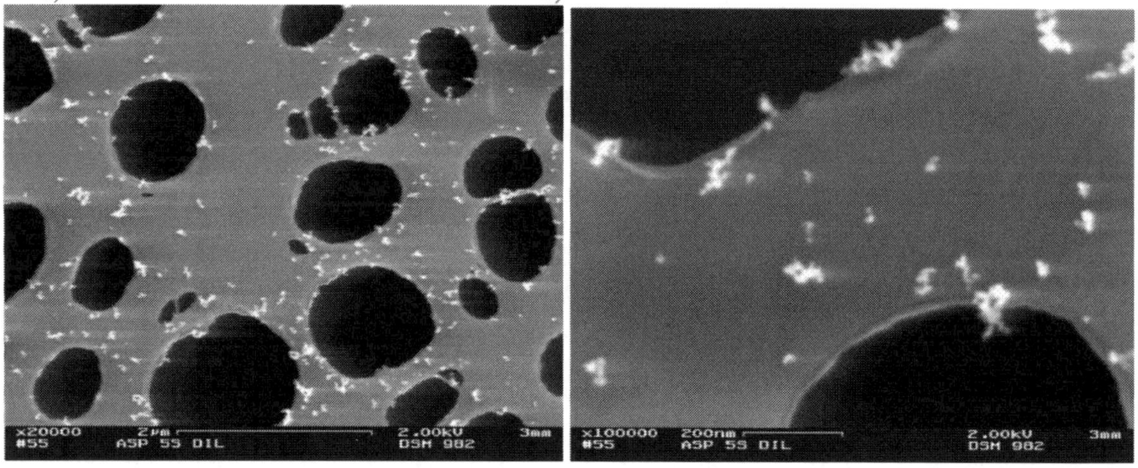

Figure 13. SEM images of TiO_2 particles collected with the aspiration EM grid sampler equipped with a pre-impactor (aerodynamic cut size $d_{ae,cut}$=600 nm), sampling time 5 s.
a) Overall view of the particles. b) Detailed image of the individual particles, sizes typically under d_p=200 nm.

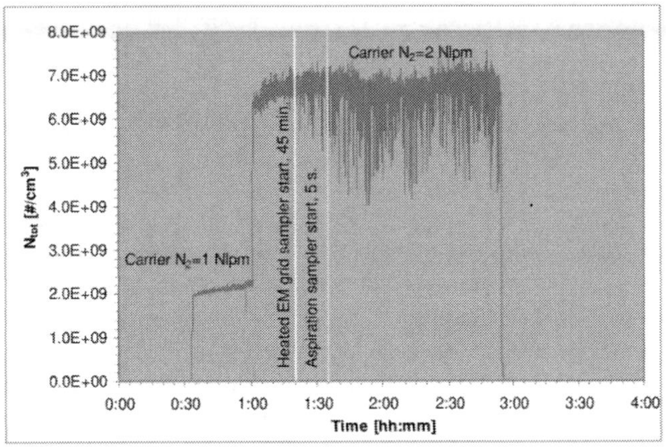

Figure 14. Total particle number concentration measured after the dilution, dilution ratio DR=1001 (Figure 4). The particle samples are collected simultaneously with the heated EM grid and aspiration sampler, grid mounted with carbon tape, aerosol flow sweeping parallel to surface of the grid. The images of the collected particles are shown in Figures 12 and 13.

4. Conclusions

An electron microscopy grid sampler based on nanoparticle diffusion and large particle thermophoretic repulsion (anti-thermophoresis) was designed. The dimensions of the current version are 79x22 mm at the largest. In addition, the performance of the heated EM grid sampler was compared to an aspiration EM grid sampler built and designed at VTT equipped with a commercial preimpactor (aerodynamic cut size approximately $d_{50,ae}$=600 nm).

Based on the experimental results a temperature difference of dT=10°C between the heated EM grid sampler and the passing parallel aerosol flow is high enough to hinder large particle (d_p>200 nm) collection on the EM grid. Mounting the grid on the sampler either with carbon tape or with a brass hat does not seem to cause any significant difference on the collected particles (d_p<200 nm; experimentally determined from the SEM micrographs) at these conditions (dT=10°C and $N_{tot} \approx 5.3\text{-}5.5 \cdot 10^6$ #/cm^3 after two-stage dilution, DR=1001). Because the brass hat is currently much easier to mount it is preferred over the carbon tape mounting. However, dimensions in nanoparticle diffusion are very small and this may cause some error for sampling, as the passing aerosol flow cannot sweep the surface of the grid (actually first sweeps the surface of the hat) when reaching the grid sampler, and the grid cannot be mounted exactly on top of the grid sampler, it is approx. 1-2 mm below the surface of the hat. This cannot be totally avoided with hat-type mounting of the grid.

In addition, when using carbon tape for mounting the EM grid, it is important to ensure that the grid has been mounted properly. If the contact between the grid and the head of the heating resistor is poor thermophoretic repulsion is very weak or even non-existent. However, there are possibilities to make the carbon tape mounting easier and better to ensure good contact between the head of the resistor and the grid. Partial solution to this problem would be to increase slightly (approximately dT=20-30°C instead of dT=10°C) the temperature difference between the passing flow and the grid.

The performance of the heated EM grid sampler was also compared to another individual particle sampler called an aspiration sampler equipped with a commercial preimpactor (orifice diameter 0.0457 mm) normally inserted in front of a TSI electrostatic classifier model 3080. The sampling times for the aspiration sampler are usually very short, usually two orders of magnitude shorter than with heated EM grid sampler. Therefore, at low concentrations when sampling times should be kept short, the aspiration sampler is a good choice. However, overloading of the sampling grid is a potential problem because of the very short collection times. For optimal operation of the aspiration sampler a more suitable preimpactor than the commercial one applied in this study could be designed.

References

[1] Oberdörster G, Maynard A, Donaldson K, Castranova V, Fitzpatrick J, Ausman K, Carter J, Karn B, Kreyling W, Lai D, Olin S, Monteiro-Riviere N, Warheit D and Yang H 2005 *Particle and Fibre Toxicology* **2**:8

[2] Oberdörster G, Ferin J and Lehnert B E 1994 *Environmental health perspectives* **102** 173

[3] Weiss M, Verheijen P J T , Marijnissen J C M and Scarlett B 1997 *J. Aerosol Sci.* **28** 159

[4] Beddows D C S and Telle H H 2005 *Spectrochimica Acta B* **60** 1040

[5] Flagan R C 2001 Electrical techniques, in: Aerosol measurement – priciples techniques and applications, P. A. Baron and K. Willeke (Eds.) Van Nostrand Reinhold, New York, 537-568

[6] Laskin A and Cowin J 2002 *J. Aerosol Sci.* **33** 405

[7] Hinds W C 1999 Aerosol technology: properties, behaviour and measurement of airborne particles, second edition. New York, John Wiley & Sons

[8] Baron P A and Willeke K 2001 Gas and particle motion, in: Aerosol measurement – priciples techniques and applications. P. A. Baron and K. Willeke (Eds.) Van Nostrand Reinhold, New York, 61-82

[9] Backman U, Tapper U and Jokiniemi J K 2004 *Synthetic Metals* **142** 169

[10] Backman U, Lyyränen J, Tapper U, Auvinen A and Jokiniemi J 2008 *International Conference on Safe production and use of nanomaterials, Nanosafe 2008*

[11] Pyykönen J and Jokiniemi J 2003 *Fuel Process.Technol.* **80** 225

Suspension characterization as important key for toxicological investigations

Tobias Meißner, Annegret Potthoff and Volkmar Richter

Fraunhofer Institute for Ceramic Technologies and Systems, 01277 Dresden, Germany.

E-mail: Tobias.Meissner@ikts.fraunhofer.de

Abstract. To assess potential health risks of nanoparticles by means of *in vitro* or *in vivo* assays and to determine dose-action curves a defined and reproducible method of particle administration is required. The interpretation of the toxicological results should be based on a comprehensive chemical-physical characterization of the particles used. Therefore, we developed a method to suspend nanoparticles stably and homogenously in physiological media. Our approach consist of three steps: (1) physical-chemical characterisation of the powders as delivered, (2) preparation and characterization of a non-physiological electro-statically stabilized nanoparticle suspension and (3) assessment of the nanoparticles behaviour in physiological media with or without proteins. This approach is demonstrated on a titanium dioxide and a tungsten carbide nanopowder. Results showed that particles agglomerate in protein-free medium within minutes, whereas in the presence of bovine serum albumin or foetal bovine serum an agglomeration is hindered.

1. Introduction

Due to the increasing use of nanomaterials, the assessment of health risks of nanoparticles become urgent. It is a central task to reveal possible risks that can emerge through the small dimension of nanostructured materials [1-3]. For a complete risk and safety assessment of nanoscaled materials, such as nanoparticles, nanotubes and nanoplates, a comprehensive chemical-physical characterization of these substances is necessary. Chemical composition, cristallinity, morphology, particle shape and size as well as the specific surface area of the powders in the as-delivered state are key properties. But, for a correct interpretation of *in vivo* and *in vitro* the particles behaviour under physiological conditions has also to be understood. Physical parameters like the size of nanoparticles, for instance, should be measured in cell culture media under conditions as close as possible to the toxicological testing [4]. Particle-particle interactions resulting in agglomeration processes and reactions between particles and serum proteins have to be taken into account. In cell culture media, proteins are rapidly adsorbed on the nanoparticle surface. The resulting protein corona modifies the appearance of the particles and their interactions with biological structures. For example, the cellular uptake of nanoparticles can be influenced by adsorbed serum proteins [5, 6].

As a result, the characterization of suspended particles in physiological media is one of the most difficult operations in a complete biological and chemical-physical assessment of nanoparticles toxicity, but it is one of the most important ones [4, 7]. Without an adequate material characterization the obtained toxicological results can hardly be interpreted correctly [8]. The toxicity of nanoparticles may be elucidated only through the combination of physico-chemical characterization and toxicological testing [9].

In this paper we describe an approach of preparing and using stable nanoparticle suspensions for toxicological testing. Instructions for a correct measurement of parameters such as particle size, size distribution and zeta potential of suspended particles in water as well as in physiological media are given. The preparation of a stable initial nanoparticle suspension is explained with the example of a titanium dioxide (TiO_2) and a tungsten carbide (WC) powder, respectively. Subsequently the initial suspensions were added to different physiological media to study the agglomeration behaviour of the nanoparticles and the influence of proteins to the stability of the suspensions. The main result of our investigations is that the agglomeration of particles in physiological media can be perfectly inhibited in the presence of bovine serum albumin (BSA) or foetal bovine serum (FBS). This enhances the knowledge of nanoparticles actions in culture media and gives toxicologist a helpful tool to work with well-defined suspensions.

2. Materials and Methods

Investigations were done using a TiO_2 (Aeroxide P25, Evonik) and a WC powder. The specific surface area was determined with the N_2-BET method. Cristallinity of both powders was analysed by X-ray diffraction (XRD). Primary particle size, morphology and aggregation degree of the particles were determined by means of scanning electron microscopy (SEM).

Before preparing a stable initial suspension, a titration of a TiO_2 and a WC suspension was performed to get the zeta potential as a function of pH-value. Suspension background was 1 mM NaCl, and HCl or NaOH at different molarity was used for titration. Resulting from the titration, WC was dispersed in pure water whereas TiO_2 was dispersed in pure water which was adjusted to pH 4 with HCl to have stable initial suspensions. All zeta potential values were obtained from electrophoretic mobility measurements using Smoluchowski equation. The initial suspensions were prepared by sonication with an ultrasound horn. Sonication duration was as long as no further de-agglomeration progress was achieved. For that, particle size and size distribution was measured by means of dynamic light scattering.

Furthermore, these initial suspensions were added to physiological media to record the agglomeration behaviour by determining particle size and zeta potential. Physiological media were phosphate buffered saline (PBS) or PBS supplemented with bovine serum albumin (BSA) as well as Dulbecco's Modified Eagle Medium (DMEM) or DMEM supplemented with 5% foetal bovine serum (FBS).

3. Results

3.1. Powder Characterization and preparation of initial suspension

Diffraction analysis shows that TiO_2 consists of anatase and rutil whereas the WC powder consists of a pure tungsten carbide phase. Both powders differ strong in their specific surface area obtained from BET and also in their size and aggregation grade which can be estimated from electron micrographs (Figure 1). The specific surface area (BET) of TiO_2 is 55.9 m^2/g and 6.9 m^2/g for WC, respectively. The primary particles of TiO_2 are below 50 nm, nearly spherical and highly aggregated, while particles of WC are in the range of approximately 50 to 200 nm, having different shapes and are more isolated.

Figure 1. SEM micrographs of TiO_2 (left) and WC (right).

In order to get stable initial suspension, high absolutes values of zeta potentials are necessary for electro-statically stabilized particles. In general, a suspension is stable when the zeta potential value is less than -30 mV or greater than +30 mV [10]. Figure 2 shows the result of the zeta potential titration of both suspensions. For TiO_2 high enough zeta potential values were obtained in the acid region at a pH lower than 4. Thus, TiO_2 was suspended in 10^{-4} M HCl which yields a zeta potential of about 39 mV. Titration of the WC suspension indicates values between -30 mV and -40 mV over the whole measured pH-range which is acceptable for an electrostatic stabilization. A pH adjustment is not necessary. WC was just dispersed in water which results in a pH of 5 and a zeta potential of -35 mV.

The volume-weighted size distributions of both sonicated suspensions are very similar, although the primary particles of tungsten carbide are significantly greater than those of titanium dioxide (Figure 3). Interestingly, the distribution of WC particles is shifted to smaller sizes compared to TiO_2. The particle size for the WC suspension agrees with the values from electron microscopy. This means that the WC suspension contains mainly primary particles after sonication. For TiO_2, sonication is not able to destroy the strong bridges between the primary particles detected by SEM. For that reason, the TiO_2 suspension includes aggregates and no primary particles.

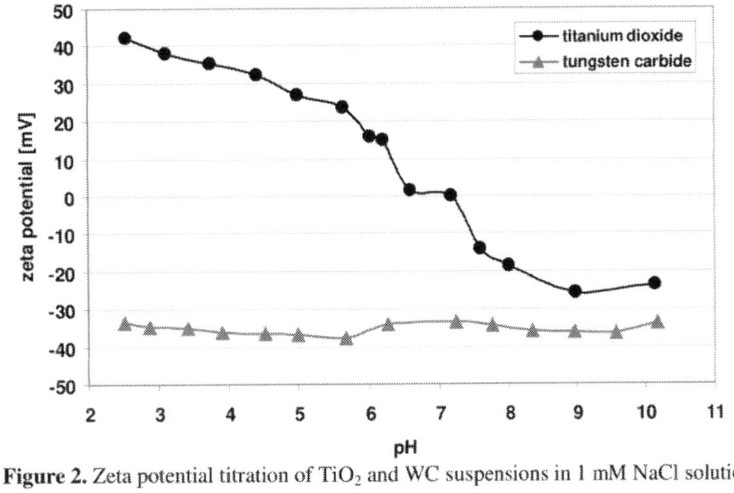

Figure 2. Zeta potential titration of TiO_2 and WC suspensions in 1 mM NaCl solution.

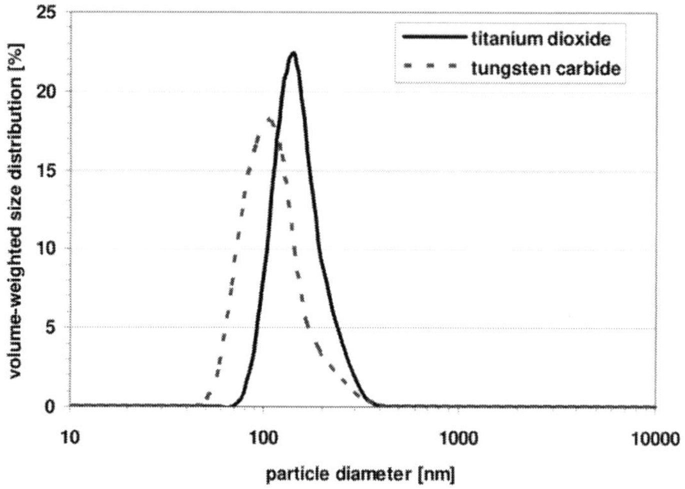

Figure 3. Size distribution of titanium dioxide and tungsten carbide particles measured by means of dynamic light scattering.

3.2. Behaviour under physiological conditions

In the next step we studied the agglomeration kinetics of the particles after adding the initial suspension (containing the particles) to the appropriate physiological media. The final mixture consists of 10% nanoparticle suspension (v/v) and 90% (v/v) media. The high percentage of media guarantees that pH and salt content of the media are not changed in a critical way. This was confirmed by *in vitro* viability tests using different cell types. The control solution which contained 90% media and 10% water or appropriate background solution without particles did not reduce the cell viability [11, 12]. Figure 4 shows the agglomeration behaviour of TiO_2 and WC at a mass concentration of 10 µg/ml. In the absence of BSA or FBS, both types of nanoparticles agglomerate quickly, whereas the agglomeration of TiO_2 is much faster compared to WC. Furthermore, the cluster formation seems to be independent of the used physiological media. The zeta potentials of the particles change radical in the physiological environment. We obtained -23 mV (TiO_2) and -22 mV (WC) in PBS and -8 mV (TiO_2) and -23 mV (WC) in DMEM. All values are outside the electrostatic stable area which gives an explanation of the agglomeration. In analogy to blood, cells are cultivated in physiological media containing proteins. So our agglomeration investigations were extended to study the nanoparticles activities in PBS including BSA (data not shown in the diagram) and in DMEM supplemented with FBS. For the tested protein concentrations tested we found that BSA (0.5 mg/ml) and FBS (5% FBS/DMEM) prevent the particles from agglomeration. BSA and the proteins from serum adsorb on the particles surface and stabilize the suspensions. BSA is the major component in FBS. We assume that the serum albumin is mainly adsorbed on the particle surface causing a stabilization of the particles. Proof can be given by the results from zeta potential measurements. We determined -11 mV in DMEM with FBS for both nanoparticle types and in PBS with BSA the zeta potentials are -15 mV for TiO_2 and -13 mV for WC. These values are very similar and close to that of free BSA with -11 mV (measured in PBS). An electrostatic stabilization of the formed nanoparticle-BSA-complexes is not possible at these very low absolute values of the zeta potential. Therefore, the stabilization effect can only be of steric or electro-steric nature.

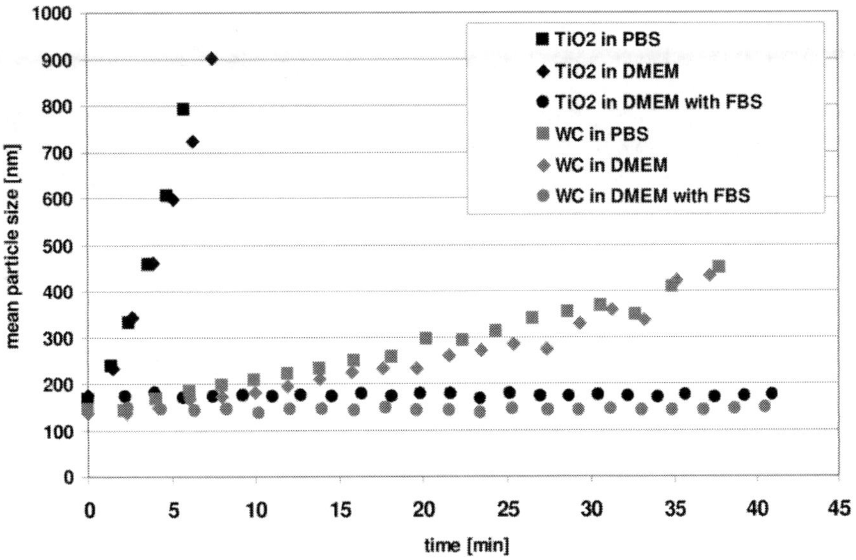

Figure 4. Agglomeration behaviour of TiO_2 and WC nanoparticles at a concentration of 10 μg/ml in PBS, DMEM and DMEM supplemented with 5% FBS. Results in PBS containing 500μg/ml BSA have the same curve shape as DMEM with FBS and are not shown for clarity reason.

One question rising from the agglomeration curves remains unanswered. Why does TiO_2 agglomerate so much faster then WC although the mass concentration is identical and the size distributions (see Figure 3 again) are very similar? The surface of nanomaterials is often considered as appropriate dose quantity in toxicological testing and research into the health effects of nanoparticles [13, 2]. Therefore, we carried out agglomeration experiments with TiO_2 and WC with the same particle surface area per unit volume instead of the same mass (Figure 5). The curves received differ only slightly. That demonstrates that the surface area offered by the particles in the suspension has a strong effect on the agglomeration rate. The exposure of nanoparticles to cells in *in vitro* tests should be compared by a surface based dosage concentration.

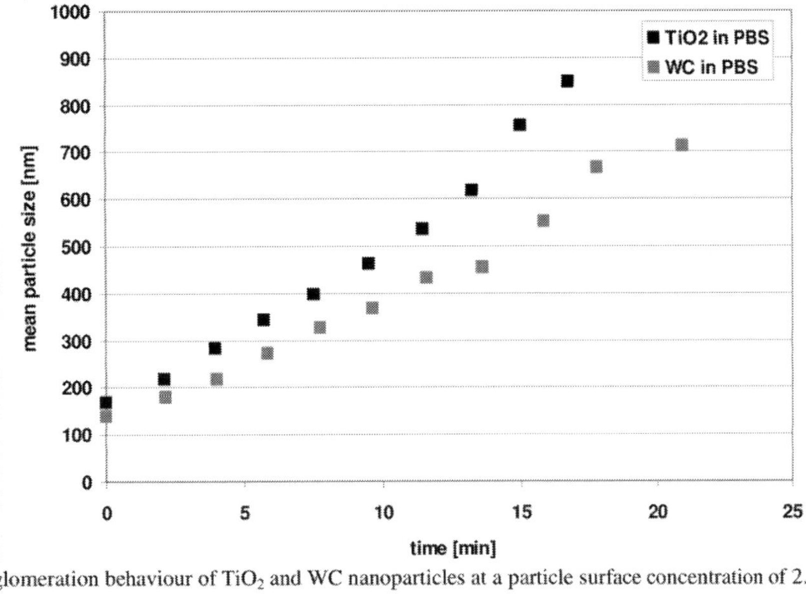

Figure 5. Agglomeration behaviour of TiO_2 and WC nanoparticles at a particle surface concentration of 2.5 cm²/ml in PBS.

4. Conclusions

In this paper the chemical-physical characterization of nanoparticles prior to toxicological studies is described. Resulting from our investigations we have derived the following guidelines for preparing a homogenous physiological nanoparticle suspension. First of all, a stable initial suspension is developed which will be later mixed with the corresponding cell culture media. Stability of the initial suspension is ensured by electrostatic forces and can be quantified by zeta potential measurements. If necessary, suspensions have to be adapted by pH adjustment or addition of a non-toxic dispersant to get a sufficiently high absolute value of the zeta potential. In protein-free cell culture medium nanoparticles agglomerate rapidly in order to reduce surface area and the surface energy. Therefore, exposure concentrations in toxicity testing should be related to the surface area of the nanoparticles. But, in many biological liquids like blood proteins are present in high concentrations. The protein BSA or BSA containing serum like FBS hinder agglomeration. They act as natural dispersant aids and stabilize the nanoparticles in their original size. Finally, the nanoparticles in physiological media are homogenous suspended, naturally stabilized through proteins and their size distribution is retained unchanged. This approach gives the toxicologists a helpful tool to assess health risks in *in vitro* investigations. The method has been successfully tested by our partners. The procedure we demonstrated for two different powders is recommended to become a standard preparation and operation procedure in nanoparticles risk assessment.

Acknowledgements
This work was partially financed by the German Federal Ministry of Education and Research within the INOS project (funding number 03X0013).

References

[1] Nel A, Xia T, Mädler L and Li N 2006 Science 311 622
[2] Oberdörster G, Oberdörster E and Oberdörster J 2005 Environ. Health Perspect. 113 823
[3] The Royal Sociecty. 2004. Nanoscience and nanotechnologies: opportunities and uncertainties. www.nanotec.org.uk/finalreport.htm
[4] Powers K W, Brown S C, Krishna V B, Wasdo S C, Moudgil B M and Roberts S M 2006 Toxicol. Sci. 90 296
[5] Cedervall T, Lynch I, Lindman S, Berggard T, Thulin E, Nilsson H, Dawson K A and Linse S 2007 Proc. Natl. Acad. Sci. USA. 104 2050
[6] Dutta D, Sundaram S K, Teeguarden J G, Riley B J, Fifield L S, Jacobs J M, Addleman S R, Kaysen G A, Moudgil B M. and Weber T J 2007 Toxicol. Sci. 100 303
[7] Schulze C, Kroll A, Lehr C-M, Schäfer U F, Becker K, Schnekenburger J, Schulze Isfort C, Landsiedel R and Wohlleben W 2008 Nanotoxicology 2 51
[8] Warheit D B 2008 Toxicol. Sci. 101 183
[9] Unfried K, Albrecht C, Klotz L-O, von Mikecz A, Grether-Beck S and Schins R P F 2007 Nanotoxicology 1 52
[10] Malvern Instruments Ltd. Zeta potential - An introduction in 30 minutes (KB000734). www.malvern.com
[11] Bastian S, Busch W, Kühnel D, Springer A, Meißner T, Holke R, Scholz S, Iwe M, Pompe W, Gelinsky M, Potthoff A, Ikonomidou C and Schirmer K 2009 Environ. Health Perspect. 117 530
[12] Kühnel D, Busch W, Meißner T, Springer A, Potthoff A, Richter V, Gelinsky M and Schirmer K 2008 Investigation of toxic potency of cobalt-doped tungsten carbide nanoparticles to rainbow trout gill cells (RTgill-W1). nanoECO Monte Verita, Switzerland. www.empa.ch/plugin/template/empa/*/69998/---/l=1
[13] Moss O R 2008 Int. J. Nanotechnol. 5 3

Dispersion and stability of TiO_2 nanoparticles synthesized by laser pyrolysis in aqueous suspensions

Carole Sentein[1], Benoit Guizard[1], Sophie Giraud[1], Chang Yé[2] and François Ténégal[1]

[1]CEA-Saclay,DEN/DMN/SRMA/LTMEx, Bât.460, 91191 Gif sur Yvette, France

[2]Laboratory for Biological Effects of Nanomaterials and Nanosafety, Institute of High Energy Physics, Beijing 100049, P. R. China.

E-mail: carole.sentein@cea.fr

Abstract. Nanoparticles of titanium dioxide (TiO_2) were synthesized by the laser pyrolysis method at pilot scale using an organometallic aerosol injected in the reactor. In order to secure the global process of nanoparticle production, we investigate the liquid recovery of the raw powders.
The goal was to recover the nanopowders directly in water at the exit of the reaction zone mainly for safety reasons. For that, the dispersion of TiO_2 nanopowder in water using a dispersing agent was optimized and comparison of the powder characteristics produced after a dry or a liquid recovery was done.

1. Introduction

In the framework of the European Project NANOSAFE2, the development of secure industrial production systems and safe applications is of major interest. To achieve this goal, we worked on the development of equipment able to collect in liquids the nanoparticles produced by gas-phase processes such as laser pyrolysis [1, 2]. Hence, synthesis and liquid collection of TiO_2 nanoparticles was developed on one hand and on the other hand, dispersion in water and stability of TiO_2 nanoparticles synthesized by laser pyrolysis was studied.

In the perspective to recover/collect in water the TiO_2 nanoparticles produced by laser pyrolysis, studies were held to optimize the dispersion and the stability of the suspensions of TiO_2 nanoparticles. Different type of dispersing agents [3-6] may be added to the suspensions to favour the slurrying of the nanoparticles in the liquid and to produce stable suspensions for storage and applications. The influence of different parameters on the dispersion and the stability was studied.

2. Effect of the Specific surface Area (SSA) on the stability

The specific surface area (S_{BET}) of as-grown nanopowders was determined by the Brunauer-Emmett-Teller (BET) method using nitrogen adsorption. A first study compared the evolution in time of suspensions for TiO_2 nanoparticles with different specific surface areas.

As it is possible to produce nanoparticles having several specific surface areas (SSA) by laser pyrolysis, it appeared interesting to study the stability of the suspensions as a function of the SSA of the powders. Consequently, two powders with different SSA have been selected (06-TiO-14 →

27 m^2/g and 06-TiO-20 → 140 m^2/g) to study the effect of the SSA on the dispersion and stability behaviour in pure water.

Stability is measured on the basis of Zeta potential measurements. The Zeta potential measurements (by means of the electrophoretic mobility) were performed on a Malvern Nano ZS equipment. A Zeta potential (ζ) of at least 30 mV (positive or negative) is normally required to achieve a reasonably stable dispersion.

The Figure 1 presents the Zeta potential (ζ) for the two powders synthesized by laser pyrolysis compared to the Degussa powder P25 (SSA = 50 m^2/g) immediately after sonication. The powder 06-TiO-14 (27 m^2/g) has a similar behaviour to that of the Degussa P25 on all the range of pH (almost the same isoelectric point and Zeta potential). The powder 06-TiO-20 (140 m^2/g) displays a different behaviour. For this powder, the isoelectric point is clearly shifted toward the low pH values.

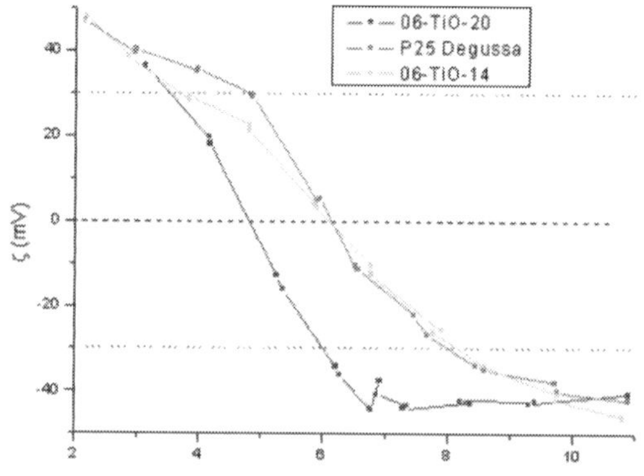

Figure 1. Potential versus pH for different TiO2 nanoparticles.

This difference could be explained by the differences of the SSA and/or of the residual carbon content. Indeed the chemical analysis revealed that the powder 06-TiO-20 contain 16 wt.% of carbon whereas its colour (between light grey and white) is close to that of 06-TiO-14 which contains only 0.7 wt.% of carbon. This surprising difference of carbon content could explain the results but it has to be confirmed by additional carbon dosages.

The Figure 2 and Figure 3 present the effect of the powder on the stability of the suspensions at a concentration of 0.02 wt %. Each measurement of ζ-potential versus pH was done after magnetic stirring for 20 minutes followed by an ultrasonic treatment of 6 minutes of the suspension. According to the Figure 2, the Degussa P25 appears to be the most stable powder in pure water whatever the pH value up to 790 h. Nevertheless, the Degussa P25 based suspension flocculates very rapidly which should indicate a strong agglomeration of the nanoparticles.

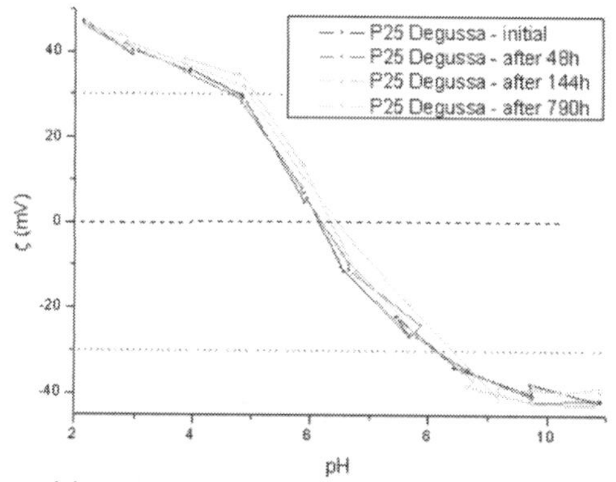

Figure 2. ζ-potential versus pH for a P25 Degussa TiO_2 nanoparticles suspension (40mg/200mL distilled water) initially and after 48, 144 and 790 hours.

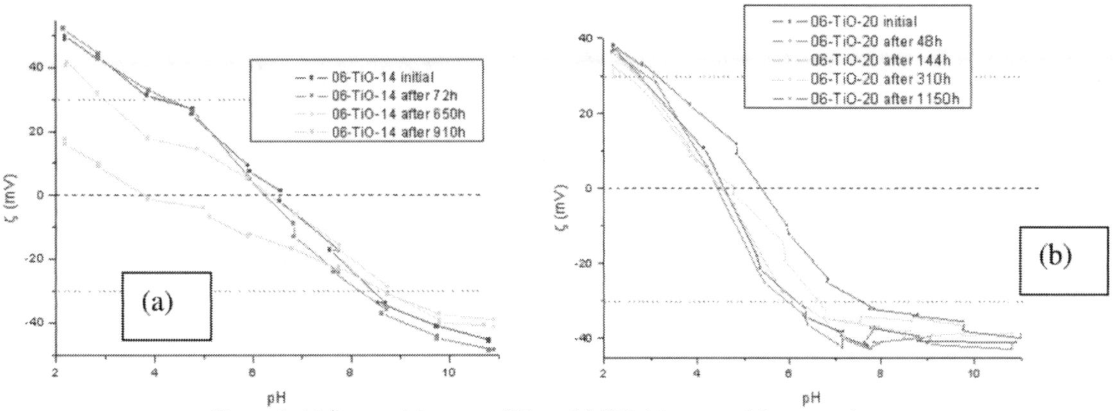

Figure 3. (a) ζ-potential versus pH for a 06-TiO-14 nanoparticle suspension (40mg/200mL distilled water) initially and after 72, 650 and 910 hours.
(b) ζ-potential versus pH for a 06-TiO-20 nanoparticle suspension (40mg/200mL distilled water) initially and after 48, 144, 310 and 1150 hours.

Figure 3 (a) shows an evolution in time of the pH stability range for the powder 06-TiO-14 (27 m^2/g). Thus, stability remains unchanged over 910 hours above pH 8.5 whereas a significant decrease of the stability is noted in the low pH range after 100 hours.

For the 06-TiO-20 powder (140 m^2/g), Figure 3 (b) shows that the stability remains unchanged over more than 1000 hours in the low pH range (below pH 3). On the other hand, a decrease of the stability range is noted in the high pH range (above pH 6.2 after 144 hours and above pH 7.5 after 1150 hours.

3. Effect of the free carbon content on the stability

We performed a study on the effect of the free carbon content upon the powder characteristics. For that, we apply to a chosen powder a thermal treatment at 460°C, which allows removing completely the free carbon. That way, we were able to compare the characteristics of two powders differing only by their free carbon content. For this comparison, we measured the Zeta potential of aqueous dispersions of different TiO_2 nanoparticles produced by laser pyrolysis, before and after thermal treatment. Result is that around pH = 7 to 8, pH of distilled water used, suspensions with no free carbon content are less stable. This result is confirmed by visual observation of the evolution in time of the suspensions. Same behaviour is observed for suspensions with a dispersing agent (10 different were tested) (Figure 4).

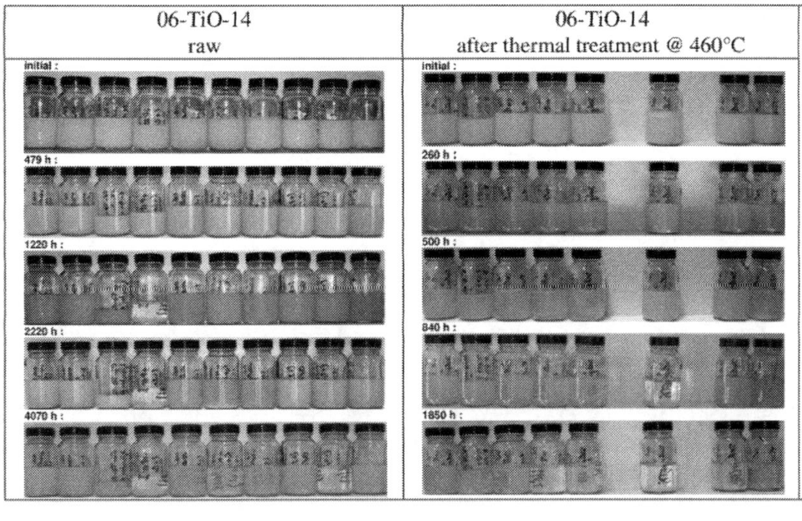

Figure 4. Stability of the 06-TiO-14 suspensions as a function of the dispersant. Left column: raw powder. Right column: after thermal treatment.
Dispersant tested from left to right: Dolapix CA, Dalais PC21, Reotan LA, PEI, Dolapix A88, KV9021, Dolapix ET85, Dolapix PC80, Dolapix CE64, Tiron.

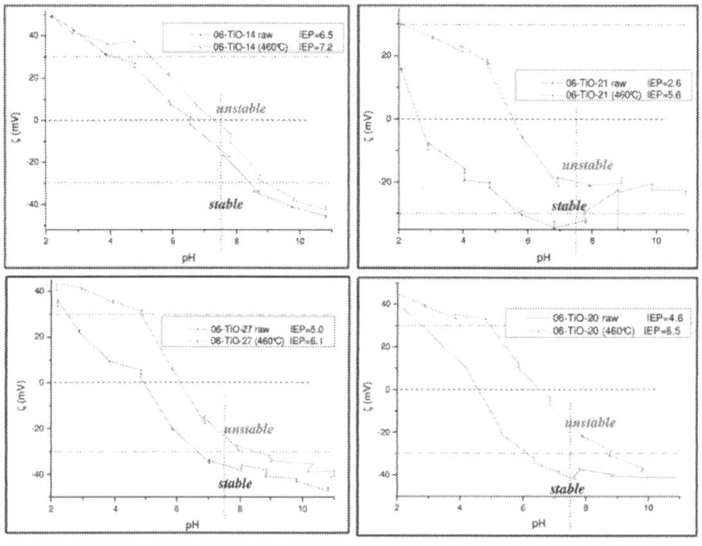

Figure 5. ζ-potential versus pH for 4 different TiO_2 nanoparticles suspensions (40 mg/200 mL distilled water). Comparison of the initial curve with raw powder (blue) and after thermal treatment at 460°C (red). Each measurement after 20' magnetic stirring and 6' ultrasonic treatment

Zeta potential was also investigated as a tool to predict dispersion stability with the same result. For the 4 different powders tested, we find again a lower stability around pH=7.5, after thermal treatment at 460°C, that is after free carbon removing (Figure 5).

Characterization of the particle size distribution was performed by Dynamic Light Scattering (DLS). DLS measures the Brownian motion of the particles and relates it to the particle size. Nanoparticles were mixed with solvent using a sonic probe. Different dispersion mediums were tested (ethanol, acetone and distilled water). The lowest mean size intensity distributions were obtained with the aqueous dispersions. The lowest values measured were around 150 to 200 nm that is much higher than individual particle size (between 12 and 80 nm, as measured by BET). We conclude then that, dispersed in a liquid, the TiO_2 nanoparticles form aggregates around 200 nm. This information may be of interest for the toxicity impact and studies. Moreover, as shown on the Figure 6, we find again a lower stability in water, around pH=7.5, for the powders treated at 460°C (particles form larger aggregates).

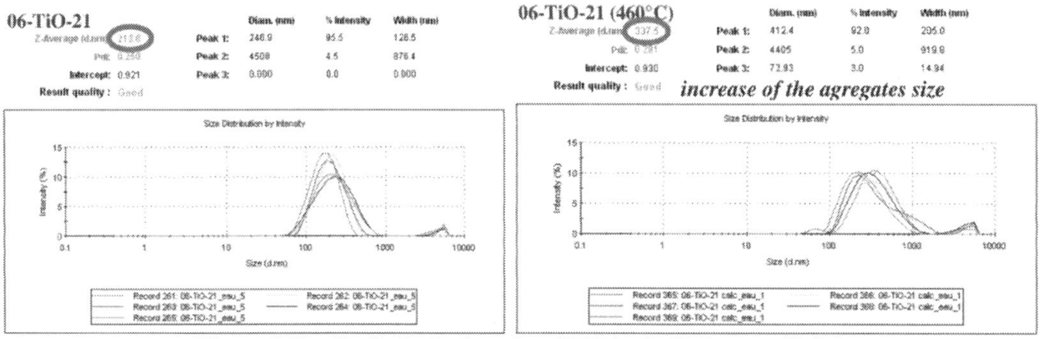

Figure 6. Size distribution by Intensity of TiO_2 in distilled water, pH=7.5. Left: raw powder. Right: after a thermal treatment at 460°C.

Microscopic studies have been performed by AFM and TEM to analyze the particle size and shape of the TiO_2 nanoparticles revealing for some the presence of free carbon originating from the organic part of incompletely pyrolysed reactant (Figure 7 and Figure 8).

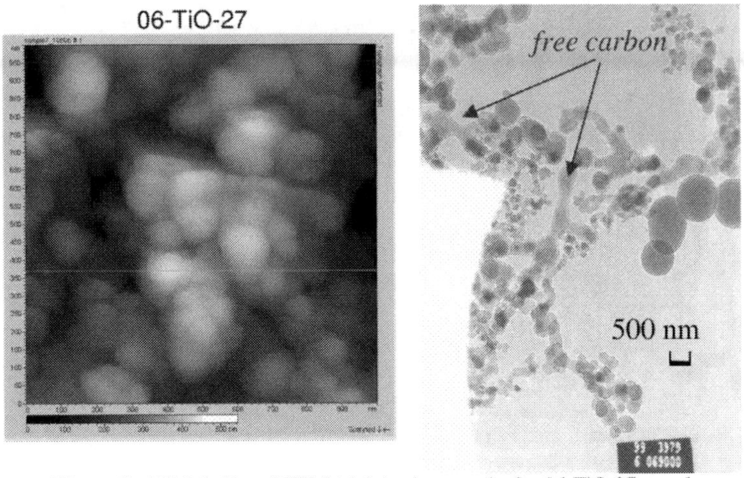

Figure 7. AFM (left) and TEM (right) micrographs for 06-TiO-27 powder.

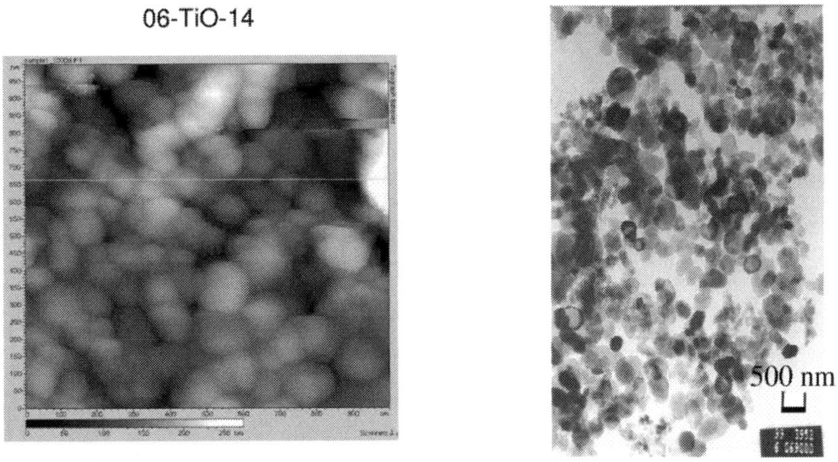

Figure 8. AFM (left) and TEM (right) micrographs for 06-TiO-14 powder.

4. Changes in the suspension stability behaviour induced by the addition of dispersing agents

Different dispersing agents were tested for 06-TiO-14, 06-TiO-20 and Degussa P25. The Figure 9 presents the Zeta potential as a function of the dispersant amount (dwb%: dry weight percentage of the TiO_2 powder basis) for the powders 06-TiO-14, 06-TiO-20 and Degussa P25. The results show that for 06-TiO-14, Reotan LA [7] and Dolapix® PC 21[8] seem to be the best dispersants. Both of them are synthetic polyelectrolytes (ammonium polyacrylate type). In this case, the deflocculation effect is a result of electrochemical interaction between the functional groups of the additive and the surface of the ceramic particles.

Reotan LA seems also to be one of the best dispersants for 06-TiO-20 and for the Degussa P25. That way, Reotan LA appears to be the more "universal" dispersant agent for the TiO_2 nanoparticles tested. The highest values of Zeta potential are obtained for the laser pyrolysis powder compared to the Degussa powder. This is clearly visible on the Figure 10 showing the Zeta potential for all the powders as a function of the amount of dispersant (% of the mass of powder). In the pH range of these measurements (pH of distilled water used between 7 and 8), it is possible to get Zeta potential values between -40 and -55 mV for laser pyrolysis powders whereas the Zeta potential value for the Degussa P25 is only between -20 and -35 mV.

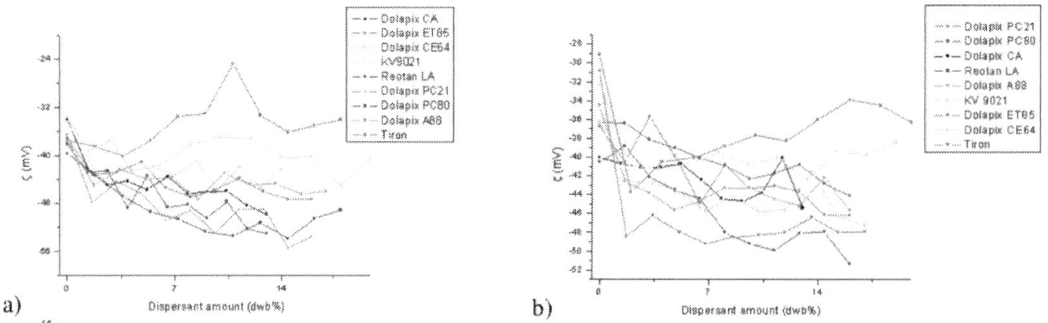

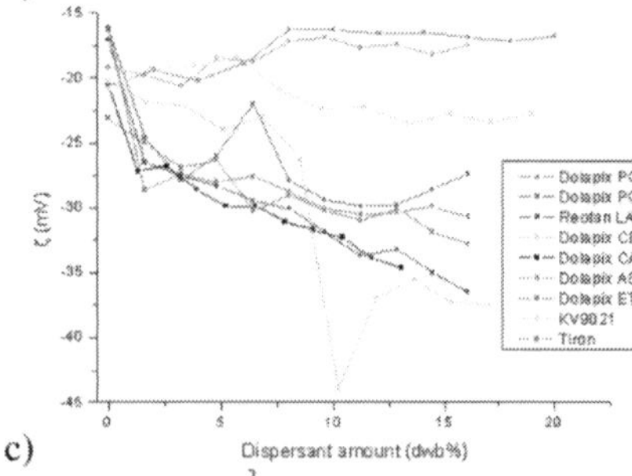

Figure 9. (a) Effect of the different dispersants tested amount on the ζ-potential of 06-TiO-14 suspensions. (b) Effect of the different dispersants tested amount on the ζ-potential of 06-TiO-20 suspensions. (c) Effect of the different dispersants tested amount on the ζ-potential of P25 Degussa suspensions.

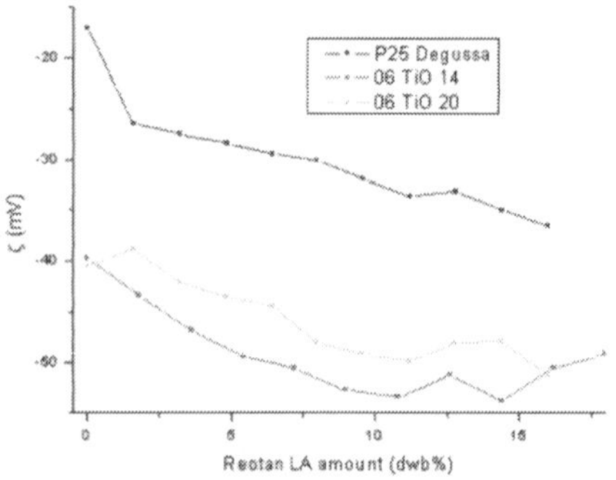

Figure 10. Comparison of the effect of Reotan LA amount on the ζ-potential of different TiO_2 nanoparticles suspensions.

As shown on Figure 11, the addition of a dispersing agent may modify deeply the pH stability range of the suspensions. On the left graph *(a)*, the 06-TiO-14 based suspension is initially stable for pH values below 4 and above 8.5. After the addition of 18 dwb% of Reotan LA, the suspension is stable for pH values above 5.3. On the right graph *(b)*, after the addition of polyethylenimin (PEI) an aqueous suspension of 06-TiO-20 become stable upon the whole range of pH studied (from 2 to 11). This behaviour was observed with different dispersants tested and could be of interest following the application.

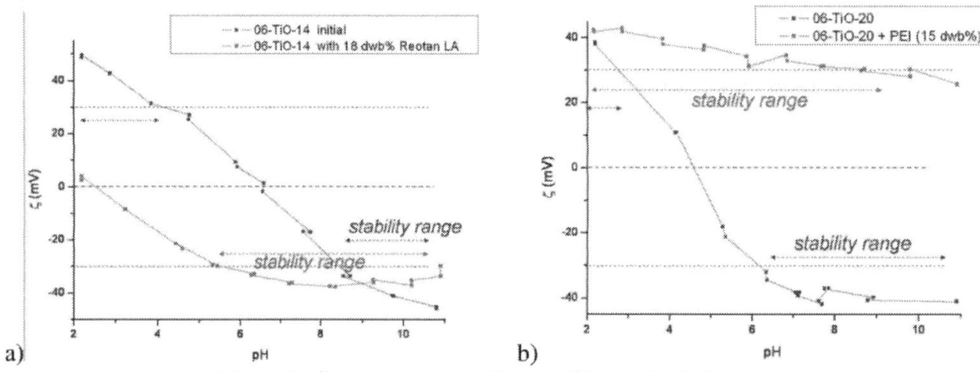

Figure 11. ζ-potential versus pH (40mg/200mL distilled water).
a) 06-TiO-14 nanoparticles suspension initially and with the addition of 18 dwb% of Reotan LA.
b) 06-TiO-20 nanoparticles suspension initially and with the addition of 15 dwb% of PEI.

5. Comparison of the powder characteristics produced after a dry or a liquid recovery

Comparison of the powder characteristics produced after a dry or a liquid recovery was performed. Specific surface areas and density are almost the same and the main difference concerns the carbon content which is much higher after a dry recovery. As the TTIP (titanium isopropoxide), reactant used for the TiO_2 synthesis, is very reactive with water, non pyrolysed part of it may hydrolyse during liquid collection and leads to different volatile sub-products which disappear during powder recovery. At last, the characteristics reproducibility for the powders issued from the liquid recovery means was tested. A good reproducibility was noticed for specific surface area, density and anatase versus rutile rate (Table 1).

Ref.	Liquid: *W = Water* *E = Ethanol* *D = Dolapix PC21*	Carrier gas flow rate Ar (l/mn)	Incident laser power (W)	S_{BET} (m^2/g)		Density (g/cm^3)		Anatase (wt%)	Rutile (wt%)
				250°C	460°C	200°C	460°C		
07-TiO-02	W + D	60	4550	74	60	3.4	3.7	93	7
07-TiO-03	W(80%) + E(20%) + D	55	4550	78	47	3.4	3.7	96	4
07-TiO-04	W(80%) + E(20%)	60	4550	60	46	3.4	3.4	97	3

Table 1. Characteristics of the powders after drying for different solvents used for liquid recovery.

6. Conclusion

Conditions have been optimized to produce stable TiO_2 nanoparticle aqueous suspensions. We determined the best dispersing agent and its best dry weight percentage of the TiO_2 basis to be added in order to obtain well dispersed suspensions and the most stable in time. Moreover, we show that liquid recovery of the nanopowders directly at the exit of the pilot reactor is possible and that powder characteristics after a liquid recovery are the same as after a dry recovery.

Acknowledgments:
The NANOSAFE2 project is partially financed by EU-FP6 program NMP2-CT-2005-515843.
The authors thank Licinio Rocha *(CEA/DRT/LIST)* for AFM scans and Marie-Pierre Hugon *(CEA/DEN/SRMA)* for TEM micrographs.

References

[1] Casey J D and Haggerty J S 1987 *J. of Materials Science* **22(12)** 4307
[2] Rice G W 1987 *J. Am. Ceram. Soc.* **70(5)** C117
[3] Heijman S G J and Stein H N 1995 *Langmuir* **11** 422
[4] Deiss J L, Anizan P, El Hadigui S and Wecker C 1996 *Colloids Surface A: Physicochem. Eng. Aspects* **106** 59
[5] Tang F, Uchikoshi T, Ozawa K and Sakka Y 2006 *Journal of the European Ceramic Society* **26** 1555
[6] Mandzy N, Grulke E and Druffel T 2005 *Powder Technology* **160** 121
[7] Reotan LA: ammonium polycarylate from Lamberti.
[8] Dolapix PC21: synthetic polyelectrolyte from Zschimmer & Schwarz.

Nanosafe 2008: International Conference on Safe production and use of nanomaterials IOP Publishing
Journal of Physics: Conference Series **170** (2009) 012014 doi:10.1088/1742-6596/170/1/012014

Measurement of nanoparticle removal by abrasion

Arnaud Guiot, Luana Golanski and François Tardif

CEA-Liten, 17 rue des Martyrs, F38054 Grenoble Cedex09, France

E-mail: francois.tardif@cea.fr

Abstract. A strong release limitation of single nanoparticles from commercial manufactured "nanoproducts" is necessary to decrease potential exposure risks of consumers and represents also a pragmatic way to facilitate acceptance for nanomaterial commercialization before obtaining definitive toxicological results. So, it is of prime importance to know how to characterize the release of small materials during usage solicitations such as mechanical, thermal, UV stress: are they single nanoparticles, aggregates or nanoparticles included in a bigger piece of the matrix?
In the frame of NanoSafe2 project, CEA developed and qualified a specific bench test where the material to be tested is mechanically solicited by abrasion using a normalized Taber equipment. The first results show that nanofillers can be released in usage by abrasion for non optimised nanoproducts.

1. Introduction

The nanoparticles are finding new industrial applications every day in fields as diverse as electronics, biomedicine, pharmaceutics, cosmetology, chemical catalysis, new materials, and others. We are about to witness the advent of a new era in the industrial history of nanoparticles. New types of nanoparticles that until now were under laboratory development are on the verge of mass-production. Economists are now talking about the dawn of a new industry for the 21st century which could rank it alongside the automobile and microelectronics industries.

Nevertheless, this new industry can develop dynamically only if the safety issues are solved during all the life cycle long of the nano products: from fabrication to the end of life through usage.

Due to the complexity of the nano toxicology in particularly due to the multitude of shapes, sizes, surface charges, etc. even for nanoparticles presenting the same chemical formula, it will probably take a long time to identify if some nanoparticles are dangerous and even a longer time to declare some nanoparticles as benign.

However the relevant parameter for safety issues is the risk. The risk depends on 2 factors: the hazard and the exposure. If the exposure is decreased toward zero, then the risk becomes low whatever the hazard (toxicity in this case). The only pragmatic way to secure the nanomaterials today consists in reducing the exposure to nanoparticles of the potentially exposed workers, the consumers and the environment.

A strong release limitation of single nanoparticles from commercial manufactured "nanoproducts" is necessary to decrease the potential exposure risks for consumers and represents also a pragmatic way to facilitate acceptance for nanomaterial commercialization. So, it is of prime importance to know how to characterize the release of single nanoparticles, aggregates or nanoparticles embeded in a bigger piece of the matrix during usage solicitations such as mechanical, thermal, UV, stress, etc.

In this paper, a new method is presented for measuring the release-ability of the nanofillers embedded in the nanoproducts under abrasion in order to mimic one of the main types of solicitation in usage. The objective of developing such a measurement tool is double: optimization of the "hooking" of the nanoparticles in the matrix and perhaps one day, qualifying the nano products for the market.

2. Experimental setup and method

In the frame of NanoSafe2 project, CEA developed and qualified a specific bench test where the material to be tested is mechanically solicited by abrasion using a normalized Taber equipment ASTM C1353-07. In this test, 2 parts of the same material are alternatively rubbed against each other to generate a dynamic friction Then, the nanoparticles are streamed into an ultra clean air flow before analysis performed by a Scanning Mobility Particle Sizer SMPS) and a Condensation Particle Counter (CPC). Removed nanoparticles can be measured from 5 to 1 000 nm.

The critical parameters to take into consideration to set up such a method for a standardization purpose is the sensitivity (detection of the first few released nanoparticles) and to get an absolute measurement.

- To get a good sensitivity of the method, it is necessary to decrease the background noise level of nanoparticles external to the system and to minimize the dilution of the flow collecting the released nanoparticles to the more sensitive counter.
- To get an absolute measurement, it is necessary to collect all the released nanoparticles and to know the deposited fraction before the counter.

As seen in Figure 1, the sample is disposed in a clean air flow filtered with an H14 HEPA filter. The collecting device has been optimized to collect the maximum of the generated particles with the minimum collecting flow in spite of the abrasion head displacement.

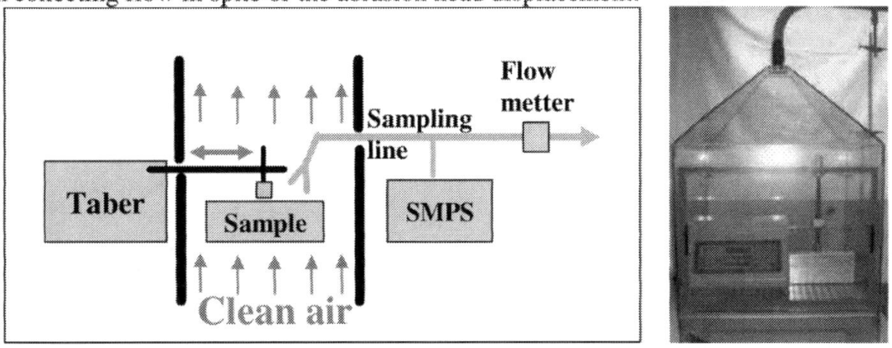

Figure 1. Schematic of the measurement tool and a picture of the equipment set up at CEA.

A 100 x 30 mm sample is placed at the bottom of the device and a 30 x 30 mm part of the same sample is placed at the moving upper part. The displacement length is 100 mm at the frequency of 60 cycles/min.

The measurement of the released nanoparticles is performed with a SMPS (Grimm 5.5-300) presenting a flow rate of 0.3 L/min and a CPC (Grimm 5.403) presenting a flow rate of 1.5 L/min. The SMPS gives the concentration of nanoparticles and their distribution which is very interesting to know if the abrasion releases single nanoparticles or bigger parts of the matrix. On the other hand the CPC gives the number of nanoparticles only but is much more sensitive. To illustrate that, we can calculate the theoretical Low Limits of Detection in the best case: when the collection flow equals the measurement flow of the counter.

- Assuming that an average of one particle in each of the 44 channels of the SMPS has to be detected during the 4 minutes measurement to get a relevant value and taking into account the flow rate of the SMPS: 0.3 L/min, calculations show that we need to release more than 1200 particles during the 4 minutes.

- For the CPC we assume that wee need 10 events in the same 4 minutes in order to get a relevant number. This means that we need to release more than 10 particles during the 4 minutes.

Measurements performed alternatively with a SMPS and a CPC are therefore complementary.

3. Qualification of the measurement tool

Thanks to the clean air flow the background noise count was measured under 5 particles @ 6 nm during 4 min whatever the TABER system is switched on or not.

As seen in Figure 2 the collection efficiency was estimated as a function of the Taber head speed by injecting intentionally a calibrated amount of Pt nanoparticles very close (1 mm) to the abrasion area. For small nanoparticles around 6 nm the collection efficiency is 100 % when the TABER system is off and decreases to 60% when the abrasion speed is maximum (0.2 m/s) due to losses induced by the head displacement. This means that it is necessary to correct the measured values for the smallest nanoparticles.

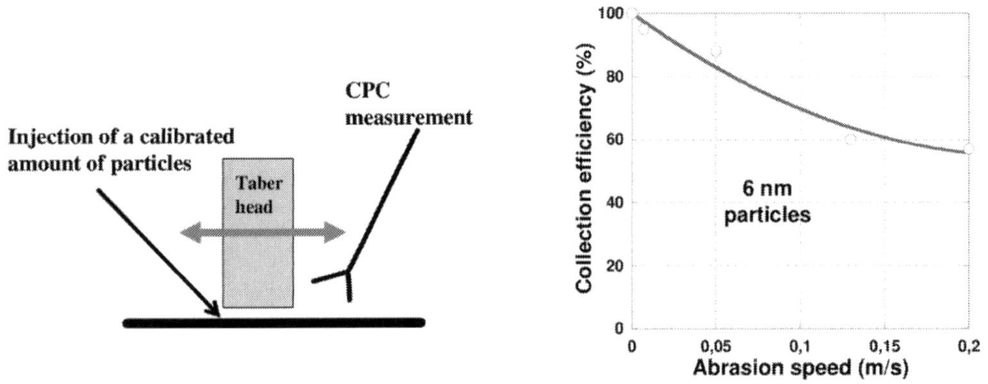

Figure 2. Schematic of the collection efficiency test and obtained results for small nanoparticles (6 nm).

Then a first test was performed on a cotton fabric sample covered with 20 nm silica particles deposited by dropping a colloidal suspension (liquid) followed by a simple air dry. First the background was quantified and estimated to be negligible. Than, the measurements performed with the SMPS show a clear release of the deposited nanoparticles. A second measurement after the first 4 minutes of abrasion indicates an important decrease of the released nanoparticles in spite the quite high level of deposited nanoparticles at the surface of the fabrics. This means that in practice it is necessary to use a new sample for each measurement in certain cases.

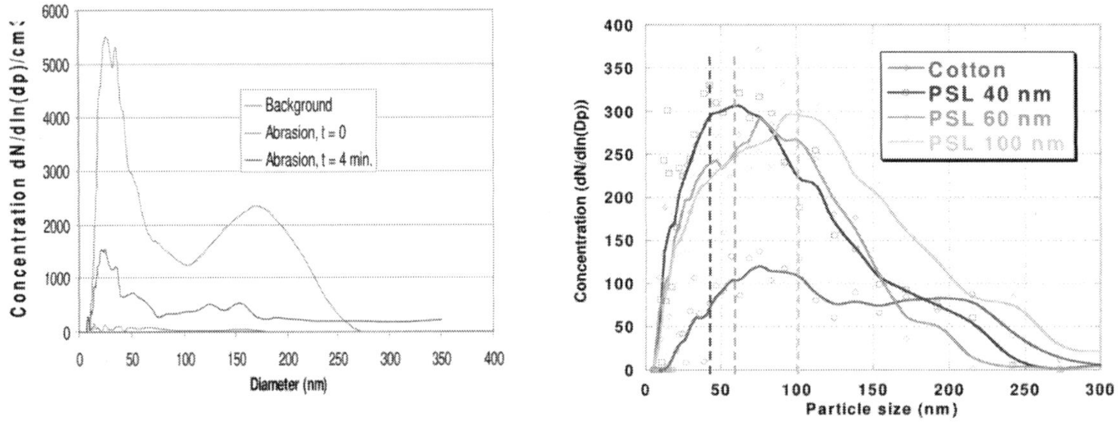

Figure 3. First results obtained on cotton fabrics containing nanoparticles. Left: results from 2 successive trials with SiO_2 nanoparticles. Right: results obtained with different sizes of PSL nanoparticles.

The method has been then also qualified with calibrated polystyrene-latex, PSL nanoparticles of size ranging from 40 to 100 nm intentionally deposited on cotton fabric pieces. The results of Figure 3 show that the abrasion aerosolizes the deposited nanoparticles as the SMPS detect the same nanoparticles sizes than deposited.

4. First measurements

Then the qualified abrasion tool was used to characterize an advanced fabric piece made of a PET layer coated with an additional PVC layer containing or not nanoclays used as nanofiller (from IFTH, Lyon, France).

Submicron particles close to 80 nm are emitted by abrasion from both PET fabric and the PVC coating. This indicates that even conventional fabrics without nanofillers can release nanoparticles. A bigger nanoparticle peak is detected when testing the PVC coating containing nanoclays. Subtracting the background contribution from the PET + PVC without nanofillers allows obtaining a peak centered at 50 nm corresponding to the nanofiller (see the right curve of Figure 4). The maximum of this peak corresponds to the nanoclays size.

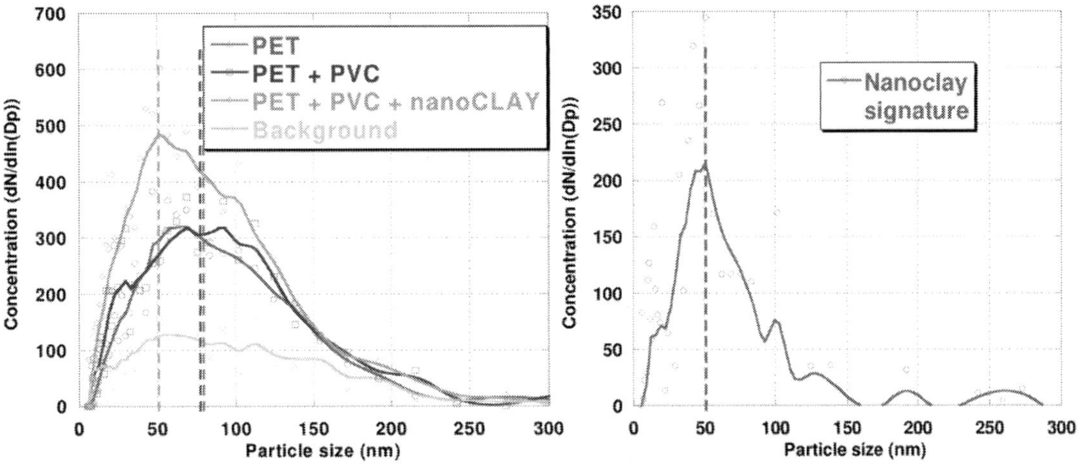

Figure 4. Abrasion test results obtained on a fabric constituted of a PET layer coated with a PVC coating containing or not nanoclays.

5. Conclusion

A new equipment to test the release-ability of nanoparticles contained in nanoproducts by abrasion is in progress at CEA to be more sensitive and accurate. The first results show that nanofillers can be released in usage by abrasion for non optimised nanoproducts. Other solicitations such as chocks and UV have also to be evaluated.

This type of equipment may be useful to protect the consumers by optimizing the hooking of nanoparticles in their matrix and perhaps one day to qualify the nanoproducts before their introduction on the market.

For this objective, a standardisation procedure at ISO is necessary in order to make this type of method recognised by all the stakeholders and regulation bodies.

Acknowledgements

This study was carried out with the financial support of the European Commission through the sixth Framework Program for research and Technology Development NMP2-CT-2005-51843 contract "NANOSAFE2."

Labelling nanoparticles by nanotracers and oligonucleotides

F. Tardif[1], O. Raccurt[1], L. Golanski[1], A. Auger[1], J. Liu[2], AG. Brachet[2], D. Jary[2], F. Vinet[2] and C. Peponnet[2]

E-mail: francois.tardif@cea.fr

Abstract. The rapidly developing field of nanotechnologies presents many opportunities and benefits for new materials with significantly improved properties as well as revolutionary applications in the fields of materials, energy, environment, medicine, etc. Nevertheless, this new industry can rapidly compete with microelectronics or automotive in terms of annual turn-over only if the nanoproducts are well accepted by consumers.
Labelling the nanoproducts as "nanomaterials" is one of the prerequisite conditions for market acceptance. One approach consists in encouraging manufacturers to insert labels since the nanoparticles synthesis, at the last step of the production reactors, in order to give consumers useful information related to nano-fillers contained in commercial products.
The first level of information could be the specification of the presence of nanoparticles in the product: a kind of tag indicating "nanomaterials inside".
The second level of labelling could supply more complex information such as: the type of nanoparticles, the manufacturer name, the batch number, the toxicity class, etc.
In European Nanosafe2 project, two main routes have been investigated: the previous one is using biomarkers based tracers. In that case the information is available thanks to conventional biological methods.
The second way requests fluorescent nanotracers which are less powerful in terms of data storage but may be more resistant against chemicals, temperature, UV aggressions and could be red in a few seconds by field operational portable decoding devices.
For economical reasons it is necessary to introduce a very small quantity of tracers in the nanoparticles. Nevertheless, for certain applications for which a high purity level of the nanoparticles is requested (*e.g.* microelectronics) this small amount can be redhibitory.

1. Labelling with biomarkers

A strategy of labelling comprising biomarkers such as oligonucleotides which can contain a very high density of coding information and also an integrated decoding apparatus has been investigated for nanomaterials traceability.

A multifunctional biomarker labelling scheme has been first designed. The nanoparticles labelling has been realised using very classical chemical grafting techniques for a covalent linkage between oligonucleotides and nanoparticles. This chemical grafting includes a generation and activation of carboxylic groups on the nanoparticle surface followed by a reaction between the biomarker and the activated carboxylic functions. The obtained solution after biomarker grafting has been tested with conventional laboratory equipments used for fluorescence detection in 10 μL tubes. It has been shown that serial dilutions can be positively detected for dilutions ranging from $1/10^4$ to $1/10^8$ corresponding to 10^{-4} to 10^{-8} mg/mL of biolabelled nanoparticles. At the lower concentration of detected nanoparticles corresponding to 10^{-8} mg/ml, only 0.1 pg of nanoparticles are present in the 10 μL test tube, which gives a detection limit much lower than what can be obtained with classical techniques for nanoparticles detection like microscopic techniques (STM, TEM, AFM for example) or spectroscopic methods like Raman spectroscopy.

ElectroWetting On Dielectric (EWOD) principle is based on the use of electrostatic forces for fluidic actuation. On an appropriate device comprising embedded electrodes, it is possible to move liquids, dispense reagents (Figure 1) and mix several droplets, thus accomplishing all basic fluidic operations necessary for performing a complex protocol by simple electrical actuations. As no moving parts, like valves or pumps, are necessary, EWOD lab on a chip are very well adapted to integration in a complete compact apparatus.

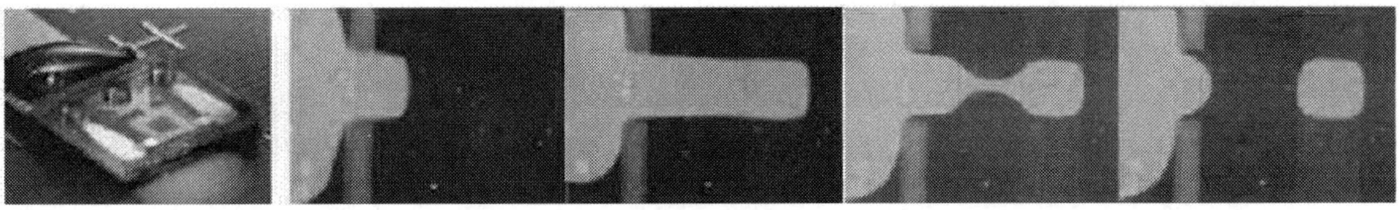

Figure 1. Droplets dispensing from a reservoir with EWOD lab on a chip by successive electrodes actuations. The obtained droplet sits on the last activated electrode.

EWOD silicon chip manufacturing is based on collective and standard micro technology processes involving successive deposition, photolithography and etching steps on a wafer. After manufacturing, the 13 mm by 13 mm silicon chips are individually packaged on a printed circuit board (Figure 2). The chip is glued on the PCB and electrical connection of the electrical pads of the chip to the PCB is obtained by wire bonding.

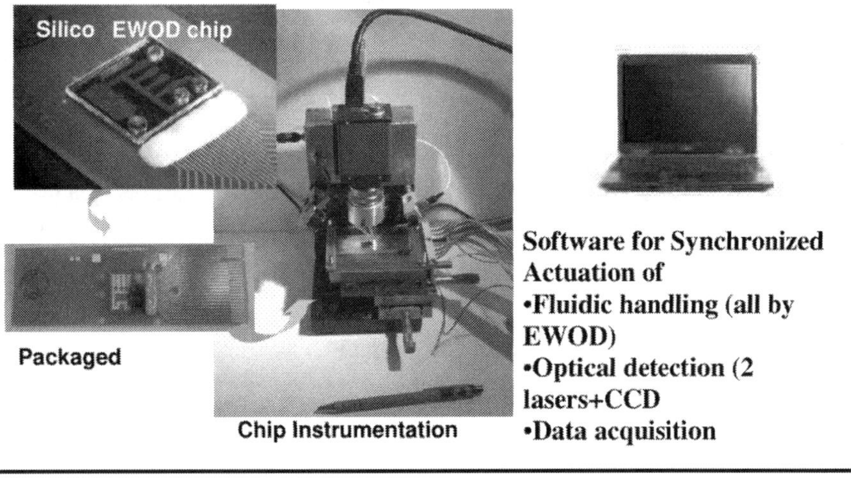

Figure 2. EWOD actuated chip and associated instrumentation.

The fluorescence detection of a diluted solution containing biomarkers labelled nanoparticles (at 0.1 pg/µL of nanoparticles) has been successfully performed in a 65 nL droplet on the lab on a chip, proving that this kind of device can be used in portable instruments for nanoparticle detection and traceability.

2. Labelling with fluorescent nanotracers

2.1. Basics of encoding

As shown in Figure 3, codes can be constituted by the relative levels of fluorescence intensities at different wavelengths. Using one of the colours as a reference, the number of possible combinations C is given by the number of colours T and the number of intensity levels N: $C = N^{T-1}$.

For 5 different intensity levels and 4 colours including one reference, the number of combination is 125. This technique is limited in practice by the pick widths and the availability of low cost dyes spread over the whole UV-IR spectrum. The example of 125 combinations given here is a good technical objective and probably sufficient to store a sufficient number of basic information to characterize nanoparticles.

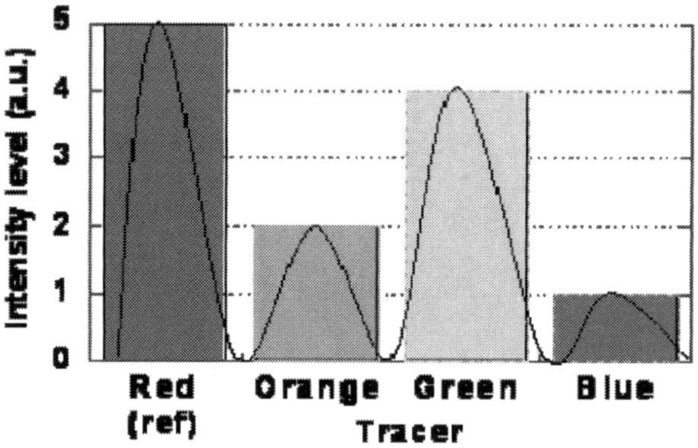

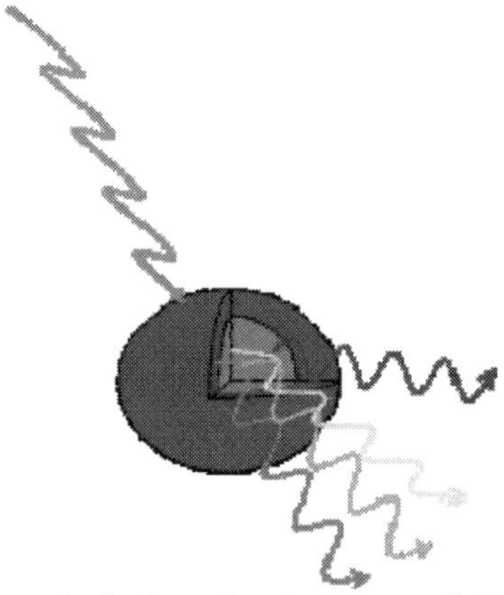

Figure 3. Code 2-4-1 constituted by the 3 colors: orange, green and blue; the red color is used as the reference (left). Schematic of a nanotracer (right).

As represented above, each nanotracer contains the different fluorescent codes which emit fluorescence under an appropriate excitation radiation.

2.2. Tests of different nanotracers

Quantum dots are potentially very interesting tracers for coding application as they present a quite sharp pick. In this case, the fluorescence is given by a quantum confinement effect: when the particle size becomes smaller than the theoretical electron-hole pair dimension. CdSe, CdS and CdZnSe were synthesized. Unfortunately the surface passivation obtained at the laboratory were not strong enough (degradation after one week) to envisage a use for nanotracer application.

Rare earth ions dispersed in oxide matrix can give even sharper picks than quantum dots. As the size is not the intrinsic parameter to obtain different wavelength, one can use bigger particles (10 - 200 nm) which present relatively less surface defects. It is quite easy to obtain phosphors by colloidal engineering. The limitation of nanophosphors for marking application comes from the quite elevated cost of raw materials.

Finally it was found that organic luminescent molecules embedded in mineral matrix represent a good opportunity for marking application as the initial materials are acceptable in terms of cost. It was shown that embedding the active molecules in polysiloxanes or organoclays to constitute the nanotracers, stabilize drastically their optical properties to temperature and UV, leading to acceptable performances for a large field of applications. Nevertheless, the drawback of using organic dyes remains their broad pick width which limits the volume of information.

A synthesis process using a sol-gel route to obtain a 20 nm organoclay with hydrophobic and hydrophilic functions was performed (see Figure 4). The incorporation of organic dye as tracers is made by dissolution in appropriate mixing of organoclays and dye. Dyes are adsorbed at the surface and in the inner porosity of nanoclays by hydrophobic functions.

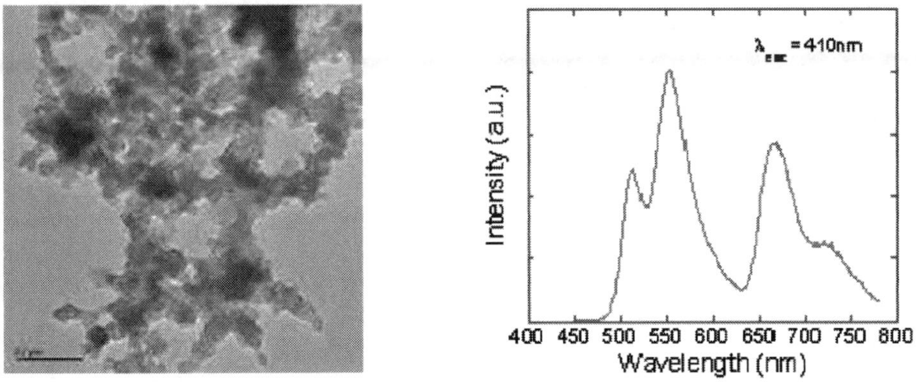

Figure 4. A three colour nanocode obtained with organic dyes inserted in organoclay (code 4-3, the first pick is used as the reference). The nanotracer size is around 20 nm. Left: TEM picture of organoclay, right: fluorescence spectra of the organoclay containing 3 fluorescent dyes (uranine, rhodamine and porphyrin) in water. Excitation wavelength 410 nm.

A 3 colour nanotracer able to contain about 50 different codes was fabricated. Figure 5 shows the spectrum of one nanotracer obtained with 3 organic dyes (code 2-1) as fabricated and then incorporated in a batch of TiO_2 nanoparticles from Degussa (P25). In spite the low concentration of nanotracers incorporated (less than 100 pp), the deconvolution of the peaks on the foot of the diffraction part of the excitation allows to recognize the initial code.

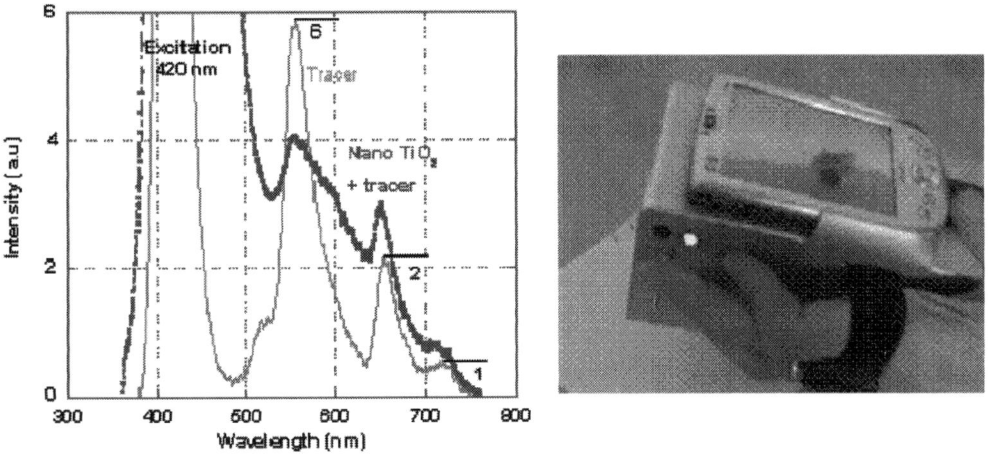

Figure 5. Left: optical spectrum of the nanotracer and residual signature when incorporated at trace level in a batch of TiO_2 nanoparticles; as measured by a spectrophotometer. Right: example of a portable decoding device able to read the nanotracers based on a micro spectrometer.

2.3. Decoding device

An example of a portable decoding device using a micro spectrometer and a PDA has been designed as a demonstrator (Figure 5). Instantaneous reading of the nanotracers can be performed.

3. Conclusion

Finally both approaches using biomarkers or nanotracers are adapted for labelling nanoparticles. The biomarker method presents the advantage to be able to carry a large number of information. On the other hand, mineral fluorescent tracers are probably more robust against temperature, UV, and chemical aggressions.

Labelling the nanoparticles as described in this work is not adapted when ultra purity is requested for nanoparticles *e.g.* microelectronic, but is suitable for nanoparticles used in a large variety of applications. A standardization initiative at ISO is needed for promoting the labelling of nanoparticles

Acknowledgements
This studied was carried out with the financial support of the European Commission through the sixth Framework Program for research and Technology Development NMP2-CT-2005-51843 contract "NANOSAFE2."

SiC nanoparticles cyto- and genotoxicity to Hep-G2 cells

Sabrina Barillet[1], Mary-Line Jugan[1], Angélique Simon-Deckers[1], Yann Leconte[2], Nathalie Herlin-Boime[2], Martine Mayne-l'Hermite[2], Cécile Reynaud[2] and Marie Carrière[1]

[1]Laboratoire Pierre Süe, CEA-CNRS UMR9956, IRAMIS, CEA Saclay, 91191 Gif sur Yvette, France

[2]Laboratoire Francis Perrin, CEA-CNRS URA2453, IRAMIS, CEA Saclay, 91191 Gif sur Yvette, France

E-mail: marie.carriere@cea.fr

Abstract. While emerging nanotechnologies have seen significant development in recent years, knowledge on exposure levels as well as data on toxicity of nanoparticles are still quite limited. Indeed, there is a general agreement that development of nanotechnologies may lead to considerable dissemination of nanoparticles in the environment. Nevertheless, questions relative to toxicity versus innocuousness of such materials still remain.
Our present study has thus been carried out with the purpose of assessing some aspects of toxicological capacities of three kinds of nano-sized particles: TiO_2 and SiC nanoparticles, as well as multi-walled carbon nanotubes (CNT). In order to address the question of their potential toxicity toward living cells, we chose several cellular models. Assuming inhalation as the most probable exposure scenario, we used A549 alveolar epithelial cells as a model for mammalian primary target organ (lung). Furthermore, we considered that nanoparticles that would deposit into the pulmonary system may be translocated to the circulatory system. Thus, we decided to study the effect of nanoparticles on potentially secondary target organs: liver (WIF-B9, Can-10, HepG2) and kidneys (NRK-52E, LLC-PK1). Herein, we will focus our attention on results obtained on the HepG2 cell line exposed to SiC nanoparticles.
Scarce literature exists on SiC nanotoxicology. According to the authors that have already carried out studies on this particular nanoparticle, it would seem that SiC nanoparticles do not induce cytotoxicity. That is one of the reasons of the potential use of these nanoparticles as biological labels [1]. We thus were interested in acquiring more data on biological effects induced by SiC nanoparticles. Furthermore, one of the particular aspects of the present study lies in the fact that we tried to specify the influence of physico-chemical characteristics of nanoparticles on toxicological endpoints (cytotoxicity and genotoxicity).

1. Materials and methods

1.1. Nanoparticle suspensions

SiC nanopowders were synthesized by laser pyrolysis [2, 3]. Their specific surface area was measured according to Brunauer, Emmet and Teller (BET) method [4]. Diameter of powdered nanoparticles could be calculated from BET results. Nanoparticles were first dispersed by sonication (Autotune 750 W, Bioblock Scientific) in ultrapure sterile water (4°C, pulsed mode: 1 s on and 1 s off, 30 min), then twice-diluted in foetal bovine serum, before being sonicated another time (same protocol). Morphology was determined by transmission electron microscopy (TEM), using a Philips EM208 microscope at 80 kV (CCME Orsay, France): a drop of each nanoparticle suspension was deposited on a TEM grid, allowed to dry and directly observed. For cells exposure, these suspensions were diluted ten times in cell culture medium (DMEM medium supplemented with 50 IU/mL penicillin and 50 µg/mL streptomycin).

1.2. Cell culture

HepG2 cells (human Hepatocellular carcinoma) were purchased from ATCC and subcultured in DMEM containing 4.5 g/L glucose supplemented with 2 mM L-glutamine, penicillin/streptomycin (50 IU/mL and 50 µg/mL, respectively) and 10% (v/v) fetal bovine. They were maintained at 37°C in a 5% CO_2/air incubator and passed at confluence.

1.3. Cytotoxicity assay

Nanoparticles cytotoxicity was assessed by using 3-(4,5-dimethylthiazol-z-yl)-2,5-dipheny-tetrazotium bromide (MTT) assay [5]. Mitochondrial dehydrogenases of viable cells reduce MTT to water-insoluble blue formazan crystals which are then solubilised by dimethyl sulfoxide (DMSO); this assay thus indicates cell mitochondrial activity impairment. After exposure, 10 µL of a 5 mg/mL MTT solution were added to each well. After 1 h at 37°C, medium was then replaced by 100 µL of DMSO and mixed thoroughly to dissolve the formazan crystals. Nanoparticles were allowed to sediment during 1 h and 50 µL of each well were then transferred to another plate. Absorbance was measured at 570 nm. Cell viability was determined as a percentage of the negative control (untreated cells).

1.4. Genotoxicity assay

Comet assay was performed under alkaline conditions. Cells were embedded in low melting point agarose (LMPA) before being deposited on precoated microscope slides and covered by a coverslip. The slides were placed on a flat tray and kept on ice for 5 min until agarose solidifies. The coverslips were removed and another layer of 0.5% LMPA was dispensed, covered by a coverslip and allowed to solidify on ice. The slides were immersed in a cold lysing solution (2.5 M NaCl, 100 mM EDTA, 10 mM Tris, 10% DMSO, 1% Triton X-100) for 1 h. DNA was then allowed to unwind for 40 min in an alkaline electrophoresis solution (300 mM NaOH, 1 mM EDTA, pH>13). Electrophoresis was performed in a field of 0.7 V/cm and 300 mA current for 24 min. Slides were neutralized with 0.4 M Tris pH7.5 and stained with 50 µL of 20 µg/mL ethidium bromide. At least 50 comets per sample were analyzed under a fluorescence microscope (Olympus BX41) equipped with a 350-390 nm excitation and 456 nm emission filter at x20 magnification. Comet parameters were analyzed by using Comet IV software (Perceptive Instruments, Suffolk, UK).

1.5. Statistical analysis

Results are expressed as mean ± standard deviation of at least three assays. Statistical evaluation was performed with STATISTICA software (Statsoft). One-way Kruskal-Wallis ANOVA was used, followed by Mann-Whitney U-tests. $p<0.05$ was considered to be statistically significant.

2. Results and discussion

2.1. Physico-chemical characterization

All these nanoparticles were spherical. A TEM picture is presented in Figure 1. After the first sonication step, in water, nanoparticles were still agglomerated into large clusters of 800-2000 nm. After the second step of sonication, suspension dispersion state was better, but nanoparticles were still agglomerated into 50-100 nm clusters. These suspensions were stable for almost 1 month.

Table 1. Nanoparticles characteristics.

	Dimensions		
	Size BET (nm)	SSA (m^2/g)	Si/C ratio
1	19.95	94	1.27
2	15.00	125	0.83
3	13.99	134	1.01
4	13.39	140	1.24
5	36.06	52	1.14
6	57.52	33	1.13

SSA: Specific surface area, measured according to Brunauer Emmett and Teller protocol.

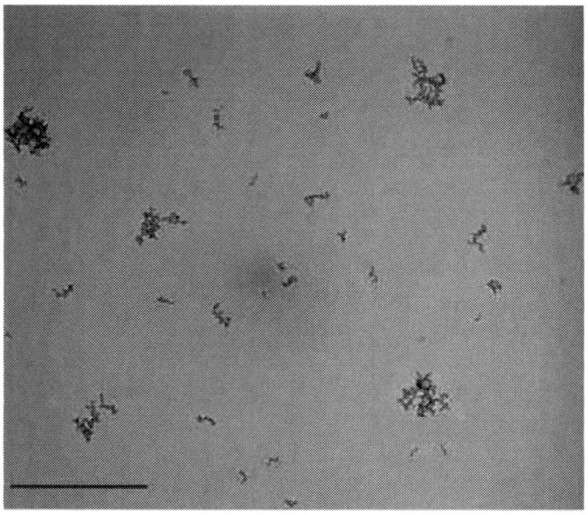

Figure 1. Transmission electron microscopy image of a nanoparticle suspension (SiC#3).
Nanopowder was first dispersed by sonication (Autotune 750 W, Bioblock Scientific) in ultrapure sterile water (4°C, pulsed mode: 1 s on and 1 s off, 30 min), then twice-diluted in foetal bovine serum, before being sonicated another time (same protocol). A drop of suspension was then deposited on a TEM grid, allowed to dry and directly observed by TEM, using a Philips EM208 microscope at 80 kV (CCME Orsay, France).

2.2. Cytotoxicity

Results relative to cytotoxicity assay are presented in Figure 2.

From a general point of view, we observed that, whatever the nanoparticle type (SiC#1, #2, #3, #4, #5 or #6), concentration (from 1 to 200 µg/mL), or time of exposure (24 or 48 h), mortality never affected more than a third of the total population. Indeed, the most drastic conditions (200 µg/mL during 48 h) only led to about 30% of mortality with SiC#1, #2, #4 and #5 nanoparticles, or to about 20% or mortality with SiC #3 and #6 (the latter one inducing the least cytotoxic effects with only 17% of cell death). This first consideration tends to demonstrate that cytotoxic effects of SiC nanoparticles are quite limited.

Considering the effects of time exposure on nanoparticles cytotoxicity, we could notice that, for some nanoparticles, major parts of the cytotoxic effects occurred during the first 24 h of exposure. The most obvious example of that trend can be seen considering results obtained after SiC#5 exposure. In this case, curves obtained after 24 and 48 h of exposure are almost perfectly superimposed. This trend was also noticed for SiC#4 (dead cells making up 19% and 29% of the total population after 24 h and 48 h of exposure respectively). Conversely, SiC#3 nanoparticle seemed to induce very weak cytotoxic effects after the first 24 h of exposure (5% cell death at 200 µg/mL nanoparticles) against more marked effects at the end of a 48 h exposure period (21% at the same concentration).

As regard the nanoparticles concentration effect on cytotoxicity, we could note that cell death is globally concentration-dependant. For some nanoparticles (#1, #2, #5), this relationship remained throughout the tested concentrations panel (from 5 to 200 µg/mL). For others nanoparticles (#3, #4, #6), cytotoxic effects seemed to reach a plateau at the highest concentrations.

Finally, nanoparticles physico-chemical characteristics (size, chemical composition) did not seem to significantly affect their impact on cell viability.

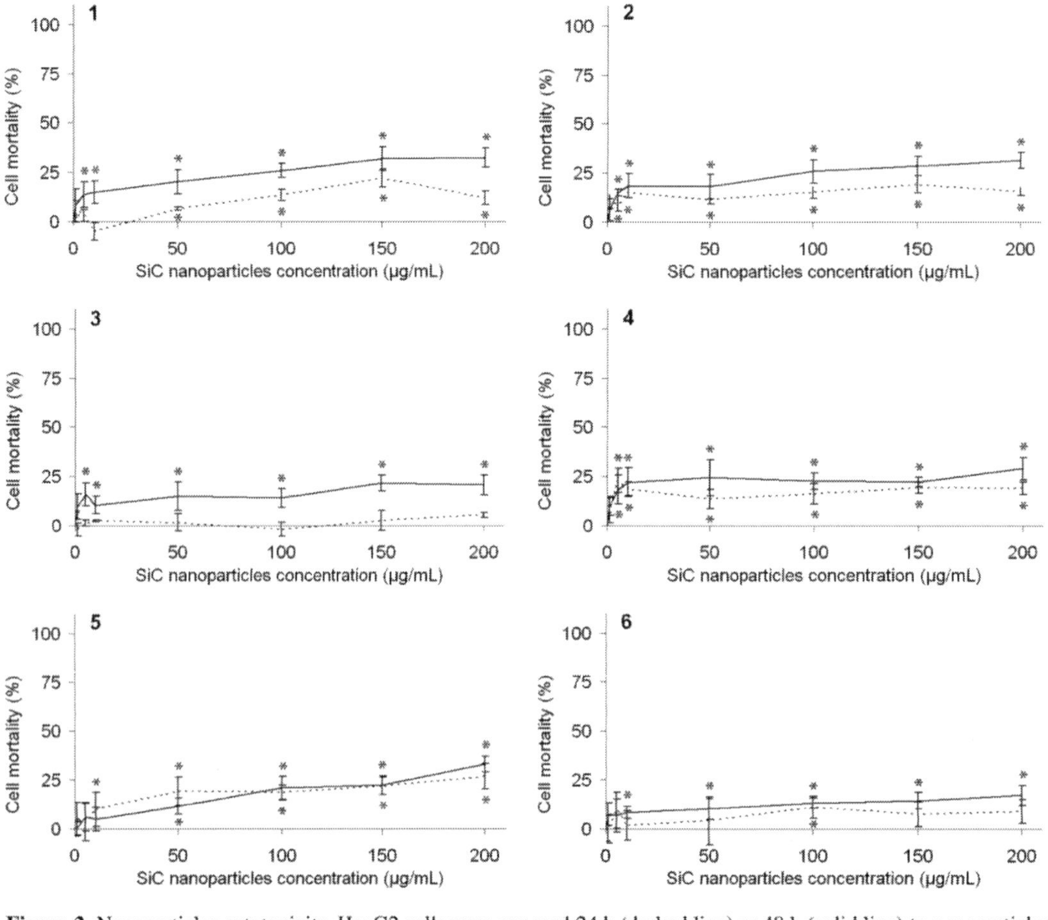

Figure 2. Nanoparticles cytotoxicity. HepG2 cells were exposed 24 h (dashed line) or 48 h (solid line) to nanoparticles and cell mortality was assessed by MTT assay (asterisks denote statistical significance, $p<0.05$).

2.3. Genotoxicity

Previous studies carried out in our laboratory tended to establish a significant relationship between nanoparticles exposure and intracellular oxidative stress (intracytoplasmic ROS overproduction, significant changes in antioxidant enzyme activities). Thus, we decided to investigate possible induction of DNA damage. However, first experimental results seem to invalidate the hypothesis of a genotoxic injury induced by SiC nanoparticles.

3. Conclusion

In this study we show that silicon carbide nanoparticles induced a slight but significant cell death. Nevertheless, it has to be noted that nanoparticles concentrations required to observe such impact are very important, and could not be appreciated as environmentally realistic. However, in case of a specific organotropism toward organs and tissues such as liver or kidney, nanoparticle concentrations could be locally disproportionately elevated and such cell response would potentially be observed.
DNA damage could not be established as a consequence of SiC nanoparticles exposure.

Considering the lack of knowledge as regards nanotoxicology, these first experimental results can be considered as a starting point for SiC nanotoxicological studies. Obviously, more complex mechanisms than just genotoxic injury need to be investigated in order to have a clearer view on nanoparticles modes of action.

Acknowledgement
This work was supported by ADEME (Agence de l'Environnement et de la Maîtrise de l'Energie), by the Région Ile-de-France in the framework of C'nano IdF. C'Nano-IdF is the nanoscience competence center of Paris Region, supported by CNRS, CEA, MESR and Région Ile-de-France. It was also funded by the French National Research Agency (ANR).

References

[1] Fan J, Li H, Jiang J, So L K Y, Lam Y W and Chu P K 2008 *Small* **4(8)** 1058
[2] Leconte Y, Maskrot H, Combemale L, Herlin-Boime N and Reynaud C 2007 *J. Anal. Appl. Pyrolysis* **79** 465
[3] Herlin-Boime N, Vicens J, Dufour C, Tenegal F, Reynaud C and Rizk R 2004 *J. Nanopart. Res.* **6** 63
[4] Brunauer S, Emmett P H and Teller E 1938 *J Am Chem Soc* **60(2)** 309
[5] Mosmann T 1983 *J Immunol Methods* **65**(1-2) 55

Biokinetics and effects of titania nano-material after inhalation and *i.v.* injection

Robert Landsiedel, Eric Fabian, Lan Ma-Hock, Karin Wiench and Bennard van Ravenzwaay

BASF SE, 67056 Ludwigshafen, Germany

E-mail: robert.landsiedel@basf.com

Abstract. Within NanoSafe2 we developed a special inhalation model to investigate deposition of inhaled particles in the lung and the further distribution in the body after. Concurrently, the effects of the inhaled materials in the lung were examined. The results for nano-Titania were compared to results from inhalation studies with micron-sized (non-nano) Titania particles and to quartz particles (DQ12, known to be potent lung toxicants). To build a PBPK model for nano-Titania the tissue distribution of the material was also examined following intravenous (i.v.) administration.

1. Material and methods

The materials tested were nano-Titania, micrometer-seized Titania and quartz dust DQ12 (Figure 1).

a) Nano-Titania	b) Micron-sized Titania	c) Quartz dust DQ 12
Primary particle 20-50 nm	Primary particle 225-500 nm	Primary particle 225-2000 nm
BET surface area 48.6 g/m^2	BET surface area 6.0 mg/m^2	BET surface area 5.9 mg/m^2

Figure 1. Transmission electron micrographs of a) nano-Titania, b) micron-sized Titania, c) quartz DQ 12.

Male Wistar (strain Crl:WI (Han)) rats (7 weeks of age) were obtained from Charles River Laboratories. All procedures for care an exposure of the animals were conducted in an AAALAC-approved laboratory in accordance with the German Animal Welfare Act and the European Council Directive 86/609/EEC.

Groups of male rats were head-nose exposed to dust aerosols for 6 h a day on five consecutive days. The amount of the test materials in the lung, mediastinal lymph nodes, liver, kidney, spleen and basal brain with olfactory bulb was determined (organ burden) in 3 rats per group and time point. The lungs of rats exposed to nano-Titania were evaluated by electron microscopy (3 rats per time point). The examinations were carried out shortly after the exposure and after an exposure-free period 14 days.

To generate the test atmosphere of nano-Titania, the material was dispersed in highly deionised water (0.5% by weight) and nebulised by a two-component atomizer (stainless steel, Schlick mod. 970). The micron-sized Titania and the quartz dust were dispersed by brush generators (developed by the Technical University of Karlsruhe in cooperation with BASF, Germany). The targeted atmospheric concentrations were 100 mg/m^3 for nano-Titania, 250 mg/m^3 for micron-sized Titania and 100 mg/m^3 for quartz dust. Technical details of the generation and characterization of the atmospheres were described elsewhere [1].

The application suspension consisted of nano-Titania dispersed in rat serum (0.5%). The Titania particle size distribution was determined by analytical ultracentrifugation of ~500 µl of a test substance application preparation that was made in parallel with that used for the i.v. administration.

A total of 12 male healthy Wistar rats, with a narrow range of body weights, were randomly assigned to three experimental groups, with one control and three treated animals in each group. The control animals were injected intravenously via the tail vein with ~1 ml of sterile saline, and the treated animals were injected i.v. via the tail vein with the test substance preparation. The dose was 5 mg/kg body weight (~1 ml of test substance preparation/kg of rat body weight). At 1, 14, and 28 days post i.v. injection rats were sacrificed by cervical dislocation under isoflurane anesthetises. Blood samples were taken, lung, liver, kidney, spleen and brain were collected and weighed immediately after sacrifice of the animals. The amounts of the test materials in these organs as well as in blood cells, plasma and popliteal lymph nodes were determined. Following this 28-day study [3], a 90-day study with i.v. injection was performed in the same manner.

To determine Ti and Si content in the tissues, wet tissue samples were digested and the solutions were ananlysed by Inductively Coupled Plasma Atomic Emission Spectrometry (ICP-AES) at wavelength of 338.376 nm for Ti and 288.15 nm for Si.

2. Results

The data of the target inhalation atmosphere concentrations, the actual analyzed concentrations and the particle size distribution measurements are presented in Table 1. The MMADs were between 1.0 and 1.2 µm, which is highly respirable for all three materials without any differences between those of the nanomaterial and the micron-sized materials. The calculated respirable fraction (MMAD < 3 µm) ranged from 81.7% to 93.1%. Only approximately 10% of the particles measured by the SMPS were smaller than 100 nm for all three exposure atmospheres (Table 1). The mass concentration of this fraction (<100 nm) was calculated (Table 1). For atmospheres with nano Titania the number concentrations of particles <100 nm represented only 0.5% of the total particle mass.

We used an analytical ultracentrifuge to measure both the highly concentrated and very turbid samples used for the i.v. injections and a sample that was diluted with FBS gold. The particle size of the Titania used for the i.v. injection was mostly in the fine fraction with components of up to 1 µm in size. Approximately 10 wt% of the particles were found in the nano-size range (< 100 nm).

After the 5-day inhalation exposure, Ti or Si was only detectable in the lung and mediastinal lymph nodes of the exposed animals (Table 2). The limit of quantification for Ti was 0.3 µg per tissue sample, corresponding to 0.5 µg Titania. The detection limit for Si was 5 µg per tissue sample, corresponding to 11 µg SiO_2.

Table 1. Particle size and concentration in the test atmospheres.

Variable	Test substance		
	Nano-Titania	Micron-Titania	Quartz DQ 12
Target concentration (mg/m^3)	100	250	100
Measured concentration (mg/m^3)	88.0 ± 6.4	274.0 ± 30.5	96.0 ± 5.4
MMAD (µm) / GSD	1.0 / 2.2	1.1 / 2.2	1.2 / 2.1
OPC (µm):count median diameter (Q_0)	0.6	1.5	0.4
SMPS (µm): count median diameter (Q_0)	0.2	0.2	0.2
Count concentration of particles in SMPS (number particle/cm^3)	88 x 10^4	13 x 10^5	33 x 10^4
Count concentration (number particle/cm^3) of particles < 100 nm	205 920	54 600	21 292
Calculated* mass fraction < 100 nm	0.5 %	0.05 %	0.03 %

* assuming all particles are spherical and have a diameter of 100 nm, the mass fraction (%) was calculated by multiplying the volume of a particle with particle count concentration (<100 nm) and physical density 4.2 g/cm^3 for Titania and 2.65 g/cm^3 for quartz dust DQ 12.

Table 2. Organ distribution after 5-day inhalation exposure
(mean values of 3 animals per group and time point in µg per organ)

Organ	n-Titania		p-Titania		Quartz DQ 12	
	Day 5	Day 19	Day 5	Day 19	Day 5	Day 19
Liver, kidney, spleen, basal brain with olfactory bulb (µg)	< 0.5*	< 0.5*	< 0.5*	< 0.5*	< 11**	<11**
Lung (µg)	2025	1547	9182	7257	2190	1975
Lymph nodes (µg)	2.2	8.5	8.2	108	19	56

* 0.5 µg/organ was the detection limit by ICP-AES for Titania.
** 11 µg/organ was the detection limit by ICP-AES for quartz.

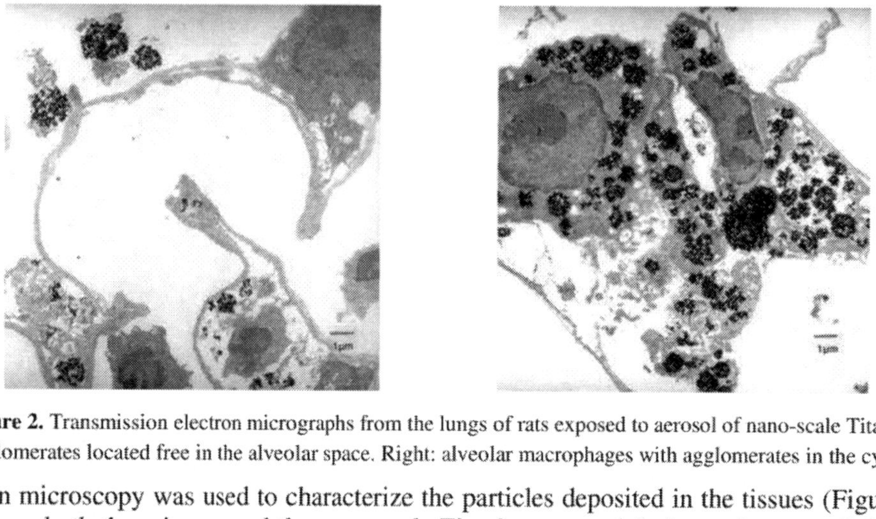

Figure 2. Transmission electron micrographs from the lungs of rats exposed to aerosol of nano-scale Titania. Left: agglomerates located free in the alveolar space. Right: alveolar macrophages with agglomerates in the cytoplasm.

Electron microscopy was used to characterize the particles deposited in the tissues (Figure 2). The particles from both the micron- and the nano-scale Titania were mainly located extra-cellularly in the lumen of the alveoli and bronchi. Moreover, particles were detected in the cytoplasm of alveolar macrophages. The particles from nano-scale material found in the lung, were mostly agglomerates of about the same size as found in the atmosphere; there were no signs of desagglomeration of the inhaled agglomerates. There was an inflammatory response in the lung and sensitive parameters in the 5-day inhalation study correlated well with results from previous subchronic inhalation study. [2]

After i.v. administration of nano-Titania, there were no detectable levels of Titania (<0.5 μg Ti per sample) in blood cells, plasma, brain, or lymph nodes (mediastenal, mesenteric, popliteal) at any of the three time points tested. The average distributions of Titania in the liver, spleen, lung, and kidney are shown in Table 3. The Ti levels were highest in the liver, followed in decreasing order by the levels in the spleen, lung, and kidney (very low levels of <0.7 μg). In kidney, a clearance could be demonstrated. The residue concentrations of Ti in liver, lung and spleen showed a high variability with a decreasing trend over time in the 28 day study, the data after 90 days, however, demonstrated that in these organs, no clearance occurred during the 3 month observation period. Ti was not detectable in urine. Ti in feces of treated animals was not significantly higher than in feces of untreated animals.

The use of this biokinetik studies for PBPK modeling and is presented at this *NanoSafe2008 Conference* [4].

There were no clinical changes and no changes in blood parameters at any timepoint, indicating that there was no detectable inflammatory response or organ toxicity.

Table 3. Titania distribution in organs after i.v. injection of 5 mg nano-Titania/kg body weight[a].

Organ	Titania on day 1		Titania on day 14		Titania on day 28	
	Control [b]	Treated [c]	Control [b]	Treated [c]	Control [b]	Treated [c]
Liver	0.5	133.8±26.0	0.5	99.5±55.8	0.5	111.3±27.3
Spleen	0.8	78.7±19.3	0.9	48.8±27.8	0.7	33.3±28.8
Lung	1.5	8.8±0.6	1.7	2.8±1.4	1.5	2.3±1.0
Kidney	< Loq	0.67±0.06	< Loq	< Loq	< Loq	< Loq

[a] Values expressed as mean ± SD (μg/g wet tissue) [b] n=1 [c] n=3.

3. Conclusion

Due to comparable aerodynamic particle size, inhaled nano- and micron-sized Titania as well as quartz (DQ12) particles deposited similarly in the lung, but had significantly different effects. Translocation to other organs than the lung-draining lymph nodes was not observed. nano-Titania administered by i.v. injection did not cause toxic effects; it was mainly found in the liver and no clearance was detected within 90 days.

References

[1] Ma-Hock L, Gamer A, Landsiedel R, Leibold E, Frechen T, Sens B, Huber G and van Ravenzwaay B 2007 *Inhalation Toxicology* **19** 833

[2] Ma-Hock L, Burkhardt S, Strauss V, Gamer A, Wiench K, van Ravenzwaay B and Landsiedel R 2009 *Inhalation Toxicology* **21** 102

[3] Fabian E, Landsiedel R, Ma-Hock L Wiench K, Wohlleben W and van Ravenzwaay B 2007 *Arch. Toxicol.* **82(3)** 151

[4] Péry A R R, Beaudouin R, Brochot C, Fabian E, Landsiedel R, Hoet P and Bois F *"PBPK models to assess nanoparticles kinetics and distribution in humans and rats"* www.nanosafe2008.org

Evaluation of MWNT toxic effects on daphnia and zebrafish embryos

Maider Olasagasti[1], Noelia Alvarez[2], Carolina Vera[2] and Sandra Rainieri[1]

[1]AZTI-TECNALIA, Parque Tecnológico de Bizkaia 609, 48160 Derio, Spain

[2]INASMET-TECNALIA, Mikeletegi pasealekua, 2, Parque Tecnológico, 20009 San Sebastian, Spain.

E-mail: srainieri@azti.es

Abstract. Organisms of daphnia (*Daphnia magna*) and zebrafish (*Danio rerio*) embryos were exposed to a range of different concentrations of COOH-functionalized MWCNT suspended in an aqueous solution of Tween 20. Immobilization of daphnia and growth retardation, inhibition and malformation of zebrafish embryos were the endpoints tested after 24 and 48 hours. Immobilization of daphnia could be observed from 3 to 16 ppm and an increasing mortality of zebrafish embryo was detected at all the concentration tested. To identify more subtle toxic effects, we took advantage of the extensive information available on the zebrafish genome and monitored by RT-PCR the expression patterns of different zebrafish genes that could act as toxicity bio-markers. At some of the concentrations tested, changes in the expression profiles of the genes examined were detected. Our results suggest that MWCNT could potentially represent a risk to human health and environment, therefore a wider range of concentrations and further testing of this molecules should be carried out to define possible limitations in their use.

1. Introduction

Carbon nanotubes are increasingly used in several industrial fields. Due to their unique electrical, mechanical and thermal properties they are currently employed in electronics, computer, biosensing and they have a potential application in the areas of food packaging for example as spoilage biosensors. In spite of the increasing popularity and vast use, their innocuity for human health still remains to be fully demonstrated. The objective of this work was to assess the potential toxic effects of multi walled carbon nanotubes (MWCNT) functionalized with a carboxylic group to facilitate dispersion in solutions. To achieve this aim we employed two well established ecotoxicological models: daphnias and zebrafish. In addition, zebrafish is an extensively used vertebrate model in developmental biology, pharmacology and medical science currently relying on a wide number of genetic tools including a nearly fully sequenced genome. In particular, in this study we employed zebrafish embryos that up to 48 h post hatching allow the performance of *in vitro* tests [1, 2]. The toxicity of carbon based nanomaterials using such model organisms has been investigated by several Authors and a number of studies have demonstrated some concerns especially in relation to C_{60} fullerenes toxicity [3- 5]. The rationale of this work is to employ an acute toxicity test with daphnia as a first broad screening method and subsequently to detect more subtle effects of MWCNT on a model vertebrate, by testing the differential expression of some selected toxicity biomarker genes on exposed zebrafish embryos.

2. Materials and methods

2.1. Pristine Test material

Commercial Catalytic Chemical Vapour Deposition Multi-Wall Carbon Nanotubes (CCVD MWCNTs), were used. The MWCNT characteristics were as follows: outer mean diameter, 10-15 nm; length, 0.1-10 μm; mean number of walls, 5 – 15; Carbon content above 90 wt% without detectable and free amorphous. To verify the presence of impurities, chemical analysis was carried out on the MWCNTs by Inductively Coupled Plasma Atomic Spectroscopy (ICP-AES). A termogravimetric analysis was also performed in diluted air in argon atmosphere using a TGA 92 16.18 SETARAM equipment applying a 2° C min^{-1} ramp up to 1200ºC.

2.2. MWCNT functionalization

Carboxyl-functionalized CNTs were prepared by oxidation of MWCNT with HNO_3. MWCNT (300 mg) and HNO_3 (65%, 200 ml) were mixed in a 500 ml flask equipped with a condenser. The flask was then immersed in an ultrasonic bath under reflux at 60°C for 2 h. After cooling down to room temperature, the reaction mixture was filtered under vacuum through a polycarbonate filter of 0.2 μm pore size (Millipore). The solid was dispersed in 500 ml distilled water, sonicated in an ultrasound bath for 30 min and filtered again, repeating this washing until complete removal of the residual HNO_3. Finally, the CNT-CO_2H samples were oven-dried overnight at 60°C. Pristine and functionalized MWCNT were also characterised by Digital Instruments Multimode atomic force microscopy (AFM) using RTESP Rotated Tapping Mode Etched Silicon Probes at 300 KHz frequency by dispersing the sample of MWNTs in methanol and then depositing it on a special holder by a drop-drying method. The functionalized MWCNT were finally suspended in a solution of Tween 20 (160 mg/l) to aid dispersion and solubilisation. Toxicity tests were carried out on different concentrations of this MWCNT test solution.

2.3. Daphnia acute toxicity test

Daphnia magna neonates were exposed to four different concentrations of the MWCNT test solution: namely, 16; 12; 6 and 3 mg/l and to a control solution of Tween 20. Experiments were carried out in triplicate exposing 20 daphnias to each of the test solutions. Daphnia immobilization was assessed after 24 and 48 h.

2.4. Zebrafish embryos exposure

Fertilized embryos were obtained from adult WIK strain zebrafish (*Danio rerio*) bred and maintained in our laboratory following standard conditions [6]. Embryos at the same developmental stage (two hours post fertilization - hpf) were collected and rinsed in embryo water. Exposure experiments were carried by placing embryos at 2 hpf into individual wells of a 96-well plate containing 100 µl of the MWCNT test solution at concentrations of 12, 6 y 3 ppm. Embryos exposed to solutions of Tween 20 (160 mg/l) were used as controls. Exposure experiments were carried out in triplicate exposing 30 embryos to each of the concentrations tested. Morphological and developmental endpoints were observed under an optic microscope in treated embryos and controls 24 and 48 hpf. Embryos 48 hpf were then used for RNA extraction.

2.5. Expression analysis (quantitative real-time PCR)

RNA extraction was carried out in groups of 30 exposed embryos 48 hpf. RNA was extracted using TRIzol (Sigma-Aldrich) following the instruction of the manufacturer. RNA quantity and quality was determined using a BioAnalyzer 2100 (Agilent). Reverse transcription reactions were carried out using

30-40 ng of total RNA using a TaqMan Reverse Transcriptase Reagents kit (Applied Biosystems). Reaction conditions were as follows: 25ºC for 10 min, 48ºC for 30 min and 95ºC for 5 min.

Specific oligonucleotides were synthesized for the zebrafish genes *Ahr2*, *iNOS* and *NMDAR* using sequences stored in the NCBI and TIGR databases. Quantitative real time PCR (qRT-PCR) was performed using an ABI Prism 7000 Sequence Detection system (Applied Biosystems), with a SYBR Green PCR master mix (Applied Biosystems). 40 ng cDNA were used as a template. Reaction conditions were as follow: 50°C for 2 min and 95°C for 10 min, followed by 40 cycles of 95°C for 15 s and 60°C for 1 min. A dissociation step was added at the end of the PCR. The β-actin gene was used as a constitutive control, to normalize all samples. The threshold cycles were calculated by the 7000 system software and expression levels of RNAs were calculated by the $2^{-\Delta\Delta ct}$ method [7].

3. Results and discussion

3.1. Characterization of the MWCNT

The pristine MWCNT was subjected to termogravimetric and chemical analysis to assess its level of purity. TGA determination allowed detecting a 2% of catalyst impurities. Chemical analysis carried out by ICP-AES showed a content of 0.56% of Al and 0.93% of Fe and a percentage lower than 0.1% of Si and Mg. AFM analysis of the pristine material in comparison to the functionalized material showed that the functionalization improved MWCNT dispersion. (Figures 1a and 1b).

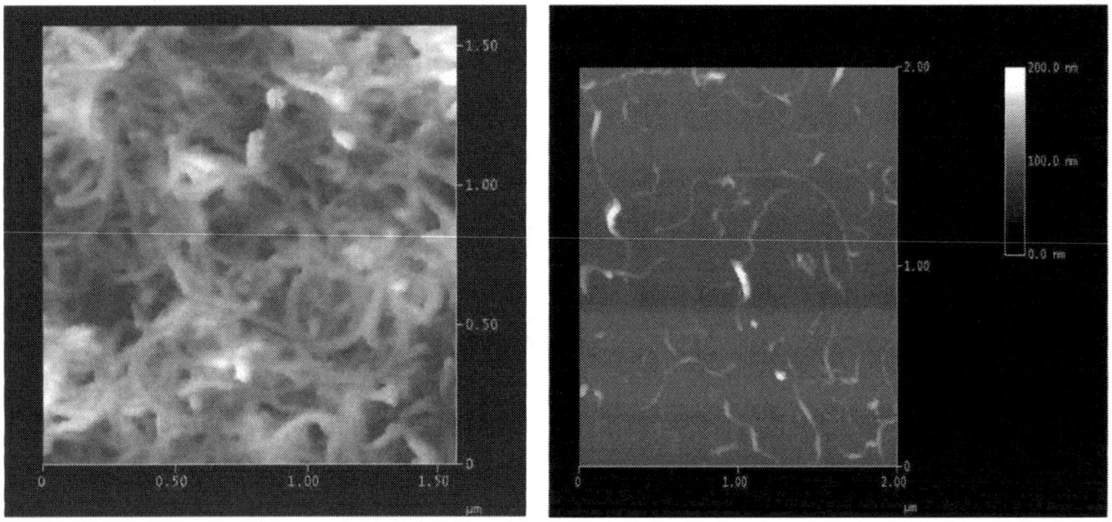

a) b)

Figure 1. Multimode atomic force microscopy (AFM) images of the MWCNT before (a) and after (b) functionalization with HNO_3.

3.2. Daphnia acute toxicity

Daphnia magna individuals were exposed to the following concentrations of functionalized MWCNT: 16, 12, 6, and 3 mg/l. As shown in Figure 2, exposure to 16 mg/l of MWCNT caused the immobilization of 95% of the daphnias after 24 h and 100% after 48 h. On the other hand, exposure to 12 mg/l caused immobilization of 15 and 10% of the animals at 24 and 48 h, respectively. Exposure at 6 mg/l only caused immobilization in 10% of the daphnias after 48 h exposure. No immobilization was detected at exposure to concentration of 3 mg/l. Exposure to the control solution of Tween 20 did not seem to cause any detectable effect on the daphnias analysed.

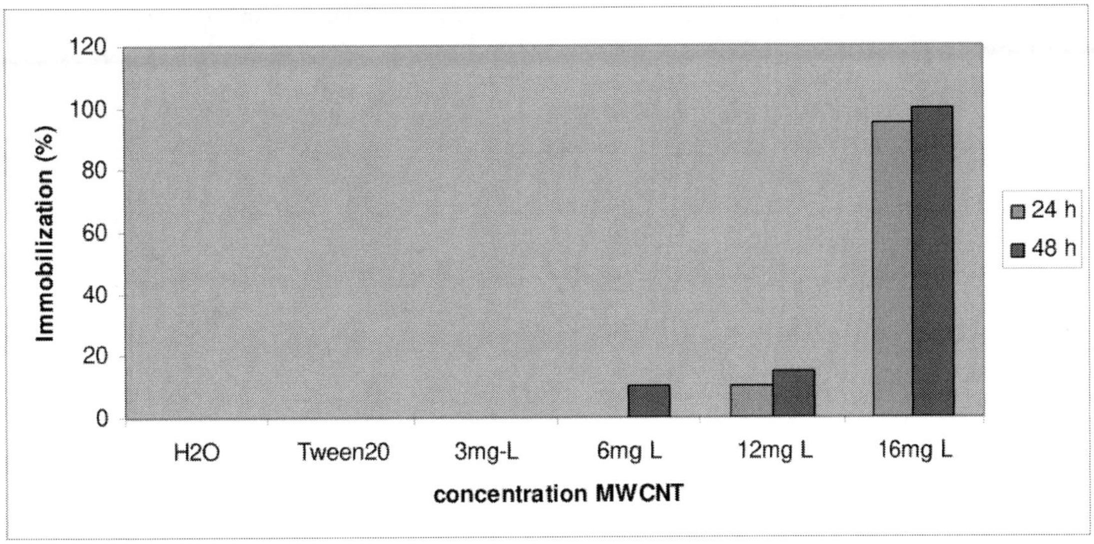

Figure 2. Immobilization % of daphnia exposed to different concentrations of the MWCNT test solution.

3.3. Zebrafish embryos acute toxicity

Zebrafish embryos were exposed to the following concentrations of the MWCNT test solution 12, 6 and 3 mg/l and to the control solution of Tween 20. The control solution showed some degree of toxicity and caused the death of 20% of the individuals already at 24 hpf. Exposure to the MWCNT test solutions caused the death of 35 to 45%, slightly increasing with the increasing of the concentration. The vast majority of the deaths occurred at 24 hpf (see Figure 3).

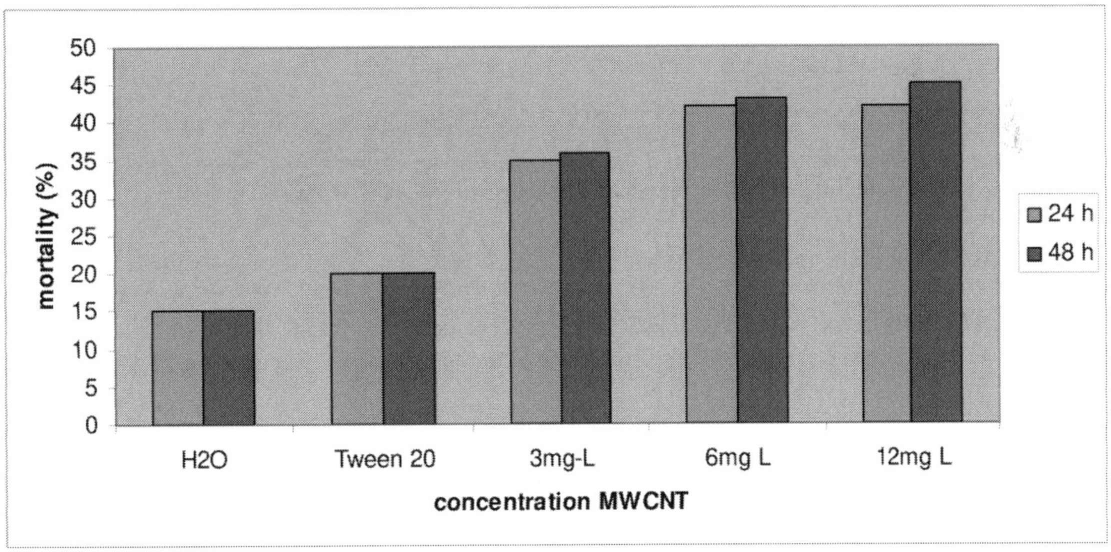

Figure 3. Mortality % of zebrafish embryos exposed to different concentrations of the MWCNT test solution.

3.4. Zebrafish embryos differential gene expression analysis

The differential expression of the following genes was tested by qRT-PCR: *Ahr2*, *iNOS* and *NMDAR*. These genes were selected as biomarkers on the basis of their function in biological processes related to the response to toxicant exposure and on the basis of previous transcription response analysis carried out with different toxicants [8]. In particular, *Ahr2* (Aryl hydrocarbon receptor 2) acts as an environmental checkpoint that senses exposure to environmental toxicants and responds by signalling cell cycle inhibition; *iNOS* (inducible nitro oxide synthase) mediates tumoricidal and bactericidal activities of macrophages; NMDAR (N-methiyl-D-aspartic acid Receptor) acts as a neurotoxicity biomarker. No changes in gene expression were detected in the embryos treated with the solution of Tween 20; however some changes were detected at some of the exposure concentrations of the MWCNT solution. In particular, as shown in Table 1, an increase in expression was clearly detected for gene *Ahr2* and this seemed to be dose responsive at the concentrations tested. *NMDAR* gene also showed an increase in the expression, but only at the highest concentration tested. These preliminary results indicate that MWCNT could potentially represent a risk to human health and environment.

Table 1. Differential expression analysis of some selected biomarker genes expressed from samples exposed to different concentration of MWCNT with respect to the control represented by the sample treated with a solution of Tween 20.

EXPOSURE SOLUTIONS	GENES FOLD INDUCTION		
	Ahr2	*iNOS*	*NMDAR*
3 mg/l	55	1	1
6 mg/l	150	1	1
12 mg/l	187	1.4	16

4. Conclusions

The results obtained indicate a possible toxic effect of MWCNT on both daphnia and zebrafish embryos for the concentrations tested. Studies carried out on SWCNT on zebrafish embryos [9] reported a delay in embryo development that the authors ascribed to the mixed effect of the CNT and of the catalyst contaminants. We do not exclude that the results obtained in the present study for both daphnia and zebrafish embryos are due to the action of MWCNTs, the catalyst contaminants as well as to the dispersant agent (in our case, Tween 20). It is indeed difficult to separate and clearly distinguish the single effect of each of these three factors; however, it should be noted that the biological application of MWCNTs is, in most cases, associated to some catalyst contaminant and to the presence of a dispersant agent. Therefore it is acceptable to consider the effect of these factors altogether. The results here presented indicate that probably the dispersant Tween 20 is not ideal for solubilising MWCNTs as it showed a degree of toxicity that was higher than expected. On the basis of the results obtained we are currently testing a wider range of dispersants, MWCNT concentrations and genes acting as toxicity biomarkers. The final aim of our research will be to define possible limitations in the use of MWCNT, especially in relation to a potential application in food packaging.

Acknowledgments
This research was supported by a grant from the Department of Industry, Trade and Tourism of the Basque Autonomous Government under the Program Etortek 2007-2009 (Project Foodbask).

References

[1] Nagel R 2002 *Altex Alternativen zu Tierexperimenten* **19** 38
[2] Scholz S, Fischer S and Gündel U 2008 *Environ Sci Pollut Res* **15** 394
[3] Oberdorster E, Zhu S Q, Blickley T M, McClellan-Green P and Haasch M L 2006 *Carbon* **44(6)** 1112
[4] Henry T B, Menn F M, Fleming J T, Wilgus J, Compton R N and Sayler G S 2007 *Environ Health Prosp* **115(7)** 1059
[5] Usenko C Y, Harper S L and Tanguay R L 2008 *Toxicol Appl Pharmacol.* **229** 44
[6] Westerfield M 2000 *The zebrafish book.* Institute of Neuroscience, University of Oregon.
[7] Livak K J and Schmittgen T D 2001 *Methods* **25** 402
[8] Voelker D, Vess C and Tillmann M 2007 *Aquat Toxicol* **86** 112
[9] Cheng J, Flahut E and Cheng S H 2007 *Environ Toxicol Chem* **26** 708

Magnetic nanoparticle effects on the red blood cells

D. E. Creangă[1], M. Culea[2], C. Nădejde[1], S. Oancea[3], L. Curecheriu[1] and M. Racuciu[4]

[1]"Al. I. Cuza" University, Faculty of Physics, 11A Blvd. Carol I, Iaşi, Romania

[2]"Babes Bolyai" University, Cluj-Napoca, Romania

[3]University of Veterinary Medicine "I. Ionescu de la Brad", Iasi, Romania

[4]"Lucian Blaga" University, Sibiu, Romania

E-mail: dorinacreanga@yahoo.com

Abstract. *In vitro* tests on magnetite colloidal nanoparticles effects upon animal red blood cells were carried out. Magnetite cores were stabilized with citric acid in the form of biocompatible magnetic fluid administrated in different dilutions in the whole blood samples. The hemolysis extent was found increased up to 2.75 in horse blood and respectively up to 2.81 in the dog blood. The electronic transitions assigned to the heme group were found shifted with about 500 cm^{-1} or, respectively, affected by supplementary vibronic structures. The Raman vibrations assigned to oxyhemoglobin were much diminished in intensity probably due to the bonding of OH group from citrate shell to the heme iron ion.

1. Introduction

The utilization of magnetite nanoparticles represents a subject of great interest in the field of multidisciplinary research since various applications in biology and medicine were carried out. The magnetite (ferric and ferrous ion oxides), in the form of functionalized and suspended nanoparticles, is used as contrast agent in Magnetic Resonance Imagery, in drug targeting techniques - by magnetic guiding of colloidal magnetite by means of adequately arrangements of magnetic fields - in cell tagging, in cancer therapy (hyperthermia), etc. The side effects induced by biological tissue contamination with magnetic compounds is also an important issue, since the magnetic field produced by them is of orders of magnitude greater than the geomagnetic field, so they can be a source of magnetic exposure, facilitating free radical formation around the magnetic particles. Until now, the consequences of nanoparticle use in medical purposes, is not fully understood.

Many reports have presented data regarding magnetic nanoparticle interaction with living tissues, especially with red blood cells. Since in most of these applications, the blood cells interaction with the magnetic nanoparticles is unavoidable, the response of erythrocytes membranes to low levels of magnetic nanoparticle concentrations remains an actual challenge. The presence of small quantities of ferromagnetic material within human body tissue was first detected while monitoring physiological magnetic fields from cardiac electric currents. Also, the liver and spleen are tissues known as able to uptake magnetic material from blood flow. Putative relaxation of magnetic nanoparticles concentrated within a tissue after intravenous injection was also reported in hamsters. The iron ions from the hemoglobin heme core might be involved in the interaction with the local magnetic field produced by the magnetic nanoparticles attached to red blood cells during fulfilling their carrier function. Generally, literature reports present positive results in pointing up the magnetic fluid's *in vitro* biocompatibility with the red blood cells, the cytotoxicity tests proving no hemolysis activity. There are a few reports mentioning the hemolytic effects of magnetic nanoparticles. For instance, the accelerated hemolysis was evidenced in the frame of an experimental study involving both electromagnetic exposure (visible radiation) and magnetic nanoparticle addition to *in vitro* red blood cells [1]. The tests described in literature as being applied to evaluate nanoparticle hemolytic effect were conducted using different methods, and even when similar approaches were used, factors such as plasma anti-coagulant, blood incubation times, centrifugal forces, assay detection wavelength varied from study to study [2-3]. Consecutively, it was really impossible to conclude what nanoparticle properties were responsible for the hemolysis – *i.e.* did it depend on size, charge, surface groups or was due to the particle absorbance.

Considering the large investigation possibilities offered by spectroscopy methods on hemoglobin behavior, the experimental investigation presented below was carried out. The present paper is focused on the magnetic nanoparticles interaction with the erythrocytes, since such situation could frequently occur during medical procedures (clinical diagnosis and therapy) or in biomedical research. The spectral behavior of animal blood in the presence of magnetic nanoparticles was investigated applying two spectral methods (electron absorption spectroscopy and vibrational Raman spectroscopy). The spectra were recorded for hemolysed blood before and after the incubation with magnetic nanoparticles, the modifications of the main spectral bands being discussed.

2. Materials and methods

The biologic material used for the experiment consists in whole animal blood (horse and dog) freshly withdrawn in the presence of heparin. The nanoparticle source was consistent with a biocompatible magnetic fluid (magnetite cores coated with citric acid), two concentrations of magnetic fluid being tested: C1 = 0.5 % and C2 = 3.0 % (v/v). The volume fraction of ferrophase was of 4.5 %, saturation magnetization of 23 kA/m, while average physical diameter was of 11.44 nm (as shown by transmission electron microscopy – Figure 1) [4]. Aliquots of aqueous magnetic fluid were brought in direct contact with the blood samples, the concentration of magnetite nanoparticles being of 10^{-12}-10^{-13}/ml. The blood samples were incubated at 37.5°C ± 0.1°C within a water bath, for 30 minutes, and then centrifuged for 5 minutes at 3 000 cycles/min, the supernatant being further investigated. The electronic absorption spectra of hemolysed blood were recorded using a Shimadzu UV 1700 spectrophotometer, the main differences, observed at the level of the two major bands in the visible domain, being discussed.

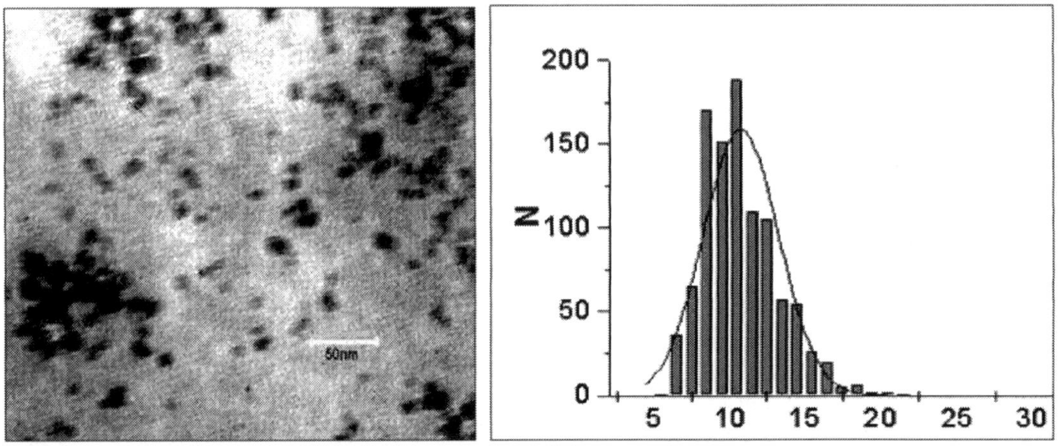

Figure 1. Left: TEM image of magnetite nanoparticles. Right: Physical diameter distribution (N: number of particles).

Hemolysis extent was assessed using the formula:

$$H.E. = \frac{A_S - A_C}{A_C} \qquad (1)$$

where A_S is the absorbance of the magnetite loaded sample at 555 nm (the most intense of the two peaks observed in the animal blood at long wavelength) and A_C the absorbance of the control sample at the same wavelength.

The absorbance ratio for the peaks having maxima at the wavelengths of 555 nm and 588 nm (18 000 cm^{-1} and respectively 16 955 cm^{-1}) was also calculated, aiming to get semiquantitative expression of the changes induced in the blood samples compared to the control:

$$R_A = \frac{A^{555}}{A^{588}} \qquad (2)$$

where A^{555} and A^{588} are the absorbance values at 555 nm and respectively 588 nm.

Raman vibration spectra were recorded at room temperature (200 Fourier transformation scans) using adequate installation provided with 400 mW laser device (Jobin Yvon T-64000 spectrometer).

Statistic significance of the investigation was ensured by working on three replies of every blood magnetite sample and controls and applying statistic t-test.

3. Results and discussion

Typical hemoglobin spectra were recorded for blood samples that were characterized by two electronic transitions in the visible range (Figure 2): relative high intensity band at 23 750 cm^{-1} (421 nm) and lower intensity and bifurcated band at 18 000-16 955 cm^{-1} (555 nm and respectively 588 nm). Significant changes were evidenced in the electronic transitions of horse blood hemolysates of magnetite loaded samples:

- the transition at 23 750 cm^{-1} was shifted toward higher wavenumber values with about 500 cm^{-1};
- the double vibronic transition at 18 000-16 955 cm^{-1} was accompanied by several new shoulders at 15 500 cm^{-1}, 19 500 cm^{-1} and 21 000 cm^{-1}.

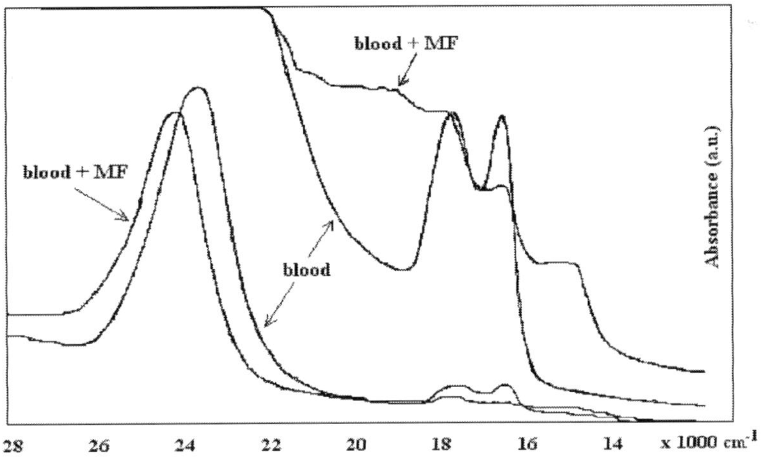

Figure 2. The electronic absorption spectra in the visible range

One could say that following the colloidal magnetite interaction with the heme group the energies of the electronic levels are changed as well as the heme stretchings, so that complex interactions are supposed to occur within the hemoglobin molecules.

In the dog blood spectra the hemoglobin bands were found at slightly shifted wavenumbers - with no more than 85 cm^{-1} toward the blue edge of the visible spectral domain. The qualitative changes, following magnetite loading were the same as for the horse blood. In all magnetite loaded samples, the hemoglobin level was higher than in the control sample, revealing the intensification of hemolysis phenomenon (Table 1) following the addition of magnetic fluid dilutions (linear approach of hemolysis extent dependence on the magnetic fluid concentration). The hemolysis extent (measured at 555 nm) significantly increased for higher concentration of magnetic fluid, following theoretical lines, with correlation coefficients higher than 0.97. Statistic significance ($p<0.05$) was revealed by t-test.

Table 1. Magnetite loading effect on hemoglobin electronic absorption spectra.

magnetite level / horse blood	hemolysis extent	absorbance ratio	magnetite level / dog blood	hemolysis extent	absorbance ratio
Control	0.00	1.045	Control	0.05	1.045
0.5 %	0.75	1.055	0.5 %	0.84	1.052
3.0 %	2.75	1.065	3.0 %	2.81	1.077

Thus, the red blood cells release significantly higher concentration of hemoglobin following the interaction with the magnetic nanoparticles, suggesting possible undesired side effects of the traces of magnetic fluid remaining for some time in the patient's blood, when used in medical applications. The iron ions of the ferrophase from the magnetic fluid could affect the relative intensity of the two electronic absorption bands studied in here, probably due to the fact that these transitions are located at the hem level - the iron ions being also involved. The possible magnetic influence of iron ions from the ferrofluid nanoparticles upon the hem structure seems to be able to modify the transition probabilities, corresponding to the main two absorption bands of hemoglobin from the visible range (the effect being also suggested by the changes induced in Raman vibration spectra).

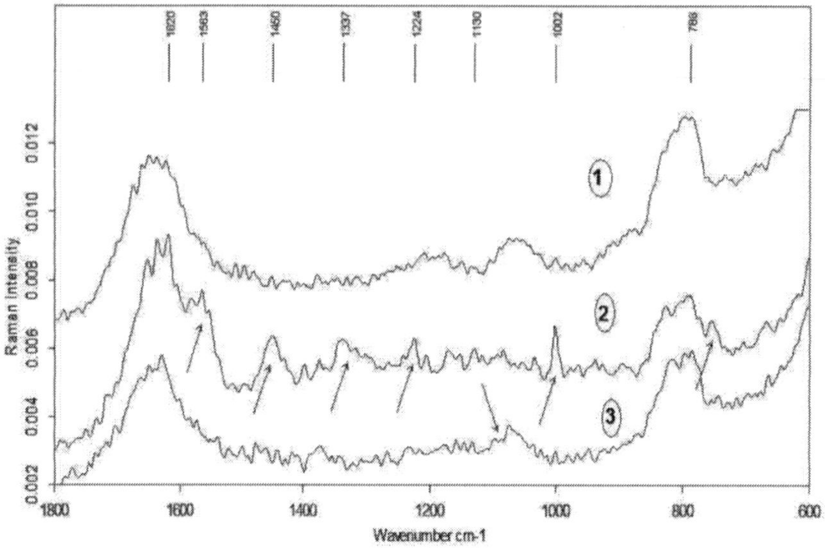

Figure 3. Raman vibrational spectra: 1) pure water; 2) horse blood; 3) horse blood incubated with magnetic nanoparticles.

In the Raman spectra of horse blood loaded with magnetic nanoparticles, several characteristic peaks at 1 563 cm^{-1}; 1 450 cm^{-1}; 1 337 cm^{-1}; 1 002 cm^{-1} and 783 cm^{-1} (Figure 3) – that are missing in the water spectrum taken as reference – are shown being diminished significantly or they disappeared after magnetite loading. Another peak, that was present at 2 950 cm^{-1} in the horse blood hemolysate, disappeared after magnetite loading. Similarly Raman peaks, corresponding to the dog blood samples at 1 543 cm^{-1}, 1 457 cm^{-1} and 1 232 cm^{-1} (Figure 4) – with lower intensity than in the horse blood – as well as that at 2,960 cm^{-1} diminished or vanished after magnetite loading; an additional peak appeared at 1 369 cm^{-1} following the interaction heme-nanoparticles. The Raman spectra allowed taking knowledge about the spectral shifts of some stretching bands corresponding to the vibration ranges, characteristic to the pyrrole ring from the heme bonded to different globins – in horse and respectively in dog blood. The behavior of the most sensitive bands - assigned to the oxyhemoglobin or to the deoxyhemoglobin - to the addition of the iron oxide particles, was interpreted taking into account the physical nature of core shell assembly, suggesting the coupling of colloidal magnetite particles to heme iron, by means of OH hydrogen bonds (OH groups being provided by the molecules of citric acid used to functionalize the magnetite particles surface within the ferrofluid composition). These results are concordant with the report of Santana [5] and Soler [6] that monitored murine hemoglobin structural changes using Raman spectroscopy, by carrying out *in vitro* biological tests using a biocompatible magnetic fluid, based on carboxymethyldextran-coated magnetite nanoparticles, emphasizing iron interaction with hemoglobin molecules. So one could say that red blood cells can be hemolysed following the magnetite loading and the released hemoglobin could interact with magnetite shell at the level of pyrrole group. It is also possible that some nanoparticles get attached to the red blood cells and exert magnetic irradiation on them, so that the electrons belonging to the heme group change their energetic levels – which can be seen in the electronic spectra behavior.

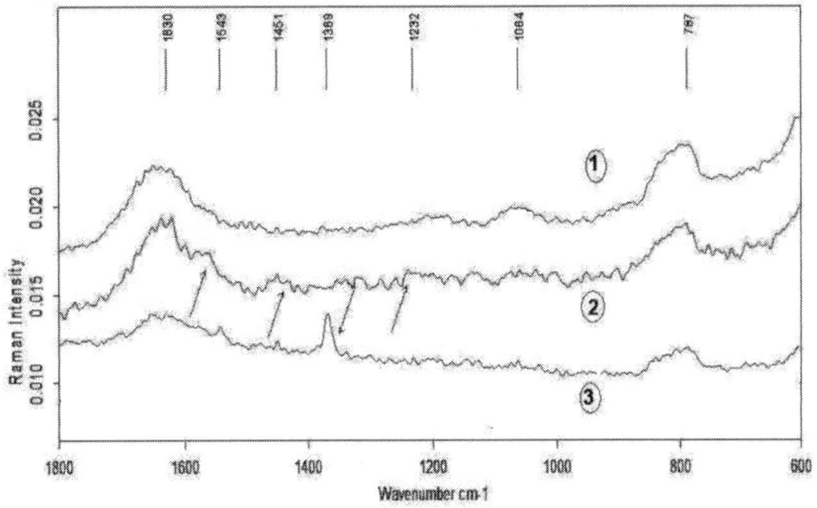

Figure 4. Raman vibrational spectra: 1) pure water; 2) dog blood; 3) dog blood incubated with magnetic nanoparticles.

4. Conclusion

The *in vitro* interaction of hemoglobin from animal blood with biocompatible magnetite nanoparticles was investigated in the frame of this study. Hemolysis extent was found significantly increased, so the hemolytic effect of magnetite nanoparticles seems to be evident. There are two possible ways that the magnetite nanoparticles influence the heme energetic levels – either electronic or vibrational ones: by interacting with heme from released hemoglobin molecules or by exerting chronic magnetic exposure on the heme iron following the addition to the red blood cell membranes. Further *in vitro* and *in vivo* investigations are planned to get detailed and deeper insight on the nature of the interactions between colloidal magnetite and heme structures.

References

[1] Valberg P.A., Burler J.P. 1987 *Biophys. J.* **52** 537
[2] Dobrovolskaia M., Aggarwal P., Hall J., McNeil Scott E. 2008 *Mol. Pharmaceutics* **5(4)** 487
[3] Dobrovolskaia M., Clogston J., Neun B., Hall J., Patri A., McNeil Scott E. 2008 *Nano Letters* **8(8)** 2180
[4] Racuciu M., Creanga D., Sulitanu N., Badescu V. 2007 *Appl. Phys. A.* **89** 565
[5] Santana J.F., Soler M.A., da Silva S.W., Guedes M.H., Lacava Z.G., Azevedo R.B., Morais P.C. 2005 *J. Magn. Magn. Mater.* **289** 452
[6] Soler M.A.G., Bao S.N., Alcantara G.B., Tiburcio V.H., Paludo G.R., Santana J.F.B., Guedes M.H., Lima E.C.D., Lacava Z.G.M., Morais P.C. 2007 *J. Nanosci. Nanotech.* **7(3)** 1069

Nanosafe 2008: International Conference on Safe production and use of nanomaterials IOP Publishing
Journal of Physics: Conference Series **170** (2009) 012020 doi:10.1088/1742-6596/170/1/012020

Antibacterial activities of PHU - $AgNO_3$ nanocomposite

Carmen Panzaru[1], M. Danciu[1], C. Ciobanu[2] and Maria-Sultana Mihailovici[1]

[1]"Gr.T.Popa" University of Medicine and Pharmacy, Iasi, Romania

[2]"Petru Poni" Institute of Macromolecular Chemistry, Iasi, Romania

E-mail: carmenpanzaru@yahoo.com

Abstract. Objective was to characterize the antibacterial action for six combination of PHU-$AgNO_3$ synthesized in "Petru Poni" Institute of Macromolecular Chemistry, Iasi, Romania. The advantages of Ag nanoparticles are durability, heat resistant, low toxicity. Silver is known for its antibacterial qualities for a long time and has been used in medicine in topical treatment.

1. Materials and methods

Three subsets of samples were tested for their antimicrobial action. The chemical characteristics for each sample are detailed in Tables 1-3. The first subset contained 5 watery solutions with Poly(hydroxyl)urethane (PHU) and silver nanoparticles (nanoAg) in different concentrations, and a witness (M_0) free of Ag nanoparticles (Table 1). The second subset of samples contained 8 solutions of Ag, but free of PHU (Table 2). The third subset contained 6 solutions of PHU and Ag nanoparticles in different concentrations, and a PHU-free witness (P-71) (Table 3). For the synthesis of metal nanoparticles we used method based on reduction process of precursor's metal ions that includes chemical reduction. PHU was used as Ag stabilizer. The morphology of the compounds was investigated with Zetasizer Nano ZS and transmission electron microscopy. The minimum inhibitory concentration (MIC) of the antimicrobial agents was determined by agar dilution (2000, EUCAST) [1]. Three reference strains and 5 wild strains (from pathologic specimens) were used. To prepare antimicrobial solution dilutions the scheme recommended by H.M. Ericsson and J.C. Sherris [2, 3, 4, 5] was used. The following reference and wild strains were used: *Staphylococcus aureus ATCC25923, Escherichia coli ATCC 25922, Pseudomonas aeruginosa ATCC 27853, MRSA (S. aureus meticilino-rezistent), Enterobacter spp., Pseudomonas aeruginosa, Acinetobacter baumannii, Candida albicans.* For testing the antifungal action we used the method recommended by CLSI (M44-A), method employing Mueller-Hinton agar supplemented with 2% glucose and 0.5% methylene blue.

Table 1. Chemical characterization (quantity in grams and concentration of PHU and Ag nanoparticles) for the solutions of the first subset.

Sample n°	H_2O (g)	PHU (g)	Concentration of PHU (%)	Sol. Ag* (g)	nanoAg^{0**} (ppm)
M_0	9.9725	0.4795	4.808	0	0
1	9.8886	0.0109	0.055	9.7958	75.6
2	10.0398	0.0402	0.265	5.0993	51.06
3	9.9826	0.0796	0.332	13.9464	88.29
4	10.0424	0.1007	0.808	2.3181	28.27
5	9.8114	0.3038	1.275	13.7042	87.65
6	10.1580	0.5030	4.045	1.7734	21.678

* - Sol. Ag – Grams of silver solution with 152 ppm Ag nanoparticles.
** - ppm – parts per million, nanoAg^0 – parts per million of silver nanoparticles less than 10 nm

Table 2. Ag^0 nanometric concentrations in the 8 polymer-free samples (second subset).

Sample n°	ppmAg
1	116.32
2	84.54
3	83
5	75.91
6	28.94
10	17.74
20	45.75
60	10.08

Tested strains: Staphylococcus aureus ATCC25923, Escherichia coli ATCC 25922, Pseudomonas aeruginosa ATCC 27853, MRSA (S. aureus meticilino-rezistent), Enterobacter spp.

Table 3. Quantities of PHU and Ag nanoparticles contained in the third subset of samples.

Sample n°	P-1	P-2	P-25	P-38	P-42	P-53	P-71
Ag. (ppm)	29.34	7.14	21.17	16	55.26	58.16	58.39
Concentration of PHU (%)	0.15	0.35	0.69	0.93	1.18	1.96	-

For the synthesis of metal nanoparticles we used method based on reduction process of precursor's metal ions that includes chemical reduction. Poly(hydroxyl)urethane (PHU) was used as Ag stabilizer. The morphology of the compounds was investigated with Zetasizer Nano ZS and transmission electron microscopy.

2. Results

Table 4 presents the minimum inhibitory concentrations of the products listed in Table 1.
No tested product acts upon *Candida albicans*.

Table 4. Minimum inhibitory concentrations of compounds 1, 2, 3, 4, 5, 6 and M_0 (first subset of samples) against the mentioned strains.

Sample	Concentration of PHU (%)	Strains*						
		1	2	3	4	5	6	7
1	0.055	75.6	37.8	37.8	-	37.8	37.8	37.8
2	0.265	-	51.06	12.51	-	51.06	51.06	51.06
3	0.332	-	-	22.07	-	-	88.29	88.29
4	0.808	-	14.13	14.13	-	14.13	28.27	28.27
5	1.275	-	43.82	21.91	-	43.82	43.82	43.82
6	4.045	-	-	-	21.678	21.678	-	21.678
M_0	-	-	-	-	-	-	-	-

(* - *1. Staphylococcus aureus ATCC25923, 2. Escherichia coli ATCC 2592, 3. Pseudomonas aeruginosa ATCC 27853, 4. MRSA (S. aureus meticilino-rezistent), 5. Enterobacter spp., 6. Pseudomonas aeruginosa, 7. Acinetobacter baumannii*)

Minus (-) signifies "no action against the species". Product 1 has the best spectrum of antimicrobial activity of all tested products (however, it does not act against MRSA); different amounts are required for inhibiting the various bacterial species. Products 2 (Fig. 1), 3, and 5 have similar spectra of activity, but at different concentrations. Product 3 acts only against *Pseudomonas aeruginosa ATCC 27853, Pseudomonas aeruginosa*, and *Acinetobacter spp.* Product 6 acts only against MRSA, *Enterobacter spp.* and *Acinetobacter spp.* Two weeks later all products were tested again and we noticed that all of them lost their inhibiting action on all strains under study. This finding suggests that solutions are unstable with respect to their antimicrobial action.

Table 5 presents the test results for the products described in Table 2. For these products, we were unable to determine the minimum inhibitory concentration (MIC) because the plates (culture medium + product to be tested) prepared by the sample provider, contained a sample at a single concentration.

Table 5. Antimicrobial action of products 1, 2, 3, 5, 6, 10, 20, and 60 (second subset of samples).

Sample No.	Concentration (Ag	
1	116.32	-
2	84.54	-
3	83.00	-
5	75.91	No action upon *S. aureus ATCC, Enterobacter*
6	28.94	-
10	17.74	-
20	45.75	-
60	10.08	Action only upon *Pseudomonas aeruginosa ATCC27853*

"-" signifies absence of growth, that is "with antibacterial activity against Staphylococcus aureus ATCC 25923, Escherichia coli ATCC 25922, Pseudomonas aeruginosa ATCC 27853, MRSA (methicillin-resistant S. aureus), Enterobacter spp".

Table 6. Antimicrobial action of products P-1, P-2, P-25, P-38, P-42, P-53, P-71(witness) (third subset of samples).

Sample no.	P-1	P-2	P-25	P-38	P-42	P-53	P-71
Antimicrobial activity	-	-	-	-	-	-	-

"-" signifies "antibacterial activity against Staphylococcus aureus ATCC 25923, Escherichia coli ATCC 25922, Pseudomonas aeruginosa ATCC 27853, MRSA (methicillin-resistant S. aureus), Enterobacter spp".

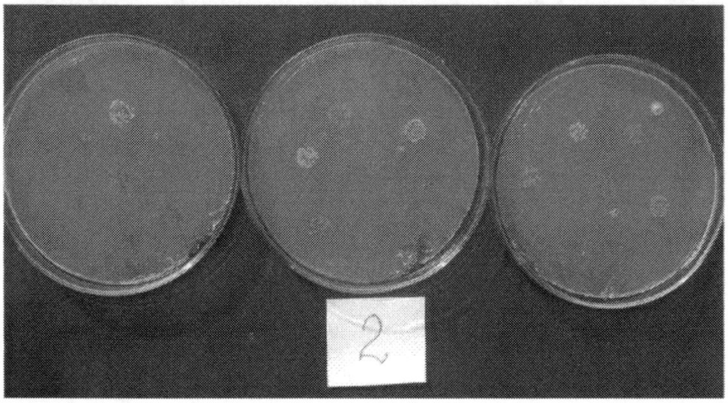

Figure 1. Antimicrobial effect of product 2 (0.0402 PHU –51.06 nanoppmAg^0): Left plate – the growth of *Staphylococcus aureus ATCC 25923* (1) and MRSA (*methicillin-resistant S. aureus*); middle plate (1/2 dilution) – only the growth of *Escherichia coli ATCC 25922*; right plate – (1/8 dilution) – no antibacterial activity at this concentration.

3. Conclusions

Product 1 ((0.0109PHU – 75.6 nanoppmAg^0) has a good spectrum of activity against: 1. Staphylococcus aureus ATCC 25923, 2. Staphylococcus aureus ATCC 25923, 3. Pseudomonas aeruginosa ATCC 27853, 5. Enterobacter spp., 6. Pseudomonas aeruginosa,7. Acinetobacter baumannii.

Products 2 (0.0402 PHU –51.06 nanoppmAg^0) (Figure 1), 4 (0.1007-28.27 nanoppmAg^0), and 5 (0.3038PHU – 87.65 nanoppmAg^0) have similar spectra of activity, but at different concentrations (*Escherichia coli ATCC 25922, Pseudomonas aeruginosa ATCC 27853, Enterobacter spp., Pseudomonas aeruginosa, Acinetobacter baumannii*).

Product 3 (0.0796 PHU – 88.29 nanoppmAg^0) acts only against MRSA, *Enterobacter spp.*, and *Acinetobacter spp.* After two weeks the products were tested again and found that all of them had lost their inhibiting action.

Of the products listed in Tables 5 and 6, except samples 5 and 60 in Table 5, all have inhibiting action against: *Staphylococcus aureus ATCC 25923, Escherichia coli ATCC 25922, Pseudomonas aeruginosa ATCC 27853, MRSA* (*methicillin-resistant S. aureus*), *Enterobacter spp.* At this stage, their MCI has not been determined.

Our results correlate the bacterial action with nanoparticle size, which is for all compounds between 1 and 10 nm.

Acknowledgment

Parts of the study have been supported by grant no. 71.146/2007 from CNMP, Romania.

References

[1] Laurian V 1986 *Antibiotics in Clinical Medicine,* Baltimore: 2nd ed. Williams and Wilkins.

[2] EUCAST DEFINITIVE DOCUMENT E.Def.1.2. Terminology relating to methods for the determination of susceptibility of bacteria to antimicrobial agents. European Committee for Antimicrobial Susceptibility Testing (EUCAST) of the European Society of Clinical Microbiology and Infectious Diseases (ESCMID) – May 2000.

[3] Buiuc D 2008 *Tratat de microbiologie clinica*. Bucuresti: 2nd ed, Ed. Medicala

[4] Durán N, Marcato P D, De Souza G I H, Alves O L and Esposito E 2007 *Journal of Biomedical Nanotechnology* **3** 203

[5] Li Z, Lee D, Sheng X, Cohen R E and Rubner M F 2006 *Langmuir* **22** 9820

Quantum Dots: proteomics characterization of the impact on biological systems

Stefano Pozzi-Mucelli[1], F. Boschi[2], L. Calderan[2], A. Sbarbati[2] and F. Osculati[1]

[1]Veneto Nanotech SCpA - European Centre for the Sustainable Impact of Nanotechnologies, Via San Crispino 106, 35129 Padova - Italy,

[2]Department of Morphological & Biomedical Sciences, University of Verona, Strada Le Grazie 8, 37134 Verona - Italy.

E-mail: stefano.pozzimucelli@ecsin.eu

Abstract. Over the past few years, Quantum Dots have been tested in most biotechnological applications that use fluorescence, including DNA array technology, immunofluorescence assays, cell and animal biology. Quantum Dots tend to be brighter than conventional dyes, because of the compounded effects of extinction coefficients that are an order of magnitude larger than those of most dyes. Their main advantage resides in their resistance to bleaching over long periods of time (minutes to hours), allowing the acquisition of images that are crisp and well contrasted. This increased photostability is especially useful for three-dimensional (3D) optical sectioning, where a major issue is bleaching of fluorophores during acquisition of successive z-sections, which compromises the correct reconstruction of 3D structures. The long-term stability and brightness of Quantum Dots make them ideal candidates also for live animal targeting and imaging. The vast majority of the papers published to date have shown no relevant effects on cells viability at the concentration used for imaging applications; higher concentrations, however, caused some issues on embryonic development. Adverse effects are due to be caused by the release of cadmium, as surface PEGylation of the Quantum Dots reduces these issues. A recently published paper shows evidences of an epigenetic effect of Quantum Dots treatment, with general histones hypoacetylation, and a translocation to the nucleus of p53. In this study, mice treated with Quantum Dots for imaging purposes were analyzed to investigate the impact on protein expression and networking. Differential mono- and bidimensional electrophoresis assays were performed, with the individuation of differentially expressed proteins after intravenous injection and imaging analysis; further, as several authors indicate an increase in reactive oxygen species as a possible mean of damage due to the Quantum Dots treatment, we investigated the signalling pathway of APE1/Ref1, a protein involved in the response to oxidative stress. Our results, although preliminary, suggest several interesting point of discussion on Quantum Dots imaging for *in vivo* diagnostic application, but also for a new therapeutic approach.

1. Introduction

Over the past few years, Quantum Dots have been tested in most biotechnological applications that use fluorescence, including DNA array technology, immunofluorescence assays, cell and animal biology. Quantum Dots tend to be brighter than conventional dyes, because of the compounded effects of extinction coefficients that are an order of magnitude larger than those of most dyes. Their main advantage resides in their resistance to bleaching over long periods of time (minutes to hours), allowing the acquisition of images that are crisp and well contrasted. This increased photostability is especially useful for three-dimensional (3D) optical sectioning, where a major issue is bleaching of fluorophores during acquisition of successive z-sections, which compromises the correct reconstruction of 3D structures. The long-term stability and brightness of Quantum Dots make them ideal candidates also for live animal targeting and imaging [1]. The vast majority of the papers published to date have shown no relevant effects on cells viability at the concentration used for imaging applications; higher concentrations, however, caused some issues on embryonic development [2]. Adverse effects seem to be caused by the release of cadmium, as surface PEGylation of the Quantum Dots reduces these issues [3]. A recently published paper shows evidences of an epigenetic effect of Quantum Dots treatment, with general histones hypoacetylation, and a translocation to the nucleus of p53 [4]. Further, several Authors indicate an increase in reactive oxygen species as a possible mean of damage due to the Quantum Dots treatment. In this study, we apply a proteomic approach to investigate the impact on protein expression and networking in the liver of mice treated with Quantum Dots, to assess potential biomarkers to be used to unravel potential biological events following the treatment; more in details, we focused our attention to the signalling pathway of APE1/Ref1. APE1/Ref-1 (Apurinic/apyrimidinic endonuclease/redox effector factor-1) is a protein involved in DNA repair [5], in particular in the base excision process; during conditions of oxidative stress, the protein migrates into the nucleus, where it works also as a transcription enhancer, strengthening the DNA binding of proteins like p53, with the effect of an inhibition of the cell cycle through the inhibition of the PI3K/AKT pathway. Our results, although preliminary, suggest several interesting point of discussion on Quantum Dots imaging for *in vivo* diagnostic application, but also for a new therapeutic approach.

2. Materials and Methods

2.1. Optical Imaging

For all the experiments a VivoVision Systems, IVIS^® 200 Series, Imaging System for small laboratory animals, Xenogen (Xenogen Corporation, Alameda USA), was used. The system is constituted by a camera sensor back thinned, back illuminated grade CCD 1 (2.7 x 2.7 cm, -90°C), with a minimal image pixel resolution of 20 µm (pixel dimension 13.5 µm, imaging pixels 2048 x 2048), quantum efficiency >85% between 500 and 700 nm, and > 30% between 400 and 900 nm. In these experiments, this fluorescent modality was chosen, with an acquisition set of excitation filters: GFP (445-490 nm), DsRed (500-550 nm), Cy5.5 (615-665 nm), ICG background (665-695 nm), ICG (710-760 nm); and an emission filter ICG (810-875 nm). A 180W Quartz halogen 3250°Kelvin lamp at high intensity; binning factor = 8; pixel resolution = 0.125x0.125; field of view = 12.8 cm; exposition time = 1 sec; opening of diaphragm (f/stop) = 2. Images acquisition protocol concerns a first "pre" acquisition and 30 "post" acquisitions after QDs800 i.v. administration. Images were acquired and analyzed with Living Image 2.6 software and Living Image 3D software (Xenogen Corporation, Alameda USA).

2.2. Animals

Athymic /nude-nude/ mice, 3-4 weeks old, were administered with a 2 µM QDs800 solution (1:5 diluted /0.01 ml/gr mouse, i.v.), and compared with control animals, injected with PBS.

2.3. Protein extraction

Animals were sacrificed by cervical dislocation, and liver was resected and snap frozen, then homogenated in 50 mM Tris-Cl pH 8.0, 100 mM NaCl, 1% NP40, and protease inhibitor. Homogenates were then incubated on a rocker 2 hours at 4°C, centrifuged at 1000 xg at 4°C for 15 minutes, and the supernatant was recovered, and stored at -20°C.

2.4. Blood fractionation assay

2 ml blood samples were collected in heparinised tubes, and centrifuged over a Ficoll gradient to separate cells from the plasma fraction. Plasma was then fractionated according to Bechtold *et al* [6]; basically, fibrinogen was precipitated overnight adding a volume of 0.5 M CaCl2, and centrifuged at 990 xg for 20 minutes; supernatant was then recovered and added with 8 ml of PBS, and then with 18 ml of a solution obtained adding 50 µl of 12 N HCl to 30 ml of Ethanol. Samples were then incubated at 37°C for 30 minutes, and centrifuged at 650 xg for 5 minutes. This step essentially leads to the precipitation of glubulins. Supernatant was finally added with 2.2 ml of 0.2 M Sodium Acetate, incubated for 15 minutes at room temperature and then centrifuged again at 650 xg for 5 minutes, to precipitate albumin. All the samples were then imaged by the mean of the Optical Imager.

2.5. Immunoprecipitation and Differential Electrophoresis

Whole protein lysate (300 µg) was incubated with 3 µg of anti-APE1/Ref1 antibody (Sigma Aldrich) 2 hours at 4°C, then immunoprecipitation was carried on by the mean of the DynaMag system (Invitrogen) according to manufacturer's protocols; electrophoresis was carried on NUPAGE 4-12% gradient gels (Invitrogen), as for all other electrophoresis analyses. Proteins were then transferred over a PVDF membrane, which was probed with an anti-p53 antibody (Invitrogen), and with a secondary Cy5-labelled antibody, and finally imaged by the mean of the Optical Imager. Differential electrophoresis was performed on 1D NUPAGE 4-12% gels, stained with Sypro Ruby; two protein bands were excised from the gels, proteins were in gel digested according to conventional protocols, and are in progress for mass spectrometry sequencing.

3. Results and discussion

Quantum Dots are semiconducting nanocrystals (2-100 nm) [6], with unique optical and electrical properties, but their more valuable property toward a biomedical application is the fluorescence spectrum, which renders them optical fluorophores for diagnostic applications. For instance, Quantum Dots can be conjugated with bioactive moieties (antibodies, ligands, oligonucleotides, ...) to target particular cell populations and structures, as for tumours [1]. Bioconjugated QDots are also being explored as tools for gene delivery, as well as for tumour photodynamic therapy. Even though several applications have been proposed for these new molecules, few data are available on the biological events that follow the use of QDots in cells and living animal models. Hence, in this preliminary study we applied a proteomic approach to identify potential alterations in the protein networks of the cells and the tissues that are site of accumulation of Quantum Dots.

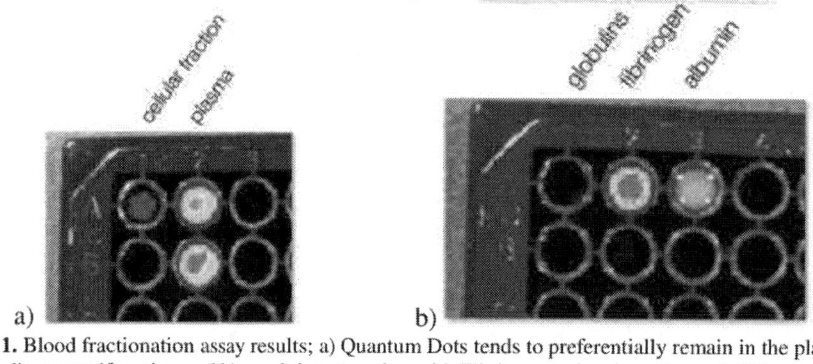

Figure 1. Blood fractionation assay results; a) Quantum Dots tends to preferentially remain in the plasmatic fraction after Ficoll gradient centrifugation and b) precipitate together with Fibrinogen, showing a low solubility or a possible binding between fibrinogen and Quantum Dots themselves.

In this work, we used untargeted Quantum Dots, commercially available from Invitrogen, with a protective layer of PEG on the surface; this layer protects both Quantum Dots from degradation and consequent release of Cadmium ions, and should reduce accumulation in particular districts of the organism, such as liver and bone marrow [3]. We first studied the behaviour of Quantum Dots in the blood flow, to investigate their partition in the various components. Using a partitioning method, we were able to understand that, at least in nude mice, QDots tend to remain in the plasma fraction (Figure 1a), without entering the cells. Moreover, as can be seen in the plasma fractionation assay (Figure 1b), they precipitate together with fibrinogen, which is the less soluble plasmatic protein; this point can be explained by a general poor solubility of QDots in the plasma, or by coverage of the QDots surface with fibrinogen. Chromatography assays are in progress to elucidate the proportions of these two events, since gel electrophoresis mobility assays are less likely to be performed, due to the size of QDots themselves (around 10 nm).

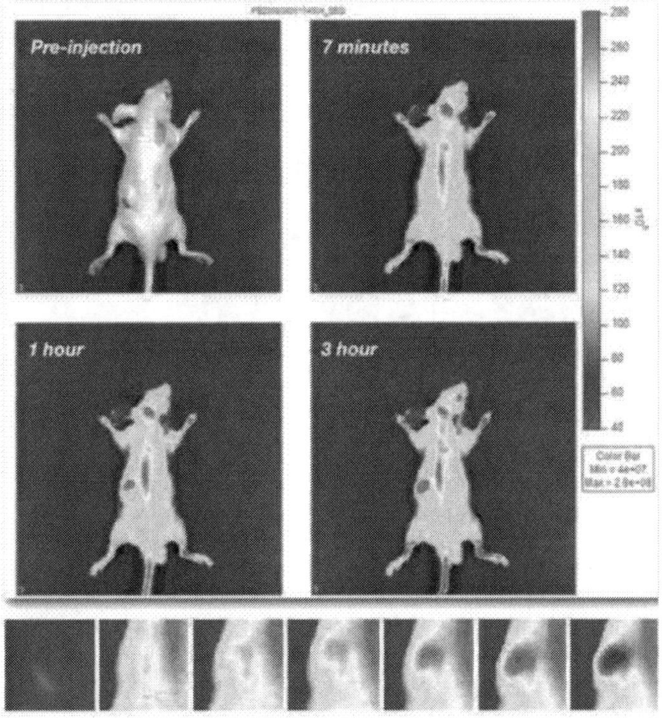

Figure 2. In vivo Optical Imaging acquisition of a mice injected with a solution of quantum dots emitting at 800 nm (Invitrogen). It is clear the accumulation in the liver area, as highlighted in the boxes below (time scale 3 hours).

To focus the molecular investigation in the actual site of accumulation, we perform a short-term distribution study of these molecules, finding that even after three hours after the intravenous injection, QDots show a strong liver accumulation, even in the presence of the PEG layer (Figure 2). We thus decide to proceed to the proteomic investigation of this organ, in a first step. Liver was resected, and processed following a non-denaturing protein extraction protocol. Proteins were separated on a 1D electrophoresis gel, stained with Sypro Ruby, and after imaging of the gels some protein bands are clearly altered (Figure 3a, red arrows), and are now in proceeding mass spectrometry investigation to sequence these proteins, to understand which molecular pathways are interested in the response to QDots treatment. As already told, a recent paper [4] showed a p53-mediated response to QDots treatment and Cadmium toxicity. Further, Fruk and co-workers [7] were able to trigger the activity of Horseradish Peroxidase by the mean of QDots light emission, proving the production of reactive oxygen species (H_2O_2, and O_2^-) during QDots light emission process, and another recent paper, showed as QDots toxicity in aquatic organism is due to ROS production [8]. We therefore decide to characterise the role of APE1/Ref1 (Apurinic/apyrimidinic endonuclease/redox effector factor-1), a protein involved in DNA repair, in particular in the base excision process. During conditions of oxidative stress, the protein migrates into the nucleus, where it works also as a transcription enhancer, strengthening the DNA binding of proteins like p53; the effect is an inhibition of the cell cycle through the inhibition of the PI3K/AKT pathway. We checked its expression levels by the mean of western blot analysis, but in the time scale of the experiments performed we weren't able

Journal of Physics: Conference Series **170** (2009) 012021 doi:10.1088/1742-6596/170/1/012021

to identify a statistically significant change; however, and surprisingly, it is important to notice the strong increase in the binding between this protein and p53. This results points out an important role of APE1/Ref-1, which is a 'sensor' of the oxidative stress in the cells, in the response to radicals produced by QDots, and can be an advantage, for instance, in the case of photodynamic therapy of p53-negative tumours, whit an incremented damage due to the reduced reactivity in this molecular pathway.

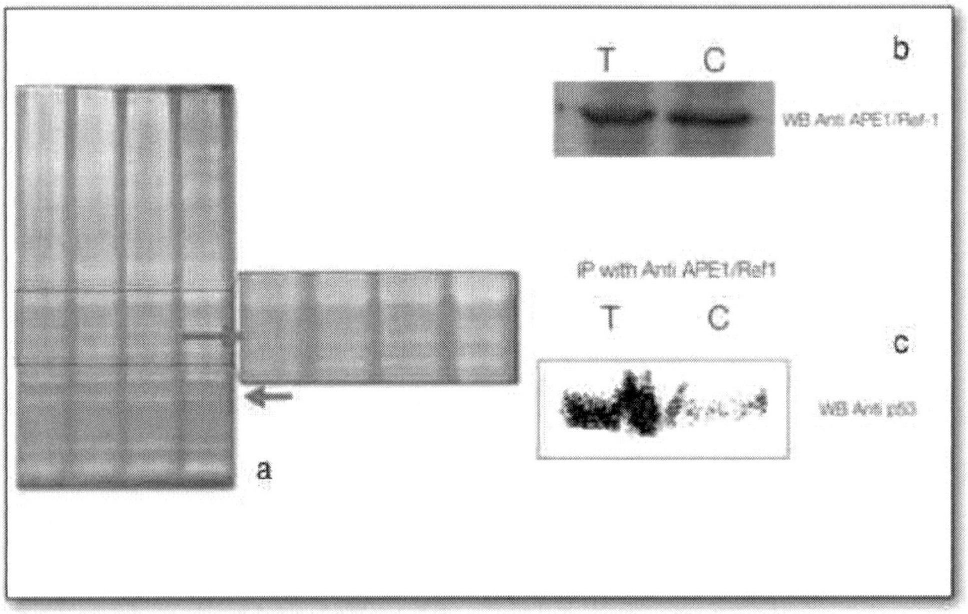

Figure 3. a) red arrows indicate differentially present protein bands in whole liver lysates, which are undergoing mass spectrometry identification; b and c) show that while the expression of APE1/Ref1 remains unchanged in the liver after the administration of Quantum Dots, its binding activity with a fundamental molecule as p53 changes dramatically, suggesting a substantial role of APE1/Ref1 in p53-mediated Quantum Dots response.

4. Conclusion

Quantum Dots are entrusted great expectations in future biomedical applications, and in particular in diagnostics. Several studies have investigated these potential applications after the injection in animal models. To date, anyway, only limited data exist about the potential side effects of Quantum Dots after the injection within the organism. We have thus applied a molecular biology approach to investigate this impact, and even though our data are still preliminary and in need of further investigation, our results point out a possible interaction with the blood flow dynamics, with a supposed binding of Quantum Dots and fibrinogen, and the activation of a p53 mediated DNA damage response pathway. In conclusion, these preliminary achievements prompt us to proceed with our approach, as it can represent an added value to conventional toxicology characterization of nanoparticles biological impact.

References

[1] Alivisatos P 2004 *Nat Biotechnol* **22** 47
[2] Duberdret B 2002 *Science* **298(5599)** 1759
[3] Cho S J 2007 *Langmuir* **23(4)** 1974
[4] Choi A O 2008 *J Mol Med.* **86(3)** 291
[5] Fung H 2005 *Mol Cell.* **17(3)** 463
[6] Bechtold W E 1992 *Carcinogenesis* **13(7)** 1217
[7] Fruk L 2007 *Chembiochem.* **8(18)** 2195
[8] Gagnè F 2008 *Aquat Toxicol* ***86(3)*** *333*

Cytotoxicity of Titanium and Silicon Dioxide Nanoparticles

Stefanie Wagner[1], Simon Münzer[2], Peter Behrens[2], Thomas Scheper[1], Detlef Bahnemann[1] and Cornelia Kasper[1]

[1]Institut für Technische Chemie, Leibniz Universität Hannover, Callinstr.5, 30167 Hannover

[2]Institut für Anorganische Chemie, Leibniz Universität Hannover, Callinstr.9, 30167 Hannover

Email: wagner@iftc.uni-hannover.de

Abstract. Different TiO_2 and SiO_2 nanoparticles have been tested concerning their toxicity on selected mammalian cell lines. Various powders and suspensions, all of which consist of titanium or silicon dioxide nanoparticles have been examined. These particles differ in the crystal structure, the size and the BET-surface area. There was also a classification in fixed particles and in particles easily accessible in solution. With focus on the possible adsorption of the nanoparticles into the human organism, via skin and via respiratory tract, the effects on fibroblasts (NIH-3T3) and on a human lung adenocarcinoma epithelial cell line were examined. Additionally, the particles were tested with HEP-G2 cells, which are often used as model cell line for biocompatibility tests, and PC-12 cells, a rat adrenal pheochromocytoma cell line.
The viability of the cells was examined by the MTT-test. The viability results were found to partly depend on the type of cells used. The experimental results show that the adhesion of the cells on the different powders strongly depends on the type of cell lines as well as on the type of powder. It was found that the lower viability of some cells on the powder coatings is not only caused by a cytotoxicity effect of the powders, but is also due to a lower adhesion of the cells on the particle surfaces. Furthermore, it could be shown that the physical properties of the powders cannot be easily correlated to any observed biological effect. While some powders show a significant suppression of the cell growth, others with similar physical properties indicate no toxic effect.

1. Introduction

The technological progress allows the assembly of new, tiny material structures having the size of only a few nanometers. In this range the materials often exhibit different physical and chemical properties as compared with their bulk counterparts and can thus be employed for specific applications. Frequently used in its nanosize is titanium dioxide (TiO_2), *e.g.* in sun creams and toothpaste. A further growing field of application for TiO_2 is photocatalysis. In this area, titanium dioxide is; for example, used for self-cleaning surfaces as well as for the treatment of polluted water or air. SiO_2 nanoparticles are used, for example, as an anti-clumping-agent in common salt and as an additive in ketchup to decrease the adhesiveness. Due to the frequent use of these materials, it is important to confirm that their biological harmlessness still exists when prepared with particle sizes in the nanometer regime.

The environmental impact of as well as the human exposure to nanoparticles is currently being widely discussed. Presently, the exposure to mostly nanomaterials at the workplace is not regulated by maximum allowable concentration (MAC) -values. One important point for any risk analysis of nanotechnological products is the mobility of these nanoparticles. Depending on their manufacturing and application, nanoparticles can eventually reach the air or the ground water.

Most nanoparticles can bind large quantities of pollutants because of their larger surface area. Therefore, more toxic substances can be transported to the ground, including excrements and pesticides, which under normal conditions are terminally mobile. Because of the high reactivity of nanomaterials, it has even been reported that naturally nontoxic occurring substances can be transformed into toxic compounds. Because of the proven bactericidal effect of some materials, the microbial composition in the ground water could thus be changed and these particles could even be absorbed by plants and living organisms. Hence, they can enter the food chain resulting in an exposure of animals and humans. In different studies the toxic effect of various nanoparticles on living organisms in aquifers has therefore been investigated. For example, an increased mortality of *daphnia magna* after the exposition to titanium dioxide nanoparticles and fullerenes (C_{60}) was observed [1]. In water containing fullerenes at a concentration of 5 ppm a significant brain damage of Juvenile Largemouth Bass was detected by the measurement of lipid peroxidation and protein oxidation. Furthermore, inflammation of the liver was observed and taken as evidence for the diffusion of fullerenes into the whole body [2, 3]. Carbon Nanotubes diminish the procreation capability of zebra fish. Hundt-Rinke *et al.* [4] have investigated the algae growth upon addition of photocatalytically active titanium dioxide under illumination with UV light. It could be shown that nanoparticles exhibit an ecotoxicologic effect depending on their dimension and crystal structure [4]. Experiments with alumina nanoparticles indicated a reduced root growth on different useful plants such as maize, cucumber, soya and carrots. Exposition to larger particles, on the other hand, had no effect [5].

There are three possible absorption paths of nanoparticles into the human body: via the skin, via the respiratory chain and via the gastrointestinal tract. By oral assimilation, nanoparticles can be transported from the intestinal tract into the lymph stream and then into the blood [6]. By breathing, the particles can reach the lung and may cause inflammatory reactions because of their insufficient elimination by macrophages [7]. Carbon Nanotubes (CNT`s), for example, can cause such reactions in lung tissue [8, 9, 10]. An additional difficulty is the elimination of fibres. Animal tests employing rats have shown a direct absorption of nanoparticles from the nose to the brain, the potential risks of which have not yet been determined adequately [11]. Once they have been incorporated into the organism, nanoparticles can reach numerous different areas of the body because of their high mobility. Nanostructured substances can pass through biological barriers such as the blood-brain barrier or cell membranes. Moreover, they can move along the nerve pathways and arrive at organs like liver and kidney. The transfer of nanoparticles from the placenta into the foetus is not only a chance for a selective therapy use but it can also be a risk. However, the potential risk of such interactions is still widely unknown. The distribution of nanoparticles within the organism depends on their dimension, form, and material properties.

Bio-degradable nanoparticles are mostly harmless, but little is presently known concerning the fate of non bio-degradable particles [12]. It can be assumed that nanoparticles concentrate in the detoxification organs such as liver and kidney. In this case the definition of the respective MAC-values will be important.

Investigations employing human lung cells (A-549) which were exposed to different concentrations of Single Wall Carbon Nanotubes (SWCNT´s) showed a very low acute toxicity effect. Transmission electron microscopy (TEM) studies confirmed that there was no intracellular localization of SWCNT´s in A-549 cells following 24 h exposure, however, an increased number of surfactant storing lamellar bodies was observed in exposed cells [13].

The role of particle size and surface chemistry for the initiation of pro-inflammatory effects in human lung epithelial cells (A-549) after treatment with different TiO_2 particles was also investigated. TiO_2 particles were rapidly taken up by the cells, however, no particles were observed inside the nuclei or any other vital organelle. TiO_2 particles even in the form of aggregates/agglomerates can trigger inflammatory responses that appear to be driven by their large surface area. It was suggested that these effects may result from oxidants generated during particle-cell interactions through an unknown mechanism [14].

In this study, nine different powders based on titanium dioxide nanoparticles and three different SiO_2 nanoparticulate preparations were investigated concerning their toxicity on four different mammalian cell lines. The particles differ in their crystal structure, their size, and their BET-surface area. Furthermore, there was a classification into fixed particles, coatings, and into particles easily accessable in their suspensions. Because nanoparticles can be incorporated into the organism via the skin, via the respiratory and via the alimentary system, the potential toxicity of the particles was investigated employing a human lung adenocarcinoma epithelial cell line (A-549), fibroblasts (NIH-3T3), a rat adrenal pheochromocytoma cell line (PC-12), and a human hepatocellular cell line (HEP-G2). The viability of the cell was examined by the MTT-test.

2. Materials and Methods

All solutions were prepared with deionised water (Arium, SARTORIUS, Göttingen, Germany). DMEM was purchased from Sigma-Aldrich (Steinheim, Germany) and foetal calf serum (FCS) and newborn calf serum from PAA (Cölbe, Germany). Horse serum (HOS) was purchased from Invitrogen (Karlsruhe, Germany).

Aeroxide TiO_2 P 25 was a kind gift from Evonik Industries (Hanau, Germany). PC 500, PC 100, PC 50, and PC 10 were a kind gift from Millennium Inorganic Chemicals (Grimsby, Great Britain) who also kindly supplied R 15, R 25, and R 34. The Silicon Dioxide Particles were synthesized by the working group of Prof. Peter Behrens from the Institute of Inorganic Chemistry in Hannover. Buffers, salts, antibiotics and other reagents were purchased from Fluka (Buchs, Switzerland) and Sigma-Aldrich (Steinheim, Germany) and are of *per analysi* quality.

2.1. Cell Culture

All cell lines were cultivated in Dulbecco's Modified Eagle's medium (DMEM) containing the according serum and 1 % antibiotics (penicillin and streptomycin) at 37°C / 5% CO_2. After 3 to 4 days the cells had grown to confluence and were then detached with trypsin and cultured in a new cell culture flask.

Table 1 shows the different DMEM additives as well as the cell number per well for the experiments.

Table 1. DMEM additives for each cell line and the number per well for the experiments.

Cell line	DMEM additives	Number per well on TiO_2 coatings	Number per well in presence of TiO_2 and SiO_2 powders in the culture medium
NIH-3T3	10 % NCS	3.000	6.000
HEP-G2	10 % FCS	5.000	10.000
A-549	10 % NCS	4.000	8.000
PC-12	10 % HOS, 5% NCS, 1 % sodiumpyruvate, 1 % L-glutamine	5.000	10.000

2.2. Cultivation on TiO_2 powder coatings

Following their autoclavation, the powders (Table 2) were suspended in the culture medium of the accordant cell line. The wells were coated with a suspension containing 0.1 % powder in the culture medium. 200 μL of the different TiO_2 suspensions were added to every well of a 96 well plate and incubated for 4 days. After 4 days the medium was removed and the wells could be seeded with a defined number of cells (see Table 1). The cells were cultivated over a period of 14 days and the viability of the cells was determined by the MTT-test on day 1, day 4, day 7, day 11, and day 14.

2.3. Cultivation in TiO_2 suspensions

Wells of a 96 well plate were seeded with a defined number of cells (see Table 1). The plates were incubated 4 days (37°C / 5% CO_2) and on day 5 the particles (Table 2) were added to the cells at a concentration of 0.1% in the culture medium of the accordant cell line. The viability of the cells was determined on day 7 using the MTT-test.

2.4. Cultivation in SiO_2 suspensions

Wells of a 96 well plate were seeded with a defined number of cells (see Table 1). 200 µL of the different SiO_2 suspensions (Table 3) in a concentration of 0.1% were added to every well. The cells were cultivated over a period of 14 days and the viability of the cells was determined by the MTT-test on day 1, day 4, day 7, day 11, and day 14.

2.5. Cell metabolism

The cells were cultivated in 96-well plates. At regular intervals the viability of the cells was analysed by MTT ((3-(4,5-dimethylthiazol-2-yl)-2,5-diphenyl-tetrazolium-bromid) test (Sigma-Aldrich, Steinheim, Germany). This assay is based on the hydrolysis of the tetrazonium ring by mitochondrial dehydrogenase enzymes to an insoluble blue reaction product.

To perform the MTT-test, at first medium of each well had to be removed. Afterwards, 100 µl of fresh medium and 10 µl of MTT solution (5 mg/ml PBS, sterile) were added to each well and incubated for 4 h at 37°C / 5% CO_2. Subsequently, 100 µl of 10% SDS in 0.01 M HCl was added and the plates were further incubated for 24 h. The transmission signals at 570/630 nm was determined using a microplate reader (Bio-Rad, München, Germany).

Table 2. Physical data (crystal structure, size and BET surface area) of the tested TiO_2 nanoparticles : PC 10, PC 50, PC 100, PC 500, UV 100, P 25, R 15, R 25, R 34.

Pulver	Kristallstruktur	Größe der Nanopartikel (nm)	BET-Oberfläche (m^2/g)
PC 10	100 % Anatas	152	10
PC 50	100 % Anatas	40	50
PC 100	100 % Anatas	26	90
PC 500	100 % Anatas	7	340
UV 100	100 % Anatas	5-13	290
P 25	80 % Anatas 20 % Rutil	37 (Anatas) 90 (Rutil)	50
R 15	100 % Rutil	20	65
R 25	100 % Rutil	27	42
R 34	100 % Rutil	36	33

Table 3. Elemental formula and structure of the tested SiO_2 nanoparticles : MFI, SOD, GOS

Particle	Elemental formular	structure
nano-Tetrapropylammonium-Silicalith-1 (TPA-MFI)	$[(C_{12}H_{28}NOH)_4 [Si_{96}O_{192}]$-MFI	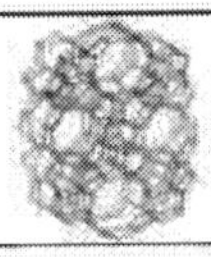
nano-Tetramethylammonium-Gismondin (TMA-GIS)	$[(C_4H_{12}N)_4 [Al_4Si_{12}O_{32}]$-GIS	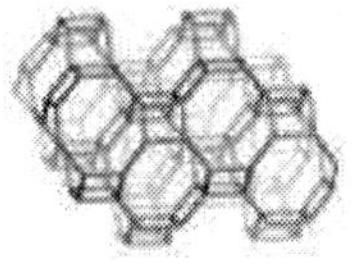
nano-Tetramethylammonium-Sodalith (TMA-SOD)	$[C_4H_{12}N)_2]$ $[Al_2Si_{10}O_{24}]$-SOD	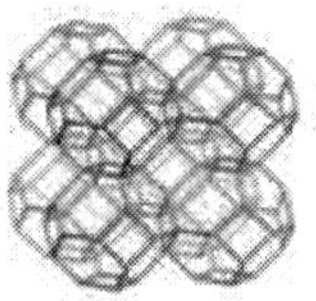

3. Results and Discussion

3.1. Cultivation on TiO_2 powder coatings and in TiO_2 suspensions

After the cells were seeded onto the different TiO_2 powders or alternatively after the cells were cultivated in presence of the powders in the culture medium, the viability was analysed using the MTT-test. Table 2 shows the different physical properties of the tested TiO_2 nanoparticles [15].

Figures 1-4 show the viability of the selected cell lines cultivated on the TiO_2 powders coatings over a period of 14 days. The NIH-3T3 cells show the lowest viability on the Aeroxide TiO_2 P 25 powder coating. On the rutile powder coatings (R 15, R 25, R 34) the NIH-3T3 cells reach a plateau and do not grow anymore (Figure 1).

As can be seen in Figure 2, the HEP-G2 cells only exhibit on the PC 10 and the PC 50 powder coatings a similar proliferation as the cells cultivated on the uncoated cell culture flask surface. No viability of the cells is observable on the R 34 powder coatings on day 7, 11 and 14. As shown in Figure 3 all A-549 cells cultivated on the powder coatings exhibit a lower viability as compared with the cells cultivated on polystyrene, with the viabilities of the cells on the rutile powders being lower than the viabilities on the anatase powders. PC-12 cells cultivated on the powder coatings as compared with all show a lower viability like the cells cultivated on PLL (Figure 4), but a higher viability than the cells cultivated on polystyrene, except for R 34.

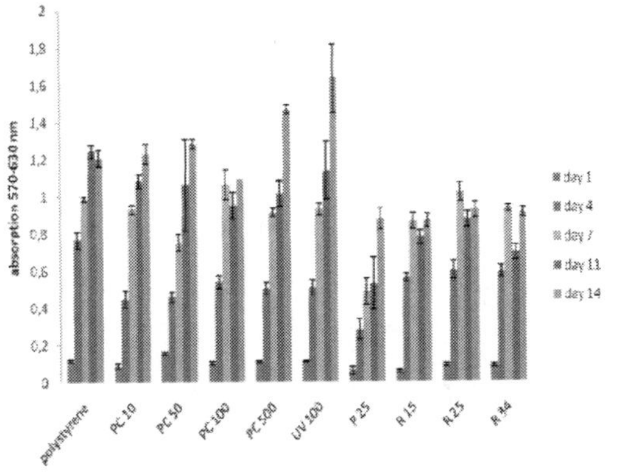

Figure 1. Viability of the NIH-3T3 cells cultivated on the different TiO_2 powder coatings over a time period of 14 days ± the standard deviation.

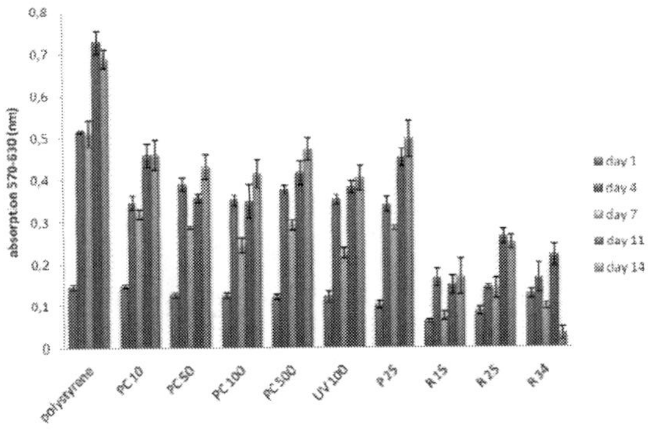

Figure 3. Viability of A-549 cells cultivated on the different TiO_2 powder coatings over a time period of 14 days ± the standard deviation.

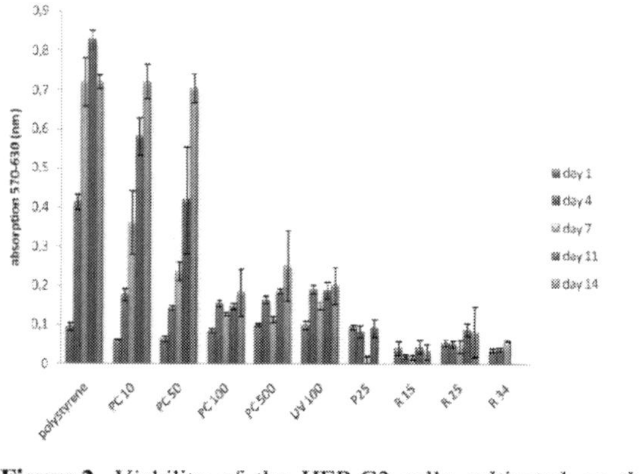

Figure 2. Viability of the HEP-G2 cells cultivated on the different TiO_2 powder coatings over a time period of 14 days ± the standard deviation.

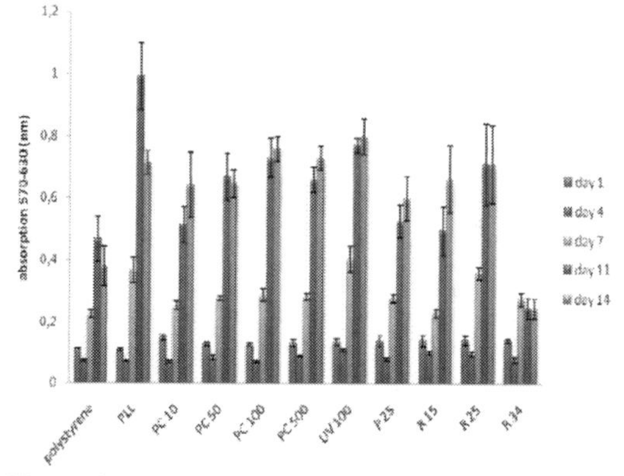

Figure 4. Viability of PC-12 cells cultivated on the different TiO_2 powder coatings over a time period of 14 days ± the standard deviation.

Figures 5-8 show the cell viability data for the cultivations carried out in the presence of the TiO_2 powders at a concentration of 0.1% in the culture medium of the according cell line. It is obvious from the results shown in these figures that the addition of the titanium dioxide powders at a concentration of 0.1% to the cell culture medium has no effect on the growth of the NIH-3T3 cells (Figure 5). The cell growth of the HEP-G2 cells in presence of the powders at a concentration of 0.1% in the culture medium was only effected by the R 34 powder (Figure 6). As shown in Figure 7 all powders at a concentration of 0.1% to the cell culture medium have an inhibitory effect on the viability of the A-549 cells. Only PC 100 and P 25 at a concentration of 0.1% have a negative effect on the viability of the PC-12 cells (Figure 8).

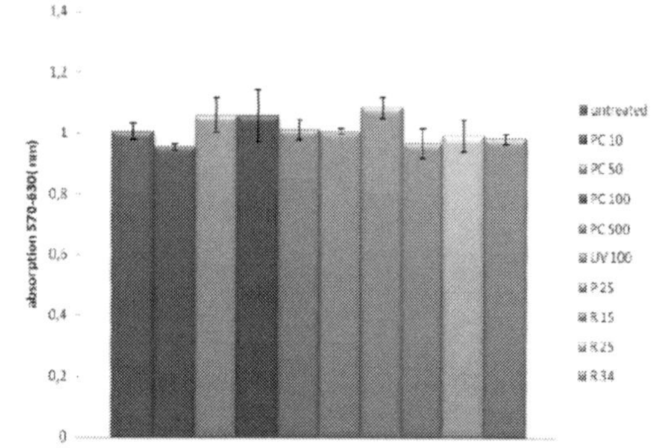

Figure 5. Viability of NIH-3T3 cells cultivated with powder presence of 0.1% in the cell culture medium ± SD.

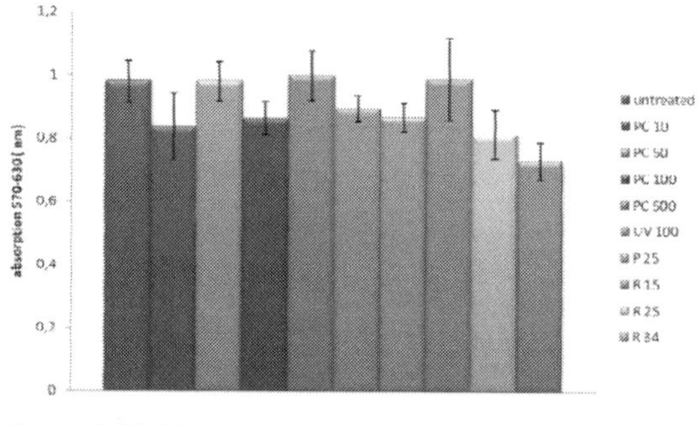

Figure 6. Viability of HEP-G2 cells cultivated with powder presence of 0.1% in the cell culture medium ± SD.

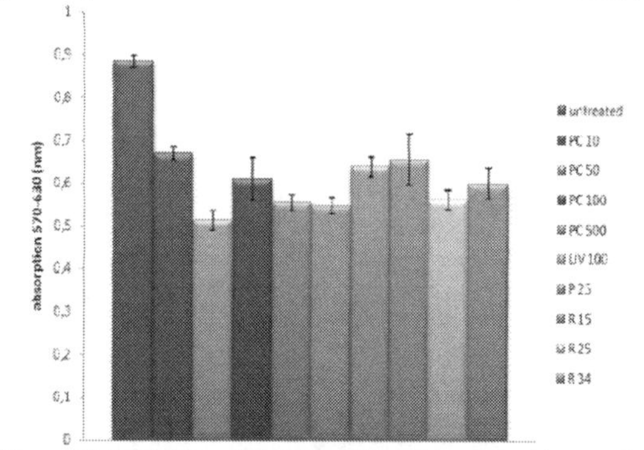

Figure 7. Viability of A-549 cells cultivated with powder presence of 0.1% in the cell culture medium ± SD.

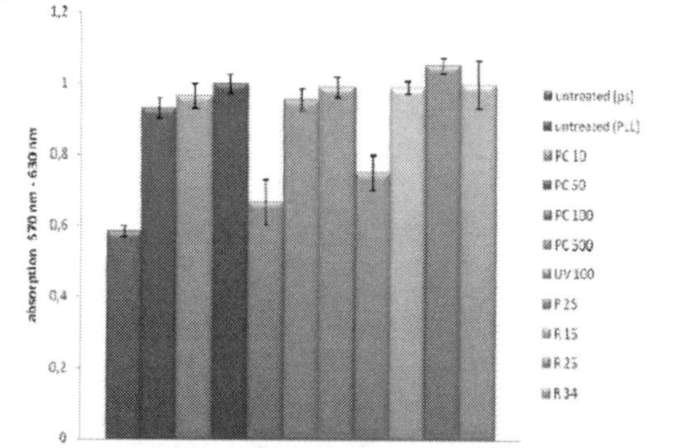

Figure 8. Viability of PC-12 cells cultivated with powder presence of 0.1% in the cell culture medium ± SD.

3.2. Cultivation in SiO_2 suspensions

Figures 9-12 show the cell viability data for the cultivations carried out in the presence of the SiO_2 powders at a concentration of 0.1% in the culture medium of the according cell line over a period of 14 days.

As can be seen in Figure 9 all HEP-G2 cells cultivated in presence of 0.1% of the SiO_2 particles exhibit a lower viability as compared with the cells cultivated in DMEM without particles. The addition of 0.1% of the SiO_2 particles to the cell culture exhibit no toxic effect to the NIH-3T3 cells (Figure 10), to the PC-12 cells (Figure 11) and to the A-549 cells (Figure 12).

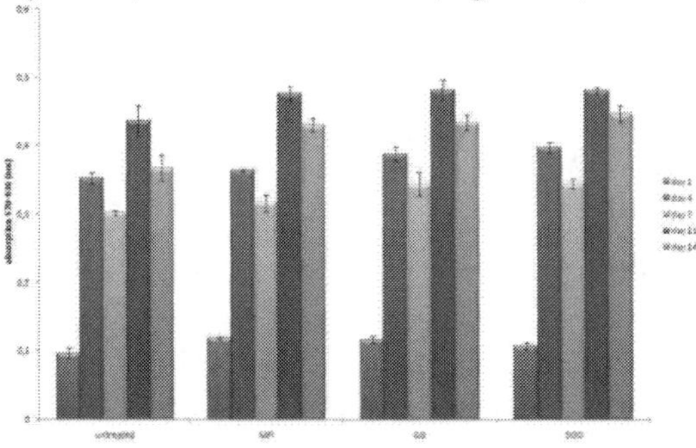

Figure 9. Viability of HEP-G2 cells cultivated in presence of SiO_2 in the culture medium over a time period of 14 days ± the standard deviation.

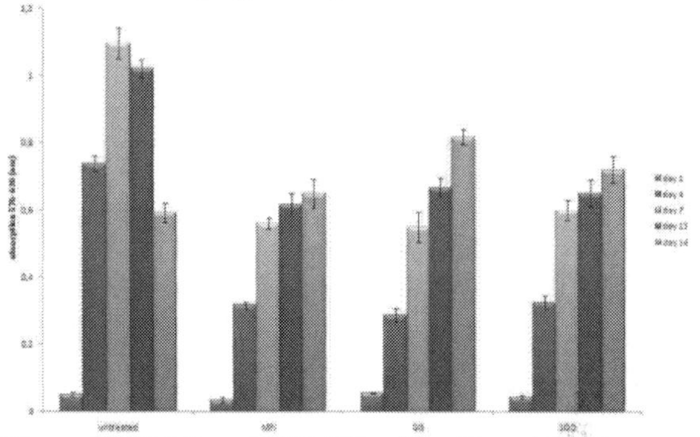

Figure 10. Viability of NIH-3T3 cells cultivated in presence of SiO_2 in the culture medium over a time period of 14 days ± the standard deviation.

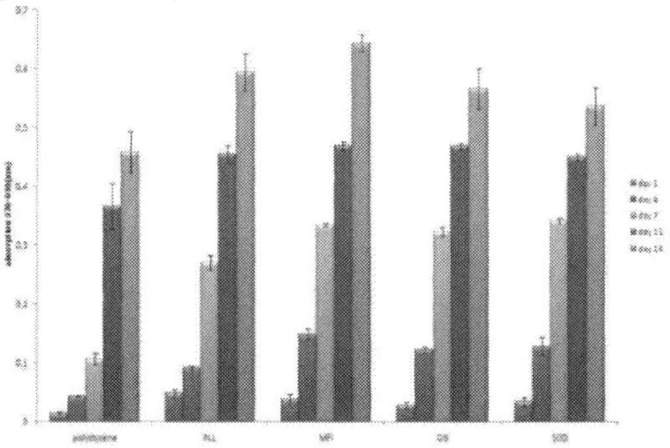

Figure 11. Viability of PC-12 cells cultivated in presence of SiO_2 in the culture medium over a time period of 14 days ± the standard deviation.

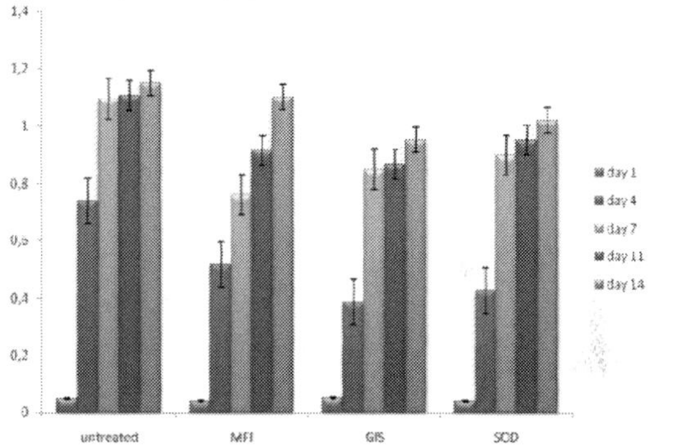

Figure 12. Viability of A-549 cells cultivated in presence of SiO_2 in the culture medium over a time period of 14 days ± the standard deviation.

4. Conclusions

Nanoparticles are less than 100 nm in size and possibly pose a health risk to humans. If they have been incorporated into the human body they can move freely because their size is similar to that of typical cellular compounds and proteins. Titanium Dioxide is because of its photocatalytic activity multifunctional applicable. Applications of TiO_2 nanoparticles are for example antibacterial and self cleaning surfaces and the treatment of wastewater and the air. Nanoparticles change their physical and chemical properties in respect to solid states, but about the change of the biological properties is so far little known. For this reason the biological harmlessness of these particles is importantly.

The objectives of this study were to assess the cytotoxic responses of four different mammalian cell lines exposed to nine unequal TiO_2 nanoparticles, which differ in their size, their BET surface area and their crystal structure and to 3 unequal SiO_2 nanoparticles. With regard to the possible absorption of nanoparticles into the human organism: via skin, via respiratory tract and via alimentary system, the toxicity was tested on connective tissue (NIH-3T3), liver (HEP-G2), lung (A-549) and kidney (PC-12).

For testing the effect of the nanoparticles on the different cells, the cells were cultivated on the powder coatings and in cell culture medium containing powders. The cytotoxic effect of the powders was determined via MTT test, a well established in vitro assay.

The experimental results have shown that the adhesion of the cells cultivated on the different TiO_2 powder coatings highly depends on the type of the cells on the type of powders. To exclude that the lower viabilities of the cells are not caused by problems in the adhesion of the cells on the TiO_2 powder coating surfaces, but due to a possible cytotoxic effect of the powders, the cells were cultivated in presence of the Titanium dioxide powders in the culture medium. These results show that the TiO_2 nanoparticles have no toxic effect on the cells. All cells are still dividable. Furthermore it could be shown that the physical properties of the particles do not refer to any observed biological effect. The tested SiO_2 nanoparticles also exhibit no toxic effect on the cells.

The concentrations of the particles tested in this study are relatively high. The TiO_2 nanoparticles concentrations in cosmetics or on self-cleaning surfaces for example are less. Consequently, he tested particles pose no danger for humans in the investigated concentration area. To get more detailed information about the possible cytotoxic effect of nanomaterials to mammalian cell lines more researches are necessary. Of great interests are researches which include the mechanism of interaction between the cells and the nanoparticles. Also important is to find out in which area of the cells when the particles are incorporated they are located.

References

[1] Lovern S B and Klaper R 2006 *Environmental Toxicology and Chemistry* **25** 1132
[2] Oberdörster E 2004 *Environmental Health Perspectives* 112 1058
[3] Oberdörster E, Zhue S, Blickley T M, Mc Clellan P and Haasch M L 2006 *Carbon* **44** 1112
[4] Hundt-Rinke K and Simon M 2006 Environmental Science and Pollution Research **13** 225
[5] Yang L and Watts D J 2005 *Toxicology Letters* **158**(2) 122
[6] Hussain N, Jaitley V and Florence A T 2001 *Advanced Drug Delivery Reviews* **50** 107
[7] Oberdörster G, Ferin J, Gelein R, Soderholm S C and Finkelstein J 1992 *Environmental Health Perspectives* **97** 193
[8] Lam C W, James J T, Mc Cluskey R and Hunter R L 2004 *Toxicological Science* **77** 126
[9] Warheit D B, Laurence B R, Reed K L, Roach D H, Reynolds G A and Webb T R 2004 *Toxicological Science* **77** 117
[10] Helland A, Wick P, Koehler A, Schmid K and Som C 2007 *Environmental Health Perspectives* **115** 1125
[11] Elder A C, Gelin R, Oberdörster G, Finkelstein J, Notter R and Wang Z 2005 *Experimental Lung Research* **31** 527
[12] Oberdörster G, Oberdörster E and Oberdörster J 2005 *Environmental Health Perspectives* **113** 823
[13] Davoren M, Herzog E, Casey A, Cottineau B, Chambers G, Byrne H J and Lyng F M 2007 *Toxicology in Vitro* **21**(3) 438
[14] Singh S, Shi T, Duffin R, Albrecht C, van Berlo D, Höhr D, Fubini B, Martra G, Fenoglio I, Borm P J A and Schino R P F 2007 *Toxicology and Applied Pharmacology* **222**(2) 141
[15] Mendive C B, Bahnemann D W and Blesa M A *Catalysis Today* **101** 237

The first equipment for protection from nanoparticles

Louis-Paul Faure, Pierre Bombardier and Jean-Marie Reinwalt

FAURE INGENIERIE S.A., Parc Technologique des Fontaines, Chemin des Fontaines, F-38190 Bernin, France

E-mail: laboratoire@faureingenierie.com

Abstract. How can we guarantee the containment of ultrafine particles but also implement the ergonomic and handling constraints voiced by researchers? This is the equation that the engineers at FAURE INGENIERIE had to resolve to develop the first barrier protection equipment for nanoparticle research.

1. Introduction

In 2005, FAURE Ingénierie (FI) decided to develop its presence in the fields of nanotechnology consulting and engineering and notably participated in the Minatec project in Grenoble, France. Initial specialized demand came mainly from research laboratories developing the first artificial nanoparticles and nanostructures. With the problems linked to asbestos fresh in everyone's memories, researchers were the first to be aware of the knowledge gap concerning the effects on health and the potential dispersal of such substances.

FI R&D department is focused on understanding and controlling indoor air quality. Amongst our priorities, controlling nanoparticles has become one our most important challenges for the management of specialized environments (laboratories and production rooms), but also for indoor air conditioning in the widest possible sense.

R&D work focused on evaluating the limits of traditional calculation tools used for designing rooms and work stations particularly for nanoparticles. Through partnerships with CEA (LITEN in Grenoble) and LEPTAB (La Rochelle) research teams, it became clear that a lot of progress was needed before the dispersal of nanoparticles could be controlled in view of their particular behavior (high dispersal rate, sensitivity to electrostatic forces, rapid development of particle size, etc.). This was the case whether nanoparticles came in the form of individual particles, aggregates or agglomerates.

As a result, FI is currently in a design and validation process that should lead to the production of the first work stations and rooms specifically designed and validated for this type of use. They present a solution for controlling risk by primary prevention.

This range of work stations called PSPN (Safety cabinets for nanostructured particles) meet the expectations of the different parties involved in the use of nanomaterials and those requesting safety cabinets. These first safety cabinets are specifically designed for research and development work.

2. Presentation

The PSPN is a reinforced safety cabinet designed to protect personnel handling harmful particles, especially nanoparticles and nanostructured particles. Made of stainless steel, this robust cabinet can be installed in a laboratory with or without dedicated air conditioning. This equipment has a unidirectional vertical airflow and controlled exhausting (like a PSM). It is easy to handle and assemble in a protected space in response laboratory requirements.

This document presents the general characteristics of the safety cabinet range which is to expand shortly. The following characteristics can be modified according to use: dimensions, filtering efficiency, chemical filtering and filter type, alarm type. Each new model will undergo the same thorough initial qualification procedure as the prototype, as containment efficiency depends on several different factors.

2.1. Functionalities

This equipment is designed for working on dry powders and or suspensions in a clean and contained environment (dynamic containment) while ensuring that the operator, handled product and environment are protected. As we are dealing with air contamination, the equipment must be effective for a wide range of particle size from 'standard' dust (D<100 μm) to nanometric particles (D<100 nm).

When in operation, the equipment must completely control the risk of particle dispersal:

- Via the equipment's front opening.
- Via leaks throughout the ductwork and air plenums.
- During disassembly, maintenance or controls on parts of the equipment that need replacing (filters, fans, joints, sensors, etc.).
- While cleaning the inside surfaces of the equipment.
- To the outside of the building (air extracted from the cabinet through the rooftop).
- To the inside room environment and possibly any areas adjacent to where the equipment is placed.

2.2. General presentation

The inside dimensions of the safety cabinet are: depth = 700 mm, width = 1300 mm and height = 950 mm. This provides enough space for laboratory equipment (scales, cyclone, mill, etc…).

The choice and preparation of the stainless steel used makes cleaning easy (minimum roughness, chemical resistance and good static dissipation) and gives it the sturdiness needed for the potentially abrasive nature of certain products.

The particle level class in the operating space is set according to the product's contamination risk level. Filtering systems can be adapted depending on the size of the aerosols to be contained (Filters for the PSPN range are manufactured especially to obtain filtering efficiency and the homogeneous pressure loss required. The homogeneity of speeds obtained guarantees a balance of flow between the different parts of the work surface). These systems concern both the quality of the operating space and the air discharged into the environment around the safety cabinet. Adsorption modules can be added to the filtering equipment to reduce any possible emission of chemical substances used in the operating space.

The work surface is perforated at the edges. It can be removed making it easier to disassemble pre-filters. The main work surface is shaped to collect liquids if these are to be used.

Fittings for glassware are included and can be modified on each cabinet. Injection and measurement points for integrity testing are situated at each filtering level to make such operations easier to carry out.

2.3. Airflow

Modeling has been used to design the cabinet's airflow to decide or analyze:

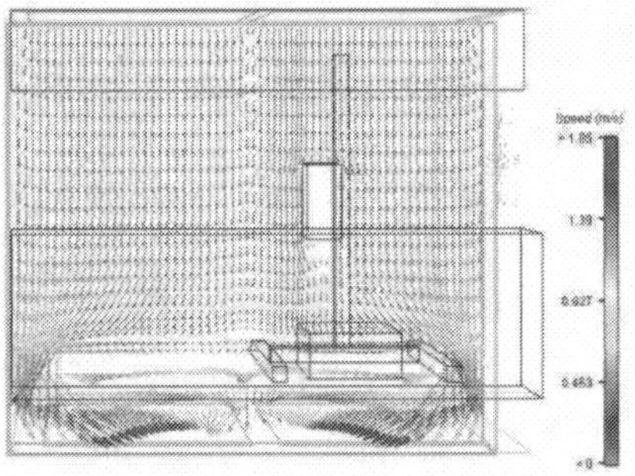

- The size and position of exhaust openings.
- The size of the work surface (non-perforated).
- The effects of having a sink.
- The disturbances caused by equipment used in the work surface area.
- The influence of external airflow conditions that can be very prominent.
- The limits for safe operation in order for level 1 (malfunction leading to corrective action) and level 2 (immediate evacuation) alarm thresholds to be set.
- The horizontal speed in the air barrier taking into account the influence of the position of the operator when working.

Particular attention was paid to the path taken by airflows under the work surface to reduce the risks of back-diffusion. The speeds at which air flows through exhaust openings must be minimal to avoid turbulence.

We also simulated operating two cabinets one next to the other to verify that there was no exchange between the two.

2.4. Ergonomics

These safety cabinets provide the universal and flexible qualities required for handling and testing operations in the context of research and development: large front opening and modular work surface in function of operational requirements. Safety cabinets can be set up side by side to ensure maximum protection during handling operations.

The sides are equipped with tracks for the front plate glass access panel, with two lateral sliding glass panels to allow communication between the two cabinets and with points reserved for passing fluids.

There are in-built fittings for glassware and other laboratory accessories.

2.5. Safety

When the cabinet operates normally, the operator is protected. If the safety cabinet malfunctions, there are two alarm levels:

1. Failure without consequence for operator protection
2. Malfunction with possibility of leakage (immediate evacuation of unprotected operators).

The operator has access to information and alarms on the front of the safety cabinet. The cabinet is also equipped for transmitting alarm information to a Building Management System if necessary.

3. Qualification procedure

A formal approach to qualification was followed for safety cabinet qualification in order to meet the requirements for minimizing risks needed by laboratories handling nanoparticles. All qualification tests include a report for each safety cabinet giving the operational capacity and the overall protection level corresponding to the primary protection from exposure criteria.

Qualification has two stages:

- Initial qualification (standard test) of a prototype
- Operational qualification when standard manufactured equipment is received.

Below is a description of the initial qualification stage which is the more interesting stage.

This consists of a thorough characterization of properties including all the latest modifications made to the system and optimal settings. This particular qualification required tools that are not normally used to qualify safety cabinets: An ISO 5 class clean room, a generator and a nanoparticle analyzer.

A few tests concerning comfort of use are carried out before the performance tests:

- Lighting efficiency in the work surface area
- Noise level at the front of the cabinet
- Vibration level of the structure measured over 3 axes

The first qualification stage then begins which corresponds to the standard tests we carry out on all MBSC (Microbiological safety cabinets) in conformity with the NF EN 12469 standard:

- Validation of modelling
- Measurement of blowing speeds
- Measurement of air barrier speeds
- Control of absence of particle collection and leakage to outside environment using smoke
- Evaluation of air movements outside of the safety cabinet (smoke)
- Absolute filter integrity testing using aerosols equivalent to the ISO particle class range.
- Inside and outside particle level testing
- Control of automatic alarm functions

This first stage corresponds to the operational qualification tests that will have to be carried out on standard manufactured safety cabinets.

The second qualification stage is only started once the preceding tests give satisfactory results (after setting blown and exhausted airflow rates).

3.1. Performance tests with nanoparticles:

This type of test needs to be conducted in a clean room as a nanoparticle source cannot be detected in an environment that is naturally full of them. As a result, tests were carried out in an ISO 4 to 5 area thus eliminating most of the standard nanoparticles found in an everyday environment.

The nanoparticle aerosol used was made up of graphite particles with a median diameter of 41nm (selected according to the characteristics required for the first experiments to be carried out in these cabinets).

Particles were measured by an SMPS (Scanning Mobility Particle Sizer) consisting of a DMA (Differential Mobility Analyser) and a CNC (Condensation nucleus counter) with up to 44 channels with a range of 5.5 to 350 nm at a rate of 0.3 per minute. This system gives a precise image of particle size but requires a large amount of aerosols (rate 100 times lower than that of a particle counter).

- Particle level control

Here, we simply needed to adapt the sample volume for measurements to be relevant. We decided to take samples every hour from each sample point when we used an SMPS.

An additional measurement made with just the CNC confirmed the results given by the SMPS (verification of accuracy of SMPS value correction algorithm whose overall measurement is only made up of some of the sampled aerosols).

- Filter system integrity test

The testing protocol is similar to a standard integrity test (see ISO 14644-3). The sensor speed just needed to be adapted to take the detector's slow rate (1.5 l/min) into account. Particle size development for such a test aerosol (high concentration = greater interactions) required controlling just above the filter to be tested (and not at generator outlet).

- Containment test with nanoparticles

We created a diffuser with the same dimensions as the work surface for this test. This diffuser was perforated at regular intervals all over its upper face to reconstitute what could be contamination occurring in an identical manner at any point on the work surface.

The test consisted of searching for any nanoparticle that could escape from the cabinet via the front opening by sweeping it with the cabinet in a standard operating mode and then deteriorating the operating conditions (opening the front panel wider, reducing the extraction rate, creating obstacles, etc.).

4. Performance results

Once the blowing and exhausting rates had been adjusted and some of the parts modified to obtain a completely airtight cabinet (particular attention was paid to all welding, the joint system used and cable installment), we concluded that the prototype was in compliance for the following criteria:

- Dispersion of blowing speeds

We obtained a less than 20% dispersion level. The aim was to reduce dispersion as much as possible. We checked that the position of the blowing fan did not have any adverse effect on the results.

- Air barrier horizontal speed dispersion

We obtained a less than 20% dispersion level.

This criterion is very important for controlling containment. We checked the influence of airflows situated in front of the safety cabinet by varying the blowing speeds in the clean room: as expected, a high outside air speed can disrupt air barrier homogeneity. This needs to be taken into account in clean room design and during qualification checks.

- Inside air flow paths

The visual test using smoke showed good exhaust speed and good containment efficiency. Tests were carried out by placing obstacles in front of the air barrier and on the work surface to test airflow behavior in unfavorable conditions.

- Filter and structure integrity

Localized leakage rate for filters and structures less than 0.001% MPPS (stricter criterion than the ISO 14644-3 standard that gives 0.01% as the maximum penetration value for absolute filters).

The acceptable leakage rate was very low as we went on the basis that nanoparticles handled could be toxic and weak in concentration.

- Cleanliness class

Class lower than ISO 5 for particles greater than 0.1 µm.
Less than 5 particles per m^3 between 5 and 100nm (measured 15 cm above the work surface).
It is worth noting that the level of cleanliness required for a product can vary immensely and that it will often be unknown for new products.

- Containment efficiency

Sweeping of the front of the cabinet showed that the leakage rate by back-diffusion to the outside space was less than 0.001% MPPS (identical criterion to integrity test).

5. Conclusion

The following conclusions were drawn from this series of tests:

- Manufacturing quality is of utmost importance to ensure that the cabinet is sufficiently airtight in view of the operational pressure attained.
- The results obtained by simulating air barrier speeds were validated by the tests.
- The environment in which the cabinet is placed must take into account any interfering air flows.

Here are some further remarks we can add concerning the installment of safety cabinets for handling harmful nanoparticles or nanostructured particles:

- It is very important for the future users to participate in the project to ensure both the ergonomic aspect of using the cabinet (whether laboratory apparatus fits into the workspace and the risks involved with the activity undertaken) and an understanding of how to use the cabinet (maintenance, monitoring of indicators, hand movements, etc.).
- The prototype stage is indispensable. This stage must include case-specific qualification with the aerosol that is closest to the materials that are to be handled in the standard manufactured safety cabinet.

FI projects in this domain are to develop the PSPN range in order to adapt cabinets to the different requirements and practical constraints of laboratories (size, price, level of risk involved). As a result, new prototypes are to be developed over the next two years and will undergo the same design procedure: Simulation/Prototype/Case-specific qualification.

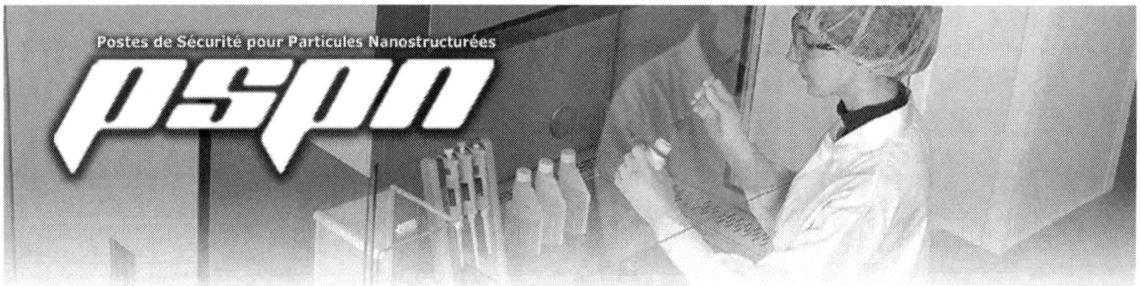

Journal of Physics: Conference Series **170** (2009) 012024 doi:10.1088/1742-6596/170/1/012024

New Methods and Standards for Fine Dust

Jürgen Spielvogel, Stefan Hartstock and Hans Grimm

Grimm Aerosol Technik GmbH, Dorfstrasse 9, 83404 Ainring, Germany

ECOMESURE, 3, rue du Grand Cèdre, 91640 JANVRY, France

E-mail: jsp@grimm-aerosol.

Abstract. There seems to be common agreement that PM10 is a suboptimal quantity for the quantification of potential dangers from fine dust due to a number of reasons, notably because the chemical composition of the particles is not considered, because the size distribution is disregarded, and because of sampling artefacts. In a first step for improving the particle measurements, the European Community has published new directives for ambient air in June 2008 (EU 2008), which as a main part included new regulations for PM2.5 measurements, in addition to the further on valid regulations for PM10. The comparison of PM2.5 and PM10 may allow a source apportionment and a better assessment of the influence of fine dust on human health. The source apportionment may allow more effective fine dust reduction strategies.

1. The coming PM2.5 regulations in Europe

In order to prepare the coming limits for PM2.5, the European directive 2008/50/EC requires new PM2.5 measurement sites in addition to the existing PM10 measurement sites. The measurements must be started at January 1st 2009 and the new directive and regulations must be part of governmental laws and process before the 11th of June 2010. The regulations state that a minimum number of PM2.5 measuring points must be installed, namely one PM2.5 site for every million people in metropolitan areas and at least one measuring point in cities with more than 100.00 people. As an additional general rule, one PM2.5 site must be installed for two PM10 sites. The regulations required also a minimum total number of measuring points (PM2.5 and PM10), which depends on PM10 concentrations and which is higher for polluted areas.

As for the data quality, the regulations state that the uncertainty must be smaller than 25 % and that the minimum data capture must not be less than 90%. The data must be reported to the European Union.

Table 1. Future limits (annual mean) for PM2.5

Year	Concentration [µg/m^3]	Limit type
2010	25	Target
2015	25	Limit
2020	20	Limit

The new regulations include future limits for PM2.5 concentrations as listed in Table1. As a rule of the thumb, PM2.5 concentrations are typically ~60% of the PM10 concentrations, and considering the present limits for PM10, annual mean < 40 µg/m³ and not more than 35 days > 50 µg/m³, these new limits for PM2.5 are hardly stronger than the old limits for PM10.

There will, however, also be exceptions for the coming PM2.5 limits. Such exceptions are permitted by the EC for a country, region, or city, if the authorities present scientific reasons based on long time measurement for the impossibility of compliance with the limits. Potential reasons could be high background values due to long-range transport or special local orographic or weather conditions.

Each candidate Method (CM), which will be used to measure PM2.5 concentrations needs an approval according to EN14907, similar to the approval for PM10 according to EN12341. For such an approval, daily values in µg/m³ measured by the CM are compared against a Reference Methods (RM).

2. Principle of optical aerosol spectrometers

Unlike nephelometers, aerosol spectrometers measure scattering intensity of each individual particle as a proxy for particle size. The light scattering process has three size regions according to the value of the optical size parameter α, $\alpha = \pi * D_p/\lambda$, with the aerosol particles diameter D_p and the wavelength of the incident light λ (Mie 1908). For $\alpha << 1$, Rayleigh-scattering occurs, where the intensity of the scattered light is proportional to the sixth power of particle diameter ($I \sim D_p^6$). The range of $\alpha \cong 1$ is addressed as Mie-scattering, where a strong interaction between the aerosol particle and the incident beam takes place. For Mie scattering, the intensity of the scattered light is a complex function of particle diameter and refractive index. The GRIMM optical spectrometers use however a special procedure to retrieve the contained size information to a large extent. The case $\alpha >> 1$, Geometric Optics, comprises reflection, refraction diffraction and absorption, and the intensity of the scattered light is usually proportional to the particle cross-sectional area ($I \sim D_p^2$).

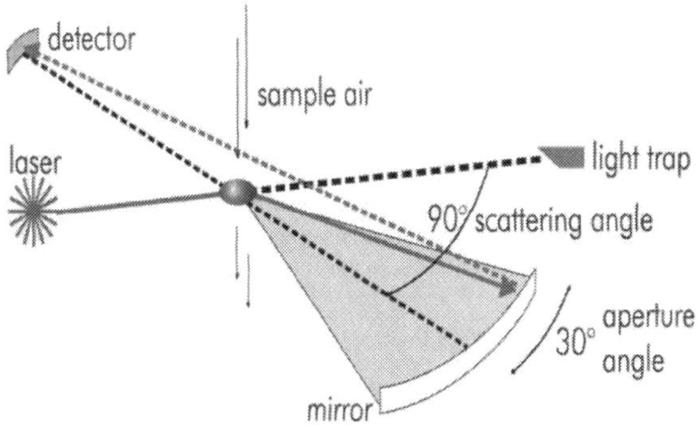

Figure 1. Principle of the GRIMM optical spectrometer.

Principle of the GRIMM aerosol spectrometers is illustrated in Figure 1. A laser-diode is used as light source (λ = 655 nm), and the beam is focused into a well defined, nearly homogeneously illuminated optical measuring volume. The scattered light from a scattering angle of 90° with an aperture angle of 30° is collected by on a photodiode. This geometry is part of the measures used to minimize the influence of refractive index on scattering intensity and to compensate the ambiguities of Mie-scattering. The detected signal is then amplified and used by a pulse height analyzer. Thus the GRIMM instruments reports particle number concentrations for 31 channels covering the size range of 0.25 – 32 µm with a time resolution of 6 s. Higher time resolution of 1 s only is possible for a reduced size range and allows the use of the instrument for flux measurements with eddy-correlation methods.

The measured number size distributions are converted into the required PM10, PM2.5 or, if desired, PM1 concentrations. The conversion is a challenge because it requires detailed investigation of aerosol properties.

These GRIMM aerosol spectrometers are very small, rugged, and require only a minimum of maintenance essentially without consumables. They can be equipped with a variety of accessories such as Nafion dryers to reduce humidity in sampling air, heated inlets to investigate the content of semivolatiles, outdoor housings, integrated filters for the collection of detected particles, or integrated batteries for mobile use. Here we describe three versions of the optical spectrometer, namely the Environmental Dust Monitor (EDM) 180 for official PM10 and PM2.5 measurements, the Wide Range Aerosol Spectrometer for measuring size distributions also of particles smaller than the lower size limit of the aerosol spectrometer (0.25 µm), and The EDM 265 for quantifying the semivolatile mass fraction.

3. The EDM 180 for official PM measurements

This 19'' rack version of the aerosol spectrometer is designed for the long term measurement in official networks, it is certified for PM10 according to EN12341 and the certification for PM2.5 according to EN14907 will soon be accomplished. The instrument includes a dehumidification system based on a NAFION tube, which is activated at an ambient relative humidity above 70%. It has a GESYTEC MODE for online data transfer.

A considerable advantage for the coming PM2.5 regulations is that the system can simultaneously measure both, PM2.5 and PM10 measurements. Thus a single instrument constitutes a PM10 and a PM2.5 site, and existing users of the EDM 180 will not need any new equipment measuring PM2.5. The system is also able to measure PM1 concentrations or total number concentrations for which, in the long run, new regulations might be established.

4. The Wide Range Aerosol Spectrometer (WRAS) to measure the full size range

The optical spectrometers can be combined with a GRIMM Scanning Mobility Particle Sizer with Condensation Particle Counter as detection system (SMPS+C). Such s SMPS+C, when operated with a differential mobility analyser (DMA) of the type M-DMA, covers the size range of 5 – 350 nm. Therefore the WRAS, i.e. optical spectrometer with additional SMPS+C, covers the size range from 5 nm to 32 µm with 72 size channels. This constitutes a comprehensive information on the particle size distribution essentially over the complete range of airborne particles. A special software accomplishes the synchronisation of optical spectrometer and SMPS+C, the time resolution is determined by the scan time of the SMPS+C and equals 4 min. If desired, the EDM data in better time resolution are still available on the internal memory of the device.

The WRAS is also available as stand-alone outdoor system, with stainless steal housing, temperature and humidity sensor, optional rain and wind sensor, suitable for long term measurement.

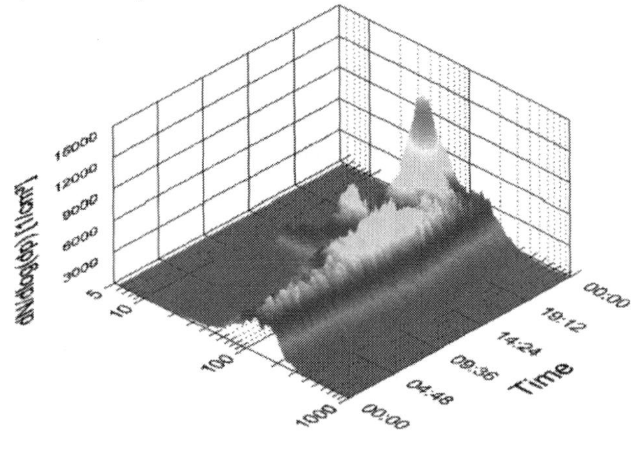

Figure 2. Example for particle size distribution measured with a WRAS system on top of the Fronau tower in Berlin at 11th of June, 2007. The high concentrations of ultrafine particles in the evening are caused by photochemistry

Figure 2 shows an example of size distributions measured with WRAS in a comprehensive study on the sources of fine dust in Berlin (Pesch 2007). In this study the excellent size resolution of the WRAS was used, together with chemical speciation and gas measurements, for source apportionment. The result was that local sources are only a small contribution to the total PM concentrations. Thus the fine dust concentrations in Berlin can only be reduced effectively if the contribution from long range transport is reduced; important sources for the long range transport are power plants in Eastern Europe. The study highlighted also a poor correlation between PM2.5 and particle number concentrations, a case for considering particle number concentrations in future fine dust regulations.

5. The EDM 265 for measurements of the semi-volatile fraction

Existing epidemiological studies require a sophisticated statistics to detect significant but small health effects from fine dust. Part of the reason for the only small effects is probably that the used PM10 or PM 2.5 data are not a good measure for adverse health effects from ambient dust since the composition of the aerosol is disregarded. It can be suspected that the health effects of ambient particulate matter are rather caused by the solid non-volatile constituents than by the semivolatile fraction, and we speculate that an epidemiological study on the connection of non-volatile particulate matter and health effects would show much stronger effects than previous results based on total PM concentrations.

Numerous studies have shown that the contribution of semivolatile compound to the PM2.5 mass concentration is typically 20 – 50% and that these semivolatile compounds are mostly nitrates, ammonium, and organic material.

The EDM 265 is devoted to the measurement of the volatile fraction, it is a stand-alone outdoor system with a heated sample pipe to evaporate semivolatile material. The sample pipe is however not continuously heated as in other instruments that just loose the semivolatile fraction, but only periodically so that the semivolatile fraction can be assessed from the concentration difference between heated and non-heated periods. The temperature of the sampling pipe can be selected between ambient and 300°C, the use of different temperatures allow to distinguish between different classes of semivolatile material.

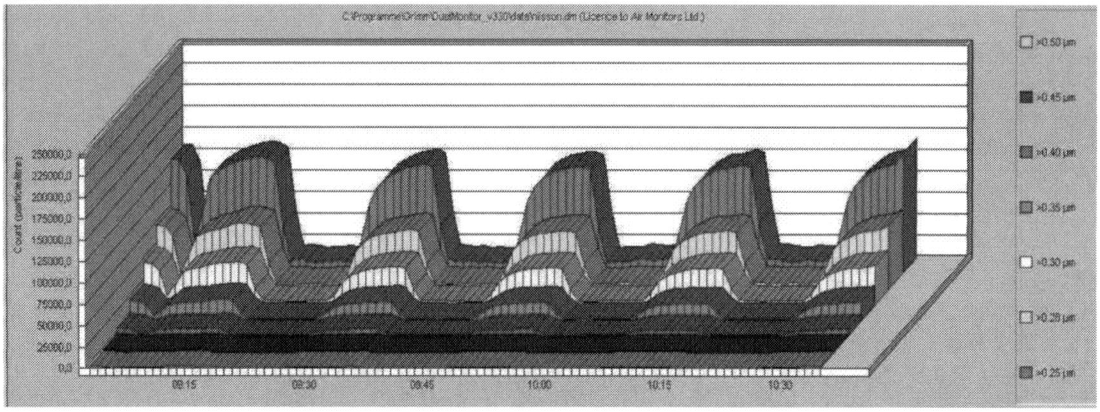

Figure 3. Data from the GRIMM model 265, shown are 7 from 31 size channels. The strong regular concentration changes are induced by heating – and non-heating periods of the sample pipe and allow a quantification of the semivolatile fraction.

6. Conclusions and Outlook

We expect that the PM values will be reduced and that future fine dust regulations will include limits for PM-1 or even for number concentrations of nano particles, similar to new EURO5 regulations for emissions from car engines. GRIMM spectrometers can already today measure PM1 values and they can be adapted to additional nano-particle measurements with accessories that fit in existing outdoor shelters.

References

[1] EN 12341 (1998). Air quality - Determination of the PM10 fraction of suspended particulate matter - Reference method and field test procedure to demonstrate reference equivalence of measurement methods; Beuth Verlag, Berlin.

[2] EU directive 2008/50/EC.

[3] http://eur-lex.europa.eu/LexUriServ/LexUriServ.do?uri=OJ:L:2008:152:0001:0044:EN:PDF

[4] Mie G 1908 *Ann. Phys.* **25** 377

[5] Pesch M 2007 2. Zwischenbericht zur Ursachenanalyse der PM 2,5-Immissionen in Berlin. Technical report, TU-Berlin Fachgebiet Umweltverfahrenstechnik

Experimental evaluation of individual protection devices against different types of nanoaerosols: graphite, TiO_2 and Pt

Luana Golanski, Arnaud Guiot and François Tardif

CEA-Grenoble, Liten, Laboratory of Tracer Technologies, France

E-mail: luana.golanski@cea.fr

Abstract. In this study different conventional individual protection devices, well-qualified for submicron particles were tested for different types of polydispersed nanoaerosols of TiO_2, Pt and graphite ranging from 10 to 75 nm (electrical mobility diameter). For that purpose two specific test benches were used: one for the filter-based devices which are tested under a controlled air flow and the other one for protective clothing and gloves under diffusion and without air flow.

1. Introduction

Nano-sized particles are defined as objects in the size range 1- 100 nanometers (reference ISO: TS 229 QG1 PG1 and ISO TS 27687, to be published in 2008). Toxicological studies have shown that some nanoparticles can potentially have an effect on human health [1]. Manufactured nanomaterials show unique properties due to their small size and therefore present a great potential for industrial applications. The number of workers exposed to the nanoparticles may increase in the future. It is today necessary to apply the principle of precaution by implementing among other, efficient personal protections against the engineered nanoparticles in order to maintain the exposure as low as possible. The chemical, physical properties of engineered nanoparticles vary widely thus, how nanoparticles in interaction with the device may behave, even under exactly the same test conditions, could be different.

Penetration of nanoparticles into the human body may occur essentially through respiration rather than by ingestion and dermal contact. Aerosol filtering is the common way to remove particles from air. Different filter types are intensively used in a variety of domains such as nuclear, microelectronics, *etc*. For uncharged fibrous filter media the Maximum Penetrating Particle Size (MPPS) is between 150 and 300 nm. According to the single fiber filtration theory for fibrous filter, particles smaller than 100 nm are trapped on the filter fibers essentially thanks to random displacements due to the Brownian motion. The particles are then in general irreversibly captured by van der Waals forces. The collection efficiency of nanosize particles through fibrous filter depends on the filter characteristics, the face velocity and on the aerosol intrinsic characteristics like particle density, particle diameter [2, 3]. Experimental works performed on the penetration of different types of nanoparticles such as silver, NaCl and DOP through fibrous filter media confirmed the expected behaviour [4-6].

Electrostatic filters are constituted of fibers electrically charged during the fabrication process by Corona or triboelectric effect. These filters show a high efficiency and a low pressure drop and are therefore used for respirator masks. For electrostatic filters exposed to neutral particles, the Maximum Penetrating Particle Size (MPPS) is expected from calculations between 50 and 60 nm [2]. The shift of the MPPS toward smaller particles for an electrostatic filter as compared to an uncharged filter is due to the effect of the fiber charge on the particle. Neutral particles passing through the media made of charged fibers are polarized by the electric field and charges are induced on the particles. Consequently, the capture of the particle is also influenced by the polarization force exercised by the fiber on the particle. Experiences performed on N95 electrostatic filters using NaCl monodispersed aerosol reported a MPPS value around 40 nm [7]. On FPP3 electrostatic filters using graphite polydispersed aerosol, the MPPS was observed to be around 30 nm [8].

In order to prevent from dermal exposure, the efficiency of protective cloths and gloves to nanoparticles penetration have to be evaluated as well. Only few works describe the influence of nanoparticle penetration through protective clothing. A previous work on protective clothing performance with regard to NaCl particle penetration has been performed in the size range of 10 to 1000 nm [8]. The results indicate that woven fabrics behave almost the same as fibrous filters with a MPPS between 100 and 500 nm.

The aim of this study was to evaluate the influence of aerosol type on the protection efficiency of commercial protective devices. In this work the efficiency of commercial protective equipments was evaluated with different types of nanoaerosols: graphite, TiO_2 and Pt. One commercial HEPA (High Efficiency Particle Air) filter and one electrostatic filter media were tested. The efficiency of three commercial protective clothing (made with woven and non-woven fabrics) was evaluated. The efficiency of three commercial gloves made of different materials (nitrile, latex, neoprene) and of various thicknesses was evaluated as well.

2. Materials and methods

2.1. Test benches

Two specific test benches were used in order to test the efficiency of personal protection devices against nanoparticles. The filter media is here tested with a controlled air flow containing a polydispersed aerosol. As shown by Japuntich *et al.* [4], the results are similar to the one obtained by the slower method using monodispersed challenge particles. The experimental set-up consists of a nanoparticle generation system, a filter support containing the filter media and a downstream pump. A detailed description of the experimental set-up is described elsewhere [8]. In this work a filter support is designed using two identical aerosol flow lines made of stainless steel: one containing the filter and the other without filter [6]. A valve allows to switch the nanoaerosol flow between the two measurement lines. The upstream and downstream particles are classified by a Differential Mobility Analyzer (DMA model 5.5-300, Grimm) and then measured by a Condensation Particle Counter (CPC model 5.403, Grimm). The filtration penetration is calculated as 1-(downstream concentration / upstream concentration).

The through diffusion bench was designed in order to reproduce the actual conditions of use of clothing (without an important air flow). The experimental set-up consists in a nanoparticle generation system, a diffusion cell containing the protective media to be tested and a CPC system (CPC model 5.403, Grimm) in order to count the nanoparticles in the down stream part of the cell. A detailed description of the experimental set-up is described elsewhere [8]. In the upper stream part of the cell, a continuous flow of graphite, TiO_2 and Pt nanoparticles is injected in order to maintain a constant concentration of nanoparticles. The differential pressure between the two parts of the cell is maintained at a few mbars in order to test the media efficiency without imposing any flow. In the down cell, the CPC takes samples of initially filtered air containing the nanoparticles which have diffused through the media. The number of particles flowing per minute in the down stream part of the cell was measured in order to estimate the efficiency of the protective clothing. For all the measurements a ^{85}Kr radioactive source is used in order to neutralize the nanoaerosol before introducing into the test filter or the diffusion cell.

Graphite nanoparticles are generated by spark discharge between two graphite electrodes. TiO_2 and Pt nanoparticles are produced by a glowing wire: the material is evaporated by passing a high current through a conducting wire, suspended in a flowing gas. The mean geometric diameter was typically about 35 nm for graphite aerosol. Smaller particles with size centered 10 nm are obtained only using the glow wire generator. The maximum mean diameter obtained by glowing wire generator was 40 nm for TiO_2 nanoparticles.

2.2. Materials

The efficiencies of a HEPA and an electrostatic commercial filter were evaluated when exposed to Pt and TiO_2 aerosols. The filters characteristics used in this study are given in the following Table 1.

Table 1. Specifications of the tested filters media given by the manufacturer.

Filtration parameters	HEPA	Electrostatic filter
Fiber material	Glass	Polypropylene
Thickness (mm/50KPa)	0,33	-
Weight (g/m2)	75±5	-
0.3 µm DOP Penetration (at 5,3 cm/s)	< 0,3%	< 1%
Solidity	0,064	-
Class given by the manufacturer	H11 according EN 1822	FPP3 according EN 149

Different types of protective clothing (thicknesses and materials) as listed in Table 2. They were tested using through diffusion method with Pt, TiO_2 and graphite nanoaerosols. The efficiency of cotton, polyester (used for protective clothing in clean room) and high density polyethylene materials was investigated.

Table 2. Specifications of the protective clothing media tested by through diffusion technique.

Name	Thickness (µm)	Type	Material
Media 1	650	Woven	Cotton
Media 2	115	Non-woven	High-density Polyethylene textile Tyvek®
Media 3	160	Woven	Polyester

The efficiency of commercial gloves made of different materials and thicknesses as listed in Table 3.

Table 3. Specifications of gloves tested by through diffusion method.

Name	Thickness (µm)	Material
Glove 1	100	Nitrile
Glove 2	150	Latex
Glove 3	700	Neoprene

3. Results and discussion

3.1. HEPA filters

The efficiency of an HEPA glass fibrous filter was evaluated for particle diameters ranging between 9 and 19 nm at a flow speed v = 5.3 cm/s for different types of nanoaerosols: Pt, TiO_2 and graphite (Figure 1). Tests were performed with graphite nanoparticles centered around 35 nm. For TiO_2 and Pt the mean diameter is around 10 nm.

The filter shows higher efficiency for graphite nanoaerosols than for TiO_2 and Pt for particles ranging between 10 and 20 nm. A difference up to more than one order of magnitude is observed for a given particle diameter. TiO_2 nanoaerosols have the highest penetration through the filter: the maximum penetration for 19 nm particles is around 0.2%. The penetration of TiO_2 and Pt through a given filter is of the same order of magnitude. The fibrous filter efficiencies for graphite, TiO_2 and Pt nanoaerosols increase with decreasing the particle size as predicted by the classical filtration model thanks to Brownian diffusion [2]. However the filtration model does not take into account the influence of the aerosol type on the filtration efficiency [2].

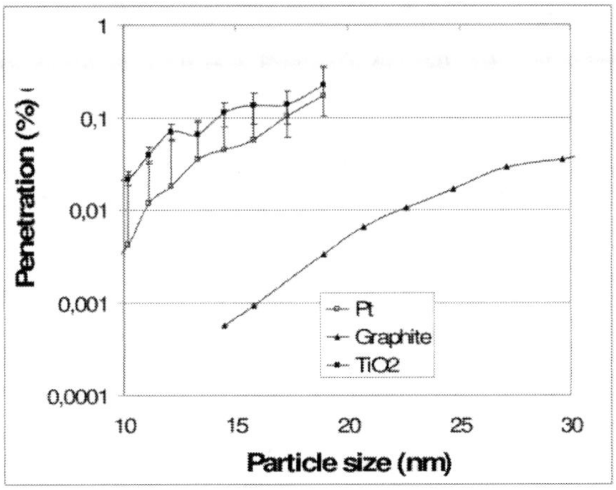

Figure 1. Evolution of graphite, TiO_2 and Pt nanoparticle penetration for the same HEPA fibrous filter as a function of particle size (nm) for a face velocity of 5.3 cm/s.

3.2. Electrostatic filters

Further, the efficiency of an FPP3 electrostatic filter was investigated for TiO_2 and graphite particle diameters ranging between 15 nm and 75 nm at the flow speed v = 5.3 cm/s. Tests were performed with TiO_2 nanoparticles centered around 45 nm and for graphite particles centred around 35 nm (as shown in Figure 2).

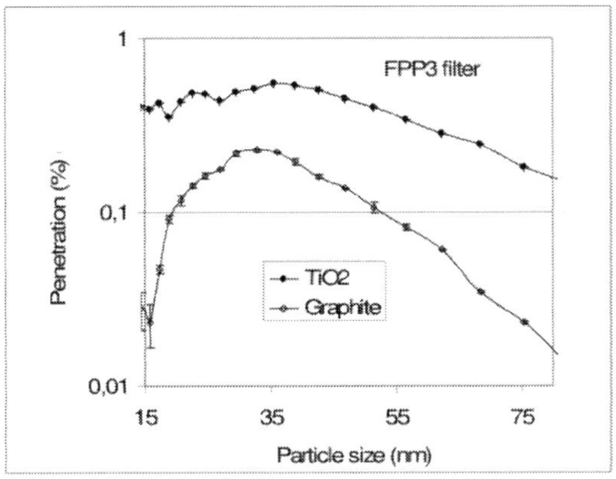

Figure 2. Evolution of graphite and TiO_2 nanoparticles penetration through the FPP3 electrostatic filter as a function of particle size (nm).

This type of filter also shows a higher efficiency for graphite nanoparticles than for TiO_2. The MPPS is around 35 nm for TiO_2 and graphite nanoparticles. The shift of the MPPS towards smaller particles is consistent with the theory: the fiber charge has an effect on the particle deposition [3]. The maximum penetration at MPPS was around 0.5% for TiO_2 and around 0.1% for graphite. The difference between TiO_2 and graphite for a given particle size increases up to one order of magnitude in efficiency when increasing or decreasing nanoparticle size around the MPPS.

3.3. Protective clothing

Different types of protective clothing, woven and non-woven were tested using through diffusion method with Pt and TiO_2 nanoaerosols.

The efficiencies of cotton, polyester (used as protective clothing in clean room) and high density polyethylene (specifications given in Table 2) were investigated. The number of particles flowing per minute in the down stream part of the cell was measured by CPC (Figure 3). The background noise of the measurement is kept below five particles/cm^3. Tests were performed first with TiO_2 and Pt nanoparticles centered around 10 nm and with a maximum concentration ranging between 7.10^5 part/cm^3 and 2.10^6 part/cm^3. The measurements, repeated three times on the same protective clothing, are found reproducible.

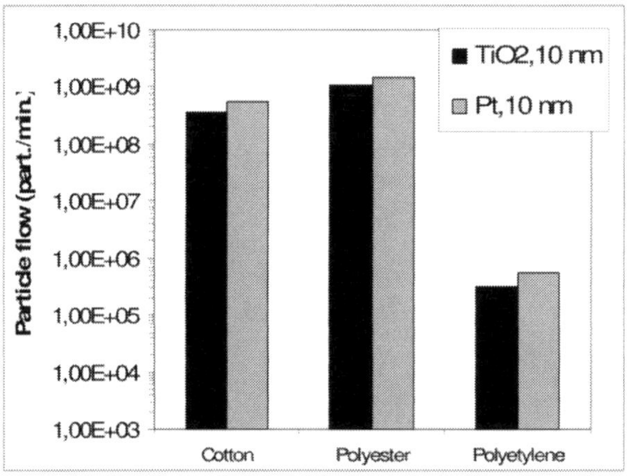

Figure 3. Evolution of clothing efficiency for TiO_2 and Pt nanoparticles as measured by through diffusion technique.

Cotton and protective clothing present almost the same efficiency when tested with TiO_2 and Pt particles centered around 10 nm. Non-woven fabrics (air-tight materials) are more efficient against Pt and TiO_2 nanoparticles centered around 10 nm than woven cotton and polyester. This is consistent with observations [8] made for graphite nanoparticles centered around 40 nm and 80 nm.

3.4. Gloves

Gloves made of three different materials *i.e.* nitrile, latex and neoprene (specifications given in Table 3) were tested with TiO_2 and Pt nanoparticles in static conditions. Gloves were exposed to graphite nanoparticles centered around 10 nm. The concentration in the up part of the diffusion cell was constant to be around 10^5-10^6 particles/cm^3. The differential pressure between up and down parts of the cell is fixed to few mbars in order to keep a diffusion regime. The background noise of the measurement is kept below five particles/cm^3. The measurements, repeated three times on the same glove is found reproductible. No TiO_2 and Pt nanoparticles centered around 10 nm penetration through the gloves is observed. Gloves are found very efficient for TiO_2 and Pt nanoaerosol penetrations. No nanoparticle penetration through the gloves is observed when exposing the glove during a few minutes. Other tests are in progress aiming at creating conditions more favorable to diffusion, although not necessarily corresponding to correct glove use: stretched gloves, long exposure time, increased differential pressure.

4. Conclusions

HEPA filters show a higher efficiency for graphite nanoaerosols than for TiO_2 and Pt nanoaerosols: a difference up to more than one order of magnitude was observed for a given particle diameter in the range 10-20nm.

The HEPA filter efficiency for TiO_2 and Pt nanoparticles is in the same order of magnitude.

At MPPS around 35 nm, the electrostatic FPP3 filter is more efficient for graphite than for TiO_2 nanoparticles. The difference in efficiency is increases, up to one order of magnitude on each side of the MPPS.

Cotton and polyester clothings showed close efficiencies when tested with TiO_2 and Pt nanoparticles centered around 10 nm. Air-tight fabrics made of non woven textile seem much more efficient to protect workers against Pt, TiO_2 nanoparticles than cotton and polypropylene.

Gloves are found very efficient for TiO_2 and Pt nanoaerosols with particle centered around 10 nm. No nanoaerosol penetration through the gloves is observed.

Acknowledgements
These results have been obtained in the frame of FP6 European program Nanosafe 2.

References

[1] Oberdörster G 2000 *Phil. Trans. Roy. Soc. London Ser. A* **358** 2719
[2] Hinds W C 1999 *Aerosol Technology: Properties, Behavior, and Measurement of Airborne particles* 2nd edition Wiley-Interscience
[3] Benoit Hervé-Bazin 2007 *Les nanoparticules: un enjeu majeur pour la santé au travail* EDP Sciences ed.
[4] Japuntich D A, Franklin L M, Pui D Y, Kuehn T H, Kim S C and Viner A S 2007 *Journal of Nanoparticle Research* **9** 93
[5] Kim S C, Harrington M S and Pui D Y H 2007 *Journal of Nanoparticle Research* **9** 117
[6] Heim M, Mullins B J, Wild M, Meyer J and Kasper G 2005 *Aerosol Science and Technology* **39** 782
[7] Balazy A, Podgorski A and Gradon L 2004 *Journal of Aerosol Science* **35(S2)** 967
[8] Golanski L, Guiot A, Rouillon F, Pocachard J and Tardif F 2009 *Human and Experimental Toxicology*

Journal of Physics: Conference Series **170** (2009) 012026 doi:10.1088/1742-6596/170/1/012026

Lessons learnt over 30 years of air filtration in the nuclear industry

J. Vendel, J.C. Laborde, N. Michielsen and F. Gensdarmes

Institut de Radioprotection et de Sûreté Nucléaire (IRSN), Service d'Etudes et de Recherches en Aérodispersion des polluants et en Confinement, F-91192 Gif-sur-Yvette, France

E-mail: jacques.vendel@irsn.fr

Abstract. The more than 30 years since the inception of the High Efficiency Particulate Air filters (HEPA) provide an incredible story. The filter's application to nuclear air cleaning and reciprocal effect on nuclear programme upon its development is even more interesting. The HEPA filter provided the capacity needed to intercept extremely small particulate matter in the airstreams of nuclear plants and laboratories. When some of the particulate matter potentially might be plutonium or other alpha radiation bearing particles in the air exhausted to the environment, the critical importance of the filter becomes obvious. From a crude and weak initial concept, the HEPA filter has developed into the backbone of particulate air cleaning for nuclear ends and has become most essential to environmental cleaning for other industrial pursuits as well.

In the nuclear industry, High Efficiency Particulate Air filters (HEPA) were the need for the containment of radioactive aerosols within the nuclear facilities. Air filtration theory has been very important in the development of HEPA filters. The early air filtration theories model air filters as the air flow around single fibres and the particle capture by these single fibres. The single fibre theories included the interference effect of neighbouring fibres by using cell flow models. Equations were derived to describe particle capture efficiency as function of system variables (air flow temperature and pressure), particle variables (size, density...) and filter characteristics (fibre diameter, fibre volume fraction, filter thickness...).

1. Mechanisms of filtration

In the viscous flow conditions which obtain in HEPA filters, the major mechanisms are considered to be diffusion, interception and inertia. The flow lines around the fibre are deflected and the three main collection mechanisms can be described as follows. A massive particle approaching the fibre deviates from the flow lines because of its inertia, the deviation being greater the more massive the particle and the higher the velocity. The particle centre needs only approach to a distance equal the particle radius from the fibre for interception capture to take place. This is more important for larger particles and does not depend on air velocity. Smaller particles are subject to molecular bombardment and diffuse across the flow lines increasing the likelihood of capture. The smaller the particle size and the lower the gas velocity, the greater is the importance of the diffusion.

The penetration of a fibre filter according to the different parameters which govern the particle filtration can be calculated with the following equation (Dorman, 1964):

$$P = exp\left(-\frac{4\alpha e}{\pi(1-\alpha)d_f}\eta_f\right)$$

where e is the filter thickness, d_f the fibre diameter, α the packing density and η_f the single fibre efficiency.

The single fibre efficiency is generally taken equal as the sum of each elementary mechanism of particle collection on a fibre:

$$\eta_f = \sum_J \eta_{f_j}$$

where J is the elementary collection mechanism.

In the case of the fibre filter, the elementary collection efficiency of a single fibre can be written as:

$$\eta_f = \eta_{f_{dif}} + \eta_{f_{\text{int}}} + \eta_{f_{ine}}$$

- $\eta_{f_{dif}}$: elementary collection efficiency by diffusion,
- $\eta_{f_{\text{int}}}$: elementary collection efficiency by interception,
- $\eta_{f_{ine}}$: elementary collection efficiency by inertia.

2. Collection efficiency

2.1. Elementary collection efficiency by diffusion

As shown on the Figure 1, the particle capture efficiency is due to Brownian motion of the particles which have a probability to reach a fibre and to stick at the surface.

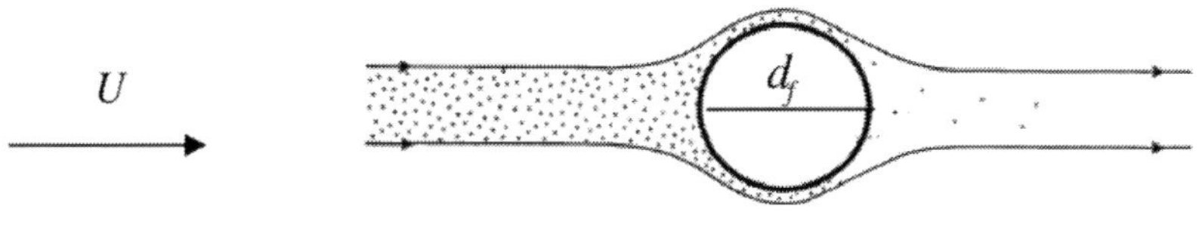

Figure 1. Collection by diffusion.

The particle collection by diffusion is governed by a characteristic number which is the Peclet number (Pe). The Peclet number is given by the following equation:

$$Pe = \frac{d_f U}{D_i},$$

where U is the flow velocity far away of the fibre and D_i the coefficient of diffusion for the particle.

The coefficient of diffusion for a spherical particle is given by:

$$D_i = \frac{kT\,Cu}{3\pi\mu\, dp},$$

where k is the Boltzmann's constant, μ the gas dynamic viscosity, d_p the particle diameter and Cu the Cunningham slip factor.

The elementary particle collection efficiency by diffusion can be written (Stechkina and Fuchs, 1966) as:

$$\eta_{f_{dif}} = \left(\frac{1-\alpha}{-\frac{1}{2}\ln\alpha - 0.75 + \alpha - \frac{\alpha^2}{4}} \right)^{1/3} Pe^{-2/3}.$$

This relation shows that if the flow velocity or the particle diameter increases the elementary collection efficiency is decreasing.

2.2. Elementary collection efficiency by interception

As shown on the Figure 2, the particle collection by interception is a purely geometrical phenomenon. The particles, which follow trajectories such as they flow nearby the fibre, will be intercepted and collected by the fibre.

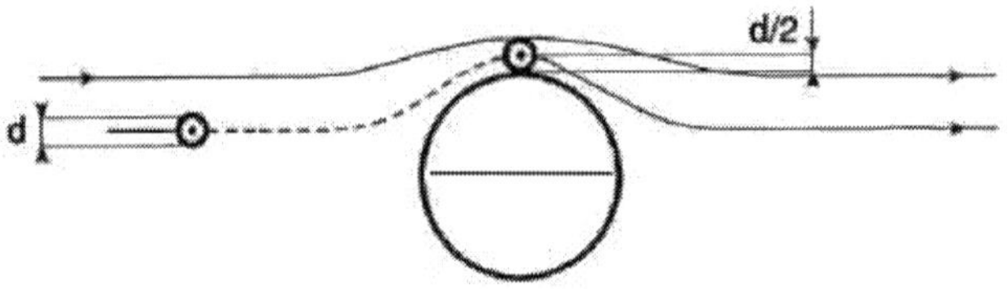

Figure 2. Collection by interception.

The particle collection by interception is governed by a characteristic number which is the interception number (R). This number is defined by the following equation:

$$R = \frac{d_p}{d_f}.$$

The elementary particle collection efficiency by interception can be written (Friedlander, 1957) as:

$$\eta_{f_{int}} = \frac{1-\alpha}{\left(-\frac{1}{2}\ln\alpha - 0.75 + \alpha - \frac{\alpha^2}{4}\right)} \cdot \frac{R^2}{1+R}.$$

This relation shows that the particle collection by interception is independent of the flow velocity and the efficiency is increasing when the particle diameter is bigger.

2.3. Elementary collection efficiency by inertia

As shown on the Figure 3, the particle collection by inertia is due to the fact that the particles, according to their mass and velocity, will be able to follow or not the streamlines around the fibre. When the particle inertia will be too high, the particles will impact the fibre.

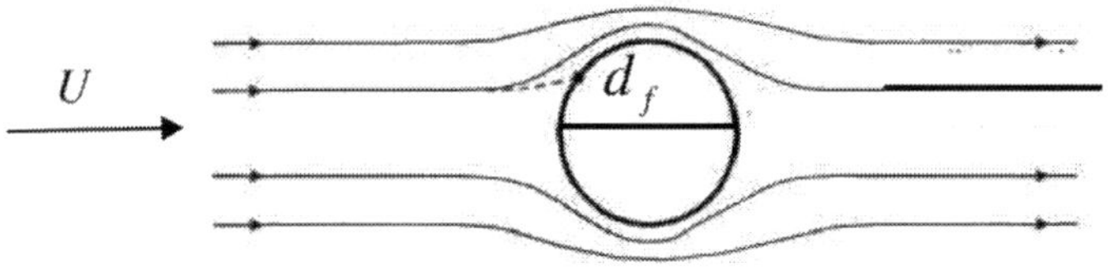

Figure 3. Collection by inertia.

The particle collection by inertia is governed by a characteristic number which is the Stokes number (*Stk*). This number is defined by the following equation:

$$Stk = \frac{\rho_p d_p{}^2 CuU}{9 \mu d_f},$$

where ρ_p is the particle density.

The elementary particle collection efficiency by inertia can be written (Davies, 1973) as:

$$\eta_{f\,int} = \frac{Stk}{4\left(-\frac{1}{2}\ln\alpha - 0.75 + \alpha - \frac{\alpha^2}{4}\right)^2}$$

This relation shows that if the flow velocity or the particle diameter increases, the elementary collection efficiency is increasing.

2.4. Total collection efficiency

As it is mentioned in the equation giving the filter penetration, the total collection efficiency of a fibre results from the sum of each elementary collection efficiency. So, according the evolution of the different mechanisms with the particle diameter, the penetration of a fibre filter corresponding to the radio between downstream and upstream concentrations, reaches a maximum as it is shown on the Figure 4.

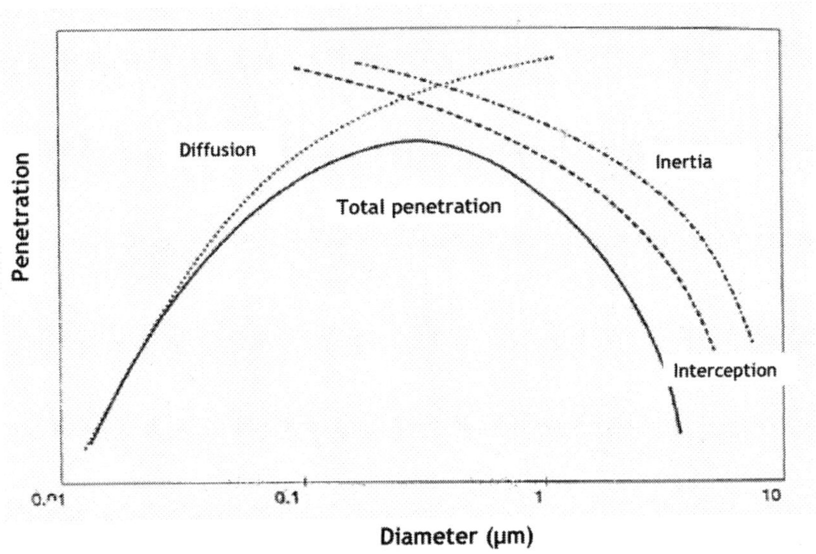

Figure 4. Filter penetration.

3. Experimental results

It is well established that at constant velocity, the penetration of particles larger than 300 nm decreases as particle size increases, while theories predict that particles much smaller penetrates less easily due to diffusion collection. Much effort has been expended on testing of the theories and on attempts to find the most penetrating particle size (MPPS). Examples of experimental results are given on figure 5 (decontamination factor is defined as the inverse of the penetration). For HEPA filters the MPPS is important, a test at the MPPS would guarantee a minimum efficiency for any unknown particle size challenge. So, the HEPA filters are tested with particles between 30 nm and 300 nm.

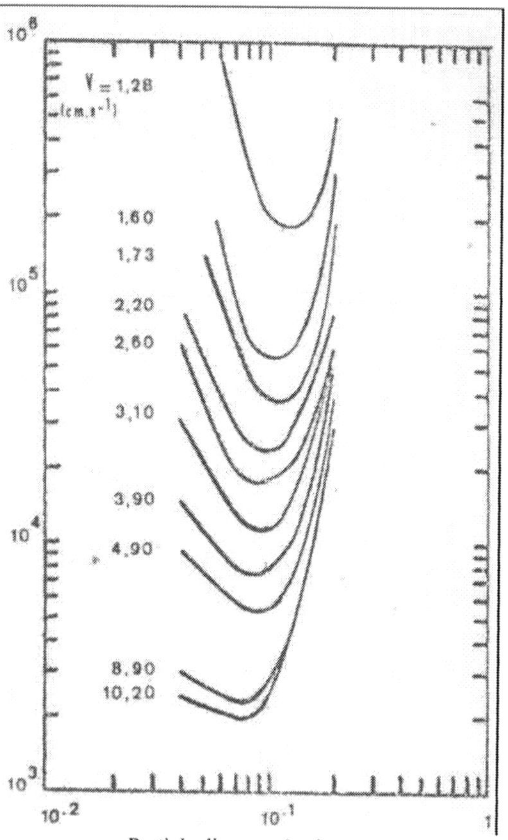

Particle diameter (μm)

J. Dupoux, A. Briand; Seminar on high efficiency aerosol filtration, Aix en Provence (France); November 22-25, 1976)

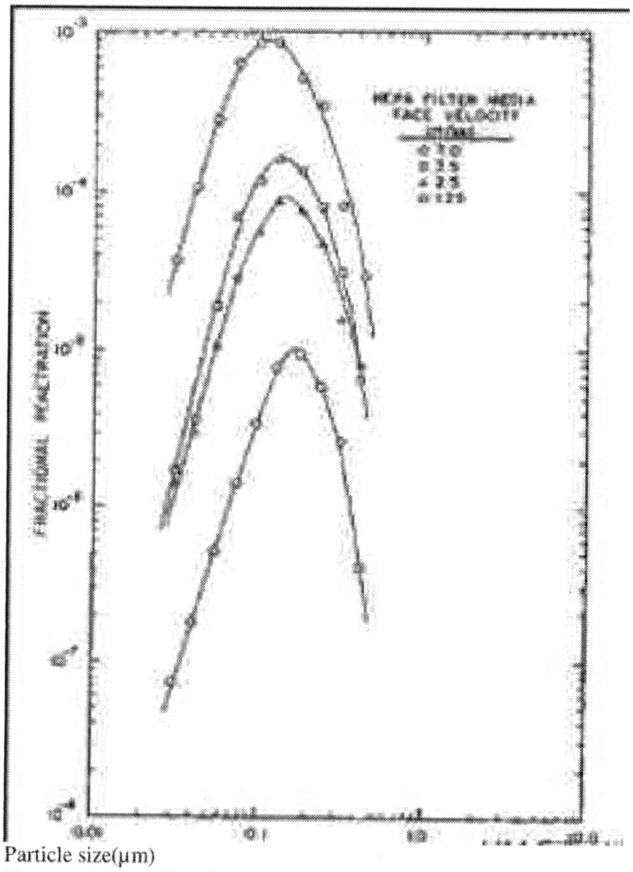

Particle size(μm)

Nanoparticle and Nanofiber Filtration: Modeling and Experiments; International Conference Hall, August 29 - September 1, 2007 Taipei,Taiwan, Tutorial IA *by* Prof. David Y.H. Pui

Figure 5. Experimental HEPA filter efficiency versus particle size.

4. Conclusion

In the frame of the development of HEPA filters a large effort has been put on the air filtration theory. It continues to be important, mainly due to the availability of CFD codes, for improving the performance of HEPA filters because of the large time and cost savings from reducing the number of tests required to optimize the filter and media design parameters.

The lessons learnt in the frame of the development of HEPA filters show that the particle collection efficiency reaches a minimum for particle diameter around 100-300 nm and demonstrate that there is no problem to filter nanoparticles which size is over 40 nm. So, efforts would be emphasized to demonstrate that the filtration theories would be extended to particle sizes of few nanometres.

References

[1] Davies C N 1973 *Air filtration*, Academic press, New York.
[2] Dorman R G 1964 *Theory of fibrous filtration in high-efficiency air filtration* White, Paf and Smith, SE Editors, Butterworths, London.
[3] Dupoux J and Briand A 1976 *Seminar on high efficiency aerosol filtration*, Aix en Provence (France) November 22-25.
[4] Friedlander S K 1957 *AIChE Journal* **3** 43
[5] Pui D Y H 2007 Tutorial IA, International Conference Hall, Taipei,Taiwan.
[6] Stechkina I B and Fuchs N A 1966 *Annals of occupational Hygiene* **9** 59.

Nanosafe 2008: International Conference on Safe production and use of nanomaterials IOP Publishing
Journal of Physics: Conference Series **170** (2009) 012027 doi:10.1088/1742-6596/170/1/012027

Spray Drying for Processing of Nanomaterials

Jesper Sæderup Lindeløv and Michael Wahlberg

GEA Niro, Gladsaxevej 305, DK-2860 Søborg, Denmark

E-mail: jesper.lindeloev@geagroup.com

Abstract. Consolidation of nano-particles into micron-sized granules reduces the potential risks associated with handling nano-powders in dry form. Spray drying is a one step granulation technique which can be designed for safe production of free flowing low dusty granules from suspensions of nano-particles. Spray dried granules are well suited for subsequent processing into final products where the superior properties given by the nano-particles are retained. A spray drier with bag filters inside the drying chamber and recycling of drying gas combined with containment valves are proposed as a safe process for granulation of potential hazardous nano-particles.

1. Introduction

Production of high-tech materials using nanoparticles is of great scientific interest as they are expected to yield products with superior characteristics. Turning a product produced on a laboratory scale into a commercial product means that process equipment needs to be available at the required capacity using suitable technologies for handling nanomaterials and their special properties. As long as health and environmental effects of nanomaterials are not fully understood the process technologies used should ensure the highest protection of operators and environment. A well established method of processing is to have the nanomaterials dispersed in a suspension. This significantly reduces the risks, however having the nanomaterials in suspension has implications due to higher transport costs of the often relatively diluted suspension and for aqueous suspensions containing perishable components the durability may restrict this method. Further, in order to produce solid materials it is required that the nanomaterials are transformed into a dry form. Spray drying produce a micronsized product from suspensions of nanoparticles, nanocapsules and other components having dimensions in the nanometer range. Hence, the spray drying technology may establish a link between the nanomaterial synthesis and the traditional processes which can produce the actual end products using micronsized materials.

Spray drying is a one step process where a liquid is atomized to micronsized droplets and dried by hot gas. The drying occurs in an insulated chamber consisting of a cylindrical chamber with a conical bottom. The hot drying gas is introduced through an air dispenser situated in the chamber roof. Atomization is performed using different techniques. Centrifugal atomizers, pressure nozzles and air blast nozzles are the techniques most frequently used but others are available. Drying time is from 1 to 5 seconds depending on the initial droplet size. Due to the energy consumption used for evaporating solvent the drying gas is cooled to an outlet temperature typically between 50°C and 120°C depending on product and application. The evaporative cooling means that the heat exposure of the product is relatively low and the final product temperature is often lower than the outlet temperature of the gas. The separation of dried product from the drying gas is done using a cyclone, a bag filter or both. Spray drying is a quick and relatively low temperature process and the promising features of the nanocomponents are preserved. By formulating the suspensions prior to spray drying re-dispersibility of the granules can be obtained.

GEA Niro is specialized in the development, design and engineering of liquid and powder processing equipments for the manufacture of products in powder, granular or agglomerated form. GEA Niro is recognized as the leading supplier of spray drying technologies for the food, chemical and pharmaceutical industry. For the past two years GEA Niro has worked on developing safe processing routes for production of non-dusty micronsized granules from formulated suspensions of nanoparticles. Through the EU founded SAPHIR (Safe, integrated & controlled Production of High-tech multifunctional materials and their Recycling) project GEA Niro has been collaborating with its 21 project partners in optimizing all processes involved in the nanoparticles synthesis through to the production of final materials of ceramic nature for use in many different applications. GEA Niro's aim in this project is to utilize the spray drying technology for production of nanostructured granules which can be handled in its dry form or redispersed. As part of this project and through other experiments performed in GEA Niro's Test Centre in Copenhagen experience in handling a range of different nanomaterials (e.g. SiC, SiO_2, TiO_2, Hydroxyapatite and carbon nanotubes) has been obtained using the spray drying process. Generally, results indicate that most suspensions of nanoparticles are easily dryable. In fact, results using the Drying Kinetic Analyzer™ (novel instrument for determining drying behaviour) have shown that the suspensions of ceramic nanoparticles tested have drying rates which are higher than other products dried in a spray drying system due to a longer constant rate drying period. This means that one stage drying can be utilized which makes up for a relative simple spray drying system. Micronsized granules may still produce dust if not treated with consideration. An area where accidental release of dust can occur from the spray drying is during changing of product containers. This risk can be removed by use of containment solutions. Buck® containment systems produced by GEA Pharma Systems are utilized in combination with the spray drying system for safe processing of nanomaterials. GEA Niro's pharmaceutical test plants are equipped with the Buck valve system to eliminate the risk of dust being released to the production environment. Especially the Buck Hicoflex® system is a containment interface for safe transfer of granules with a suitable containment level for use in nanomaterial processing systems.

This paper presents results obtained during the SAPHIR project where spray drying has been utilized for production of a non-dusty product from a suspension of nano titanium dioxide. The purpose has been to produce granules which can be used for coatings where the photo catalytic behaviour of the anatase phase of the titanium can be utilized. To ensure easy handling of the coating equipment a mean granule size below 50 μm was required and the product should be free flowing. Different spray drying configuration and drying parameters have been tested to find the optimum conditions for continued production of the product.

2. Materials and Methods

2.1. Materials

Titanium dioxide (Degussa Aeroxide TiO_2 P25) having a specific surface area of 50 m^2/g was dispersed in an aqueous solution of Dispex N40 (Ciba). Through vigorous stirring a suspension containing 37% dry matter was obtained. 2-Octananole was used as foam reducer. Sieving of the suspension was conducted to remove lumps which otherwise would block the atomization nozzle. The suspension was continuously stirred until spray drying.

2.2. Nozzle characterisation

The nozzle used for atomizing the suspension was a GEA Niro two fluid nozzle. Nozzle characterization was done using a Malvern Spraytec (Malvern Instruments) with a setup shown in Figure 1.

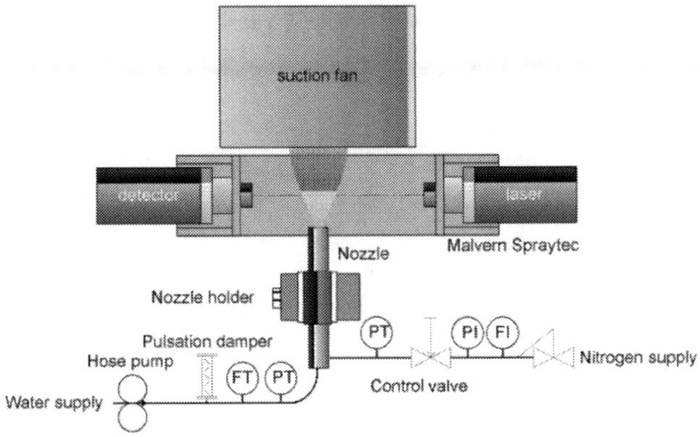

Figure 1. Schematic illustration of the set-up for droplet size measurements by laser diffraction.

Results of droplet size were recorded during atomizing water at a flow rate of 1,5; 2,5 and 4,0 kg/h using atomization gas flow rates between 1 and 20 kg/h. Knowledge of the spraying behaviour of the nozzle was used in selecting atomization conditions during spray drying.

2.3. Spray drying

A Mobile Minor™ (GEA Niro) spray dryer was used in four different configurations as illustrated in Figure 2.

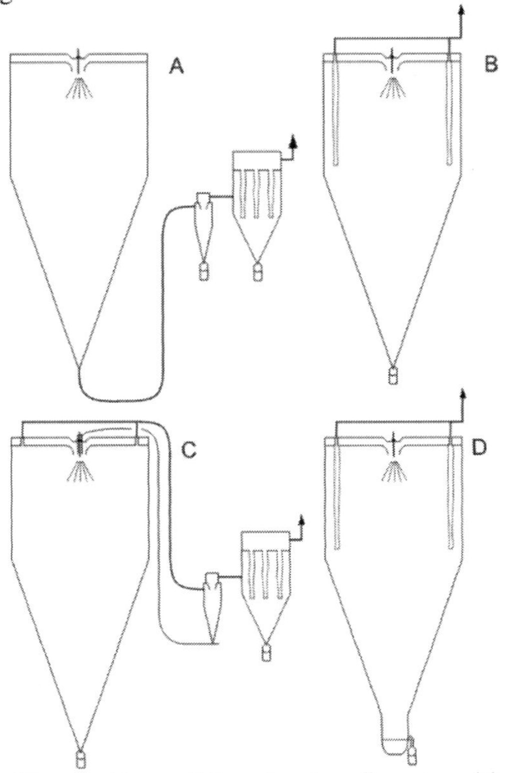

Figure 2. Different configurations of the Mobile Minor (MM) used in the experiments.
A: SD-MM configuration with product outlets from cyclone and bag filter;
B: SD-IF-MM configuration with bag filters placed inside the drying chamber and with product outlet under the chamber;
C: SD configuration with fines return from the cyclone to the atomization zone and main powder outlet under the chamber;
D: IFD-MM configuration with bag filters placed inside the drying chamber and with powder outlet from the integrated fluid bed.

The plant has a 0.8 m chamber diameter with a cylindrical height of 0.65 and a 40° cone angle. Co-current drying was conducted with the atomization nozzle positioned in the centre of the hot gas dispenser. A pneumatic hammer was mounted on the conical part of the chamber and was in operation during the experiments. Nitrogen was used as drying gas. The inlet drying gas rate was 80 kg/h at temperatures between 155°C and 195°C, and the outlet temperature was between 70°C and 90°C giving a feed rate between 2 and 5 kg/h. In using the IFD-MM configuration the inlet temperature to the fluid bed was 60°C at a rate sufficient to fluidize the granules. The bag filters were cleaned during operation by back flush purging using nitrogen gas at 2 bar(g). Process parameters were continuously recorded during operation.

2.4. Granule characterisation

The granule size was measured using a Mastersizer 2000 (Malvern Instruments) equipped with a dry feeding unit (Scirocco) dispersing the sample at 0,5 bar. Three size values are reported; D10 is the size where 90% (volume) of the population are larger, D50 is the median size and D90 is the size where 10 % of the population is larger.

Residual moisture content was measured using a Halogen Moisture Analyser (HR73, Mettler Toledo) at 105°C.

Granule morphology was investigated at Technological Institute, Center for Micro Technology and Surface Analysis using a scanning electron microscope (Ultra55, Zeiss).

3. Results and discussions

Initial test using the Drying Kinetic Analyzer™ (DKA) was performed. The DKA is a novel instrument enabling real time drying kinetic measurements and morphology development observations on a levitated drop drying at carefully controlled conditions. The drying behaviour determination of the nano-TiO_2 suspensions at 37% solids showed that the drying rate was relatively high (data not shown) and that a spherical granule was produced. The volume reduction during drying of the suspension droplets was 45%. The final size of non-agglomerated granules produced by spray drying is determined by the droplet size obtained in the atomization process and the volume reduction occurring during drying. The volume reduction is often dependent on the drying temperatures. Hence, the drying behaviour determination should be done at the same temperature as during spray drying.

Knowledge on the performance of the atomization system is important for control of granule size. Figure 3 shows the spraying performance of the two fluid nozzle atomization water at three different flow rates. The droplet size is dependent on the liquid flow rate and the atomization gas flow rate. The ratio between the atomization mass flow rate and the liquid mass flow rate (ALR) is normally used to illustrate the performance of two fluid nozzles. Knowledge of nozzles performance can be utilized in adjusting atomization conditions during spray drying experiments as suspensions will be influenced by changing liquid flow rates and ALR in a similar way as found for water.

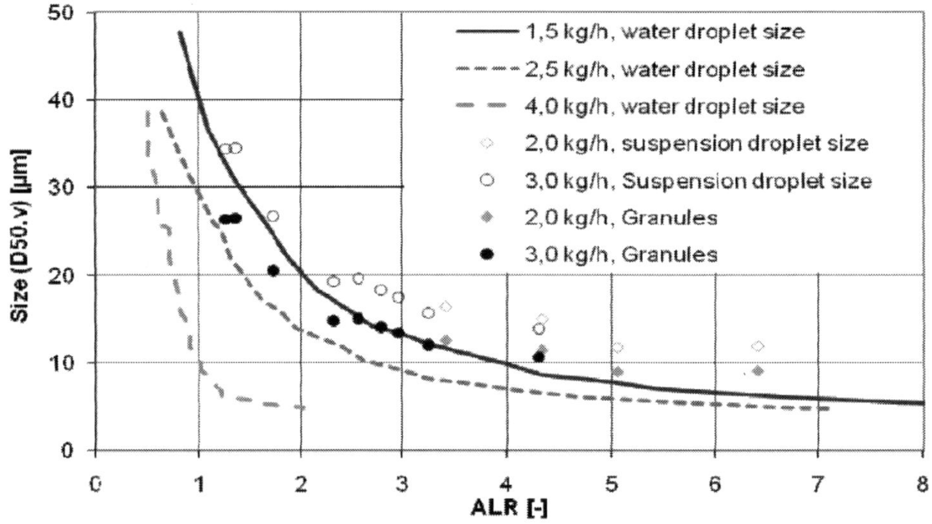

Figure 3. Graph shows droplet sizes produced by the GEA Niro two fluid nozzle operating at three different water flow rates and granule size produced in the Mobile Minor spray dryer (using the A configuration as show in Figure 2) at two different suspension flow rates. Suspension droplet sizes are calculated from the measured granule sizes using the volumetric reduction determined using the Drying Kinetic Analyzer™. Absissa is the Ratio between Atomization gas mass flow rate and the Liquid mass flow rate (ALR).

The size of spray dried granules obtained at a feed capacity of 2.0 kg/h and 3.0 kg/h are shown in Figure 3. The calculated droplet sizes of the suspensions follow the same tendency as that found for water atomization. In producing a non-dusty product the granules should be as large as possible. In this case, however, the mean size (D50) should not exceed 50 µm. Different spray drying configurations have been utilized in producing a product which has a D10 (granules size where 10 % (volume) of the population having a smaller size: i.e. dusty part of the population) as high as possible while keeping the D50 under the 50 µm. Table 1 shows process parameters, granule size and residual moisture content from spray drying experiments using the four different configurations.

Table 1. Selected results from spray drying in four different configurations

Description		Unit	Plant configuration (see figure 2)			
			A	B	C	D
Feed rate		kg/h	3.1	2.9	2.8	5.0
ALR		-	1.4	1.3	2.6	3.5
Inlet gas temperature		°C	154	182	160	195
Outlet gas temperature		°C	82	79	84	71
Granule size						
	D(10; V)	µm	9[1]	11	14[2]	17
	D(50; V)	µm	26[1]	32	39[2]	43
	D(90; V)	µm	75[1]	87	91[2]	120
Residual moisture		%	0.9[1]	0.5	0.4[2]	0.3

[1] Data from cyclone fraction
[2] Data from chamber fraction

By using spray drying configuration A the finest part of the population is not collected by the cyclone. Hence, two fractions are obtained. The cyclone fraction has a suitable granule size but the bag filter fraction contains a very dusty product with a D50 of only 5 µm. However, only 5% of the total mass produced was collected under the bag filter. Having two fractions of a product always gives rise to questions on what to do with the non-spec fraction. Often it can be redispersed and mixed with the next production; however, avoiding two fractions is preferable. Figure 4 shows SEM pictures of the granule morphology obtained using the four different configurations. In plant configuration A individual spherical granules were produced as expected from the drying behaviour analysis. Slightly higher tendency for agglomerated granules was obtained for plant configuration B. The reason is that the fine part of the population will stay in the SD-IF chamber until agglomerated with bigger wet droplets and become sufficiently heavy to fall to the product container at the bottom. Agglomeration is an efficient way to remove the finest granules from a population by agglomerating them to bigger granules. This agglomeration process is forced using plant configuration C where the granules collected by the cyclone were returned to the atomization zone. Examples of such agglomerated particles are seen in Figure 4 C. Table 1 also shows that even though atomization has occurred at a higher ALR in the plant configuration C compared to configuration B the mean granule size was larger. Again, using configuration C gives rise to a non-spec fraction collected under the bag filter containing very dusty product and this is not optimal even though that the amount was limited. Agglomeration is very distinct for granules produced using configuration D. The agglomeration occurred in a random way as in the B configuration but, the fluid bed function as a classifier by blowing off the finest granules. Control of the granule size using configuration D was very difficult and having a mean granule size below 50 µm was only obtained for a relatively short time.

Figure 5 shows a SEM picture of granules produced using configuration A. One of the granules has become damaged in the process by either high shear or impact with the production equipment. No binders were used in the suspension preparation, and this limits the granule resistance to shear and impact that may occur especially in the cyclone or in the duct from the chamber. In the cases where binders need to be avoided a low shear process should be considered to ensure that granule damage is kept at a minimum.

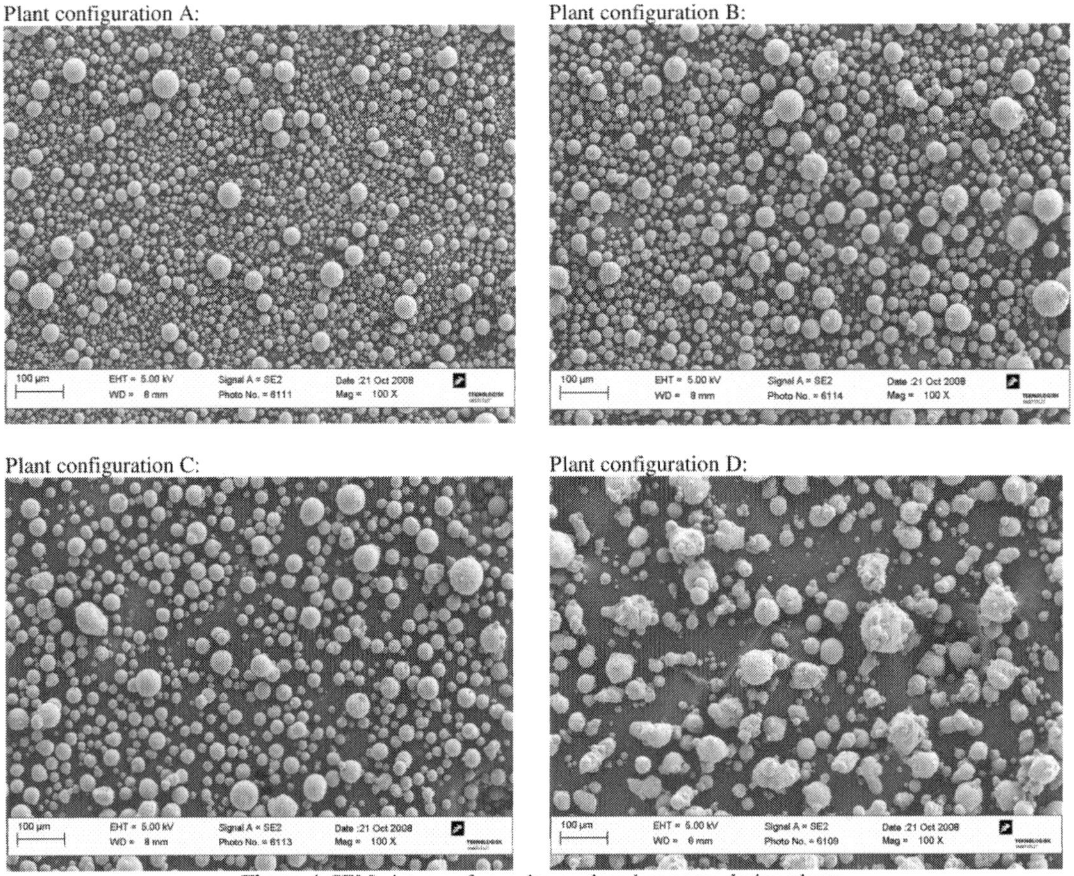

Figure 4. SEM pictures of granules produced on spray drying plants in four different configurations (A, B, C and D as shown in Figure 2).

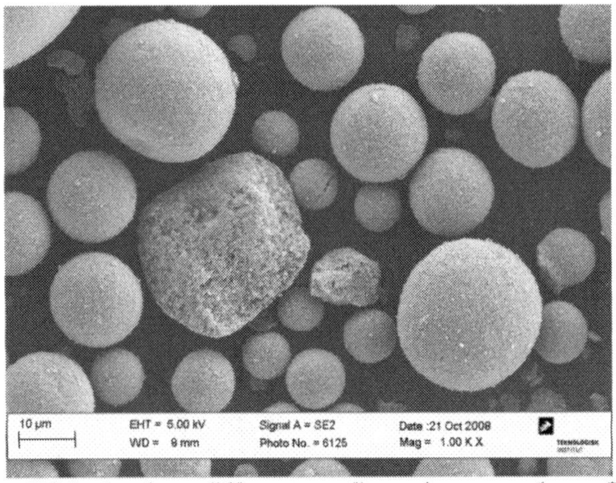

Figure 5. SEM picture of granules produced on the spray dying plant in configuration A as shown in Figure 2.

With the four different configurations tested, configuration B appears to be the most attractive solution as it is a simple low shear system, it produces only one fraction, it is easy to control and it confines all granules inside the chamber. This configuration has been utilized on other suspensions of nanomaterials where it also proved to be a suitable configuration utilising that the suspensions of nanoparticles dry relatively easily.

Safety issues in spray drying involves a number of individual risks which need to be tackled, *e.g.* dust explosion, solvent explosion, hot surfaces and emission of granules to the production environment. These risks are to be considered both with the plant in operation and during cleaning. The risks relating to explosions and hot surfaces is relatively trivial and are routinely tackled doing GEA Niro's design of spray drying equipment. Also risk of emission of granules is routinely tackled, however with the health concerns associated by dealing with nanostructured granules special emphasis should be laid on this risk. The atomization step produces micronsized droplets which dry to granules still in the micronsize range. This means that the granules produced during spray drying are much easier to control compared to nanopowders as they will remain in the area of where emission has occurred. Even though, release of granules from the spray drying plant should be avoided. This risk can be eliminated using a closed circuit plant as the one shown in Figure 6. This plant operates as the plant configuration B, but with a HEPA filter placed after the bag filters to ensure complete confinement of the nanomaterials. Further, a condenser is used to condense the vapours in the outlet drying gas which means that the gas can be reused after reheating it to the inlet temperature. The exhaust flow is very small compared to a normal open plant. Loss of granules during change of product container is eliminated through the use of the Hicoflex® system from Buck® (Figure 6 b, c). The proposed plant is a Mobile Minor sized plant having a capacity up to 5 kg/h. The plant can be scaled to the full size range that GEA Niro supplies, but in each case the design and level of safety should be selected on the basis of a risk evaluation based on the environmental and health implications with the products to process.

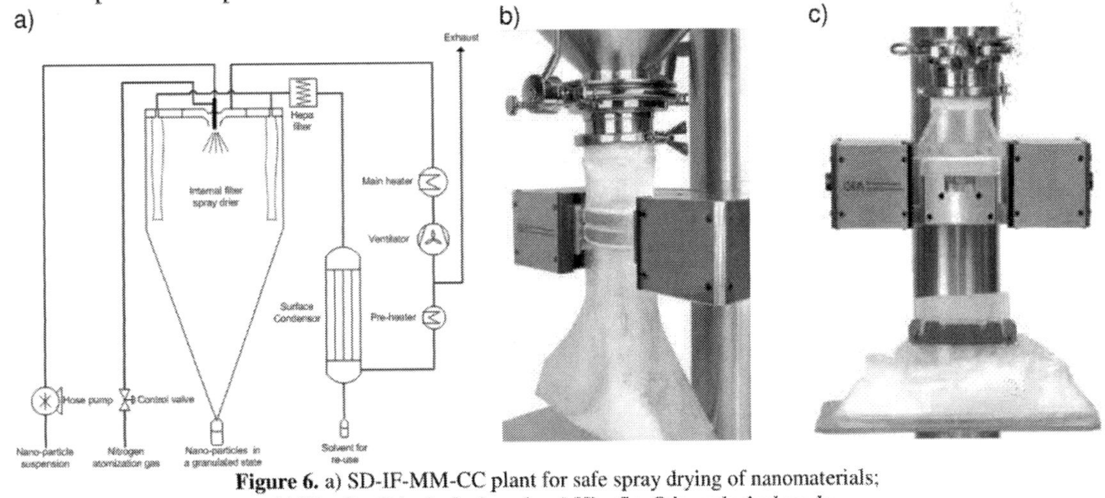

Figure 6. a) SD-IF-MM-CC plant for safe spray drying of nanomaterials; b) Hicoflex® in docked mode; c) Hicoflex® in undocked mode.

4. Conclusions

This study demonstrates that spray drying can successfully produce a non-dusty, micronsized granulated product from suspensions of nanomaterials. Four different configurations were tested for spray drying a suspension of nano-TiO_2. It was concluded that the SD-IF-MM configuration was the optimal configuration for this particular product. A closed circuit configuration of this plant equipped with the Hicoflex® system would enable safe processing of nanomaterials.

Acknowledgement

We gratefully acknowledge founding obtained from the European Commission through the 6th framework programme as part of the SAPHIR project (http://www.saphir-project.eu/). SAPHIR partner CERAM is acknowledge for developing the recipe for TiO_2 suspension preparation

Nanoparticles Filtration by Leaked Fibrous Filters

Guillaume Mouret[1], Sandrine Calle-Chazelet[1], Dominique Thomas[1], Denis Bemer[2] and Jean-Christophe Appert-Collin[1]

[1]Nancy-Université/LSGC/CNRS - 1 rue Grandville - BP 20451 - F-54001 Nancy Cedex, France

[2]INRS - Avenue de Bourgogne - F-54501 Vandoeuvre les Nancy Cedex, France

E-mail: sandrine.calle@ensic.inpl-nancy.fr

Abstract. The aim of this work is first to measured nanoparticles penetration through three different fiberglass filters intentionally-pierced with calibrated needles at different filtration velocity. Then a semi-empirical model based on the air flow resistances of the new and perforated filter media and on the mechanism of Brownian diffusion for the collection of ultrafine particles by the media enables to well predict the efficiency observed for all tested operating conditions. Results show that the increase of particles penetration is all the more important that the pinhole is large and that the particle diameter is low. Another result is that the filtration efficiency of the new filter media controlled the penetration. A high efficiency filter with a high resistance to air flow will be more damaged than a low efficiency filter when being perforated.

1. Introduction

For the last few years, numerous works have been carried out dealing with the efficiency of undamaged filters in the ultrafine particle size range [1- 3]. In contrast, very few have been done about the performances of media presenting pinhole towards the filtration of particles under 30 nm. It is however known that the performances of filters can be dramatically changed because of leakages.

So, we measured nanoparticles penetration through three different fiberglass filters intentionally-pierced with calibrated needles, at 5 and 15 cm/s. The test monodisperse aerosols of copper nanoparticles, the electrical mobility diameter of which was varied from 4 to 30 nm, were produced from a commercial spark-generator and a TSI® nano-DMA 3085 for diameter selection. The effect of the filtration velocity, the particle diameter, the properties of the filter media are studied experimentally.

On the other hand, we developed a semi-empirical model based on the aeraulic properties of the filter media which are determined experimentally. The assumption that the only mechanism which occurs during the capture of the nanoparticles by a fibrous filter is the Brownian diffusion allows the development of an expression of the penetration of nanoparticles through a filter media with a pinhole.

2. Materials and Methods

2.1. Experimental Apparatus

In front of the lack of legislation about nanoparticles handling, the precautionary principle has to be applied. That is why the experimental setup of this study was put in an air hood and a part of it was integrated inside a gloves box (schematized in dotted lines on Figure 1).

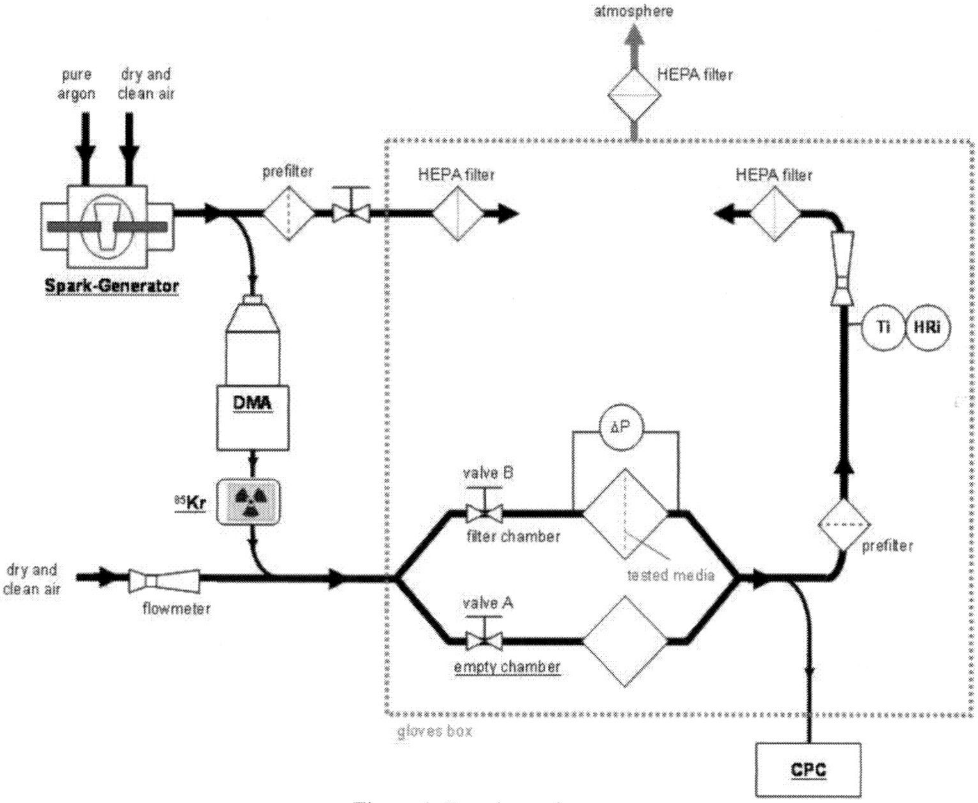

Figure 1. Experimental setup

This test bench can be classically divided into three main stages.

The first one consists in generating a polydisperse aerosol of nanoparticles. A Palas® GFG-1000 spark-generator was used for that. Developed by Helsper et al. (1993), this device uses high voltage to produce sparks between two electrodes. The resulting sublimed material condenses into very fine particles, which leave the electrodes chamber thanks to a flow of pure argon inside which they grow in size. An additional flow of air then dilutes the particles. With copper electrodes, the GFG-1000 generator produces an aerosol with a mean diameter of 7 nm.

As a second step, a part of this aerosol is sampled into a TSI® nano-DMA 3085. The particles are separated thanks to an electric field between two concentric cylinders according to their electrical mobility, which is inversely proportional to their diameter. This creates at the outlet of the classifier a monodisperse unipolarly-charged aerosol, which is then globally neutralized – Boltzmann equilibrium – thanks to a ^{85}Kr ionizing source [4].

The last step is the filtration of the selected particles carried by a dry and clean air stream, the flowrate of which is set according to the desired filtration velocity, through each of the two grounded stainless steel branches of the parallel chambers device alternatively, according to the valves A and B respective positions. Both the empty and the filter chambers are identical and have an internal diameter of 60 mm. Particle concentration at the outlet of any chamber is measured by a TSI®

CPC 3022. The 1.5 L/min it aspires are isokinetically sampled. Then, the ratio between the particle counts at the outlet of both the filter and the empty chambers respectively is equal to the penetration P of the media put inside the filter chamber.

2.2. Filter media properties

The medias tested in this study are three non woven fiberglass filters. Their physical properties are listed in Table 1.

Table 1. Physical properties of the non-woven media A, B and C

	media A	**media B**	**media C**
Mean geometric fiber diameter d_f (μm)	4.5	3.3	1.3
Thickness Z (μm)	440	420	409
Solidity α	0.060	0.065	0.078

These medias were perforated with calibrated needles of 400, 1320 and 2000 μm diameter. Pressure drop of these medias were recorded as a function of the air flow rate on the range 0-100 NL/min in both cases: without pinhole and with a pinhole. The curves obtained are linear in each case and resistance to air flow can be deduced from these experiments. Their values are listed in Table 2.

Table 2. Values of resistance to air flow of the filter medias (K and K')

	Resistance values (× 10^5 cm^{-1}) ± 3 × 10^3		
pinhole diameter d_h (mm)	**media A**	**media B**	**media C**
0	1.57	4.24	35.41
0.4	-	4.22	35.28
1.32	1.55	4.18	33.22
2	1.52	3.96	-

Then the filters were loaded by copper nanoparticles the electrical mobility diameter of which was varied from 4 to 30 nm.

3. Penetration model

3.1. Description of the modelled system

A schematisation of the filter media with a pinhole is given on Figure 2.

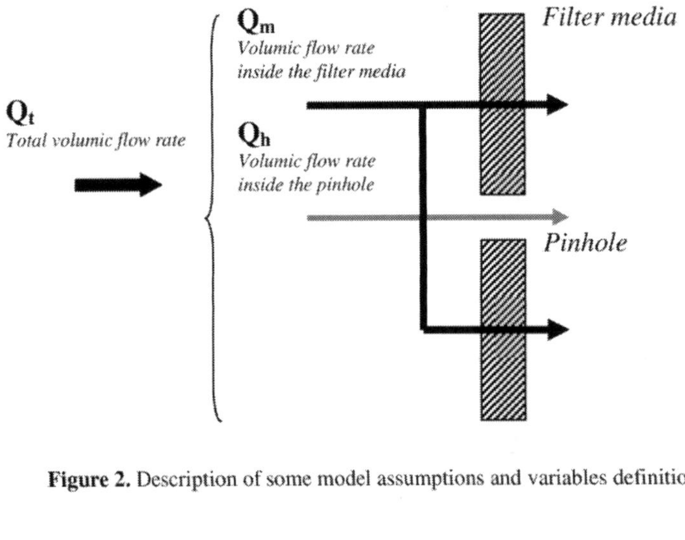

Figure 2. Description of some model assumptions and variables definition.

3.2. Expression of the total penetration

A filtration efficiency model that includes the effect of a pinhole inside the media can be derived by differentiating the leakage flow from the one passing through the remaining bed of fibers of the filter. Then, from the mass balance of the system, it can be written:

$$P_f = \frac{Q_b}{Q_f} P_b + \frac{Q_h}{Q_f} P_h = 1 - E_f \qquad (1)$$

where the subscripts f, b and h are for filter, bed and hole respectively. P_i is the penetration through the object i and Q_i is the corresponding flowrate. The total flowrate Q_f is equal to the sum of Q_b and Q_h. E_f is the global filtration efficiency. The additional subscript 0 is used to represent the values for the leakage-free filter.

From equation *(1)* and according to the respective values of $Q_b \times P_b$ and $Q_h \times P_h$, the total penetration of the filter is to be directed either by the remaining bed of fibers or by the pinhole or by the both in an intermediate case. Thus and as a first point, Q_h and Q_b have to be estimated. It can be done from the measurement of the pressure drop of the filter with and without any hole inside.

The evolution of the pressure drop $\Delta P_{f,0}$ of the three intact filters are linear as a function of the filtration velocity U_f (see Table 2). In laminar flow, $\Delta P_{f,0}$ can be mathematically expressed as:

$$\Delta P_{f,0} = K\,\mu\,U_f = K\,\mu\,\frac{Q_f}{S_f} \qquad (2)$$

with K, the flow resistance (Table 2) and μ, the air viscosity (1.8×10^{-5} Pa.s about in normal conditions). The surface of the filter S_f is equal to 2.83×10^{-3} m^2.

In addition, ΔP_f is proportional to the filtration velocity U_f by a factor $\mu \times K'$:

$$\Delta P_f = K'\,\mu\,U_f = K'\,\mu\,\frac{Q_f}{S_f} \qquad (3)$$

The values of K' are listed in Table 2.

The percentage of leak flowrate increases as the hole size increases. As a consequence, the actual velocity U_b at which particles pass through the remaining bed of fibers of the filter decreases and is less and less close to U_f as d_h is bigger (K'/K ratio decreases). This difference of real filtration velocity between the filter in intact and leaked state has to be taken into consideration in the determination of the penetration fraction P_b, through the single-fiber efficiency term η_b:

$$P_b = \exp\left(-\frac{4\,\alpha\,Z}{\pi\,(1-\alpha)\,d_f}\,\eta_b\right) = 1 - E_b \qquad (4)$$

where E_b is the particle removal efficiency of the bed of fibers. When the filter is undamaged, E_b is equal to E_f, which can also be noted $E_{f,0}$.

Then, equation *(1)* resumes to the general expression:

$$P_f = \frac{S_b\,K'}{S_f\,K}\exp\left(-\frac{4\,\alpha\,Z}{\pi\,(1-\alpha)\,d_f}\,\eta_b\right) + \left(1 - \frac{S_b\,K'}{S_f\,K}\right)P_h \qquad (5)$$

It is assumed in the following of this study that the particle penetration through the hole P_h is total. In other words, $P_h = 1$ whatever the particle size d_p and the flowrate Q_h. Moreover and in regard to the aerosol size range covered by this paper (d_p < 30 nm), it can be considered that the unique mechanism responsible for particle collection is the Brownian diffusion. Then, different models are available in the literature for the calculation of the single-fiber efficiency η_b [5-7]. It was chosen here to use the recent correlation of Wang *et al* [8] since it well fits the nanoparticles penetration values measured for the unleaky filters (Figure 6):

$$\eta_b = 0.84\ Pe_b^{-0.43} \qquad (6)$$

where Pe_b is the Péclet number of the bed of fibers:

$$Pe_b = \frac{d_f\,U_b}{D} = \frac{d_f\,Q_b}{D\,S_b} = \frac{K'}{K} Pe_f \qquad (7)$$

with D, the particle diffusion coefficient.
This equation was validated experimentally on the tested filters without pinholes.
Incorporation of equation *(6)* into equations *(5)* finally leads to:

$$P_f = 1 - \frac{S_b\,K'}{S_f\,K}\left(1 - \exp\left[-\frac{3.36\,\alpha\,Z}{\pi(1-\alpha)d_f}\left(\frac{K'}{K}Pe_f\right)^{-0.43}\right]\right) \qquad (8)$$

4. Experimental Results and Discussion

4.1. Media Penetration

Each point presented in this section is the average of three series of measurements at least. The aerosol concentrations were always lower than 10^4 part.cm^{-3} in order to detect particles in the most accurate counting mode of the CPC.

Figures 3 and 4 show the effect of calibrated holes inside media A and media B on their penetration by nanoparticles. It can be seen that the agreement between the experimental points and equation (8) is very good.

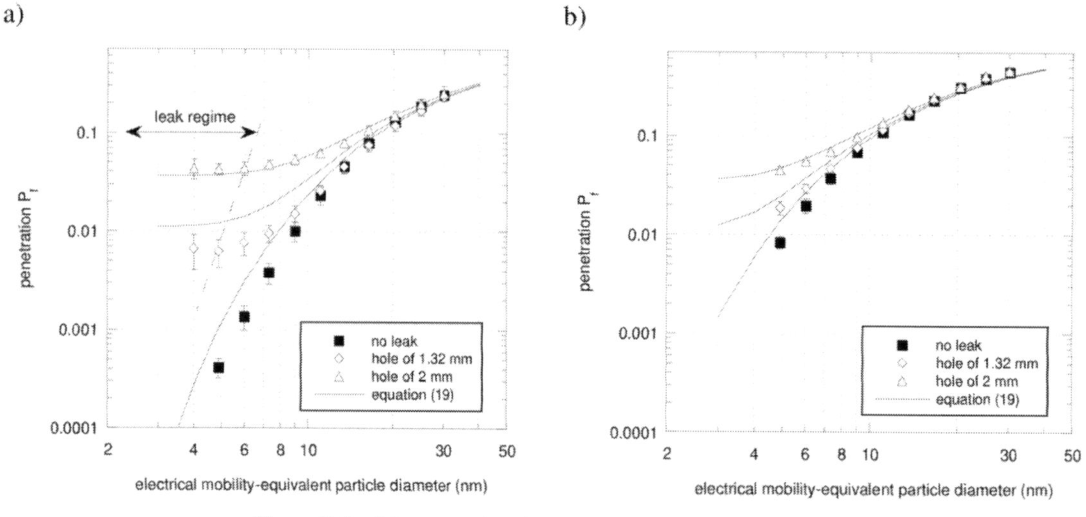

Figure 3. Particle penetration through media A as a function of both particle size and leak hole diameter (a): 5 cm/s and (b): 15 cm/s.

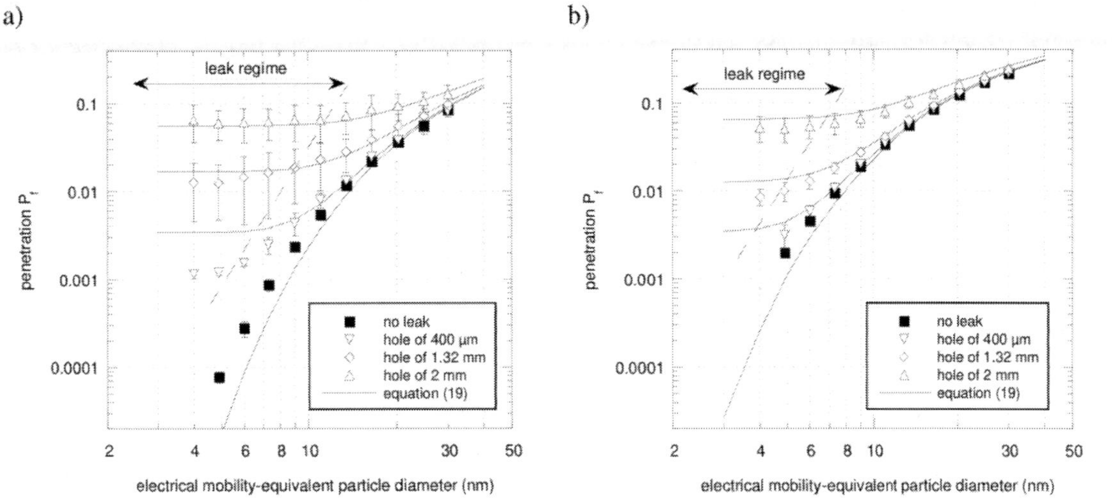

Figure 4. Particle penetration through media B as a function of both particle size and leak hole diameter (a): 5 cm/s and (b): 15 cm/s.

As it could be expected, the global efficiency of a filter decreases when this latter is damaged, all the more as the leakage is bigger. It is however notable that penetration increase is more important for fine particles than for larger ones, even for the smallest holes.

Lastly, it is well seen on Figures 3 and 4 that below a given aerosol diameter, which depends upon the leak size, the penetration of particles tends to reach a constant value, like for media C (Figure 5). The aerosol penetration is no longer determined and controlled by the capture through the bed of fibers, but by the leak (leak regime).

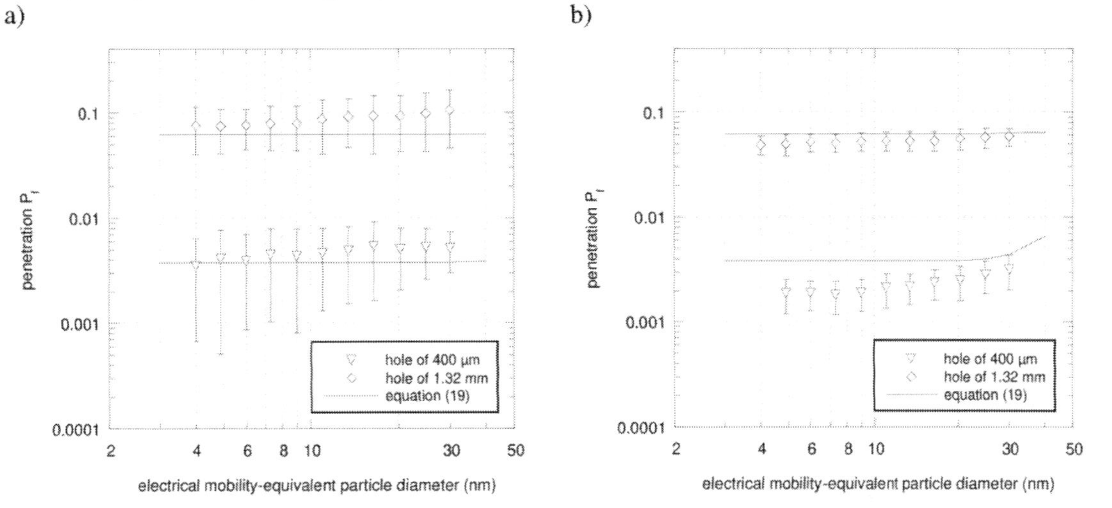

Figure 5. Particle penetration through media C as a function of both particle size and leak hole diameter (a): 5 cm/s and (b): 15 cm/s.

Whatever the particle size, media C is always in leak regime, even for the tiniest hole of 400 µm. It can be explained by the fact that this filter is so much efficient in leakage-free state that the slightest leak is enough to significantly increase the particle penetration.

Figures 3, 4 and 5 show that while media C is the most efficient of the three tested filters when they are intact at 15 cm/s, it becomes less effective than media B below 13 nm and than media A below 7 nm when they are 1.32 mm leaked. This point can be explained by the fact that the air flow resistance of media C is higher than the resistance of the two other filters. Then, at a given total flow and for a same hole size, the percent of leak flow is likely to be all the more important as the filter is higher flow-resistant. Thus, the lower the pressure drop due to the fibers is, the less significant a leak becomes in allowing particles to penetrate the media. The model well describes this phenomenon.

For a given particle size and as a well-known result, penetration through intact filters increases when the velocity increases, as a consequence of the reduction of diffusional deposition (the Péclet number is higher). In contrast, the penetration through damaged filters progressively becomes less and less velocity-dependent as the hole size increases. This confirms that in leak regime, the particle penetration is no longer influenced by the filtration velocity. Then, the gap between the penetration values for leaked and intact filters decreases as the filtration velocity increases. In a sense, the effect of a leak inside a media is reduced at high velocity.

5. Conclusion and Perspectives

The present study showed that perforation in a filter can result in a strong efficiency decrease, even though the opening percentage is small. On the one hand, these results confirm the important point that air leakage through a hole is all the more consequential as the air flow resistance of the filter is higher (i.e. for filters with higher efficiency). On the other hand, the effect of air leakage is more significant for smaller particles, in the region of 10 nm and below. This might call into question the use of very high filtering facepiece respirators for protection against nanoparticles, since the higher the media efficiency, the higher the leak consequence.

Moreover, it is shown that knowledge of air flow resistance of both new and damaged filter can be used to predict penetration increase. It must be noted that for filters with low flow resistance, air leakage has very little effect on pressure drop, resulting in potential undetected leak. Though, particle penetration can have been dramatically increased and causes the filter to fail efficiency requirements, especially for the finest particles. Consequently, this limits the use of indirect, pressure-based, methods of leakage measurement.

Finally, a study on nanoparticles penetration through leaks into respirators, taking into account the inspiration/exhalation cycle, should be soon carried out to complete this work.

References

[1] Heim M, Mullins B J, Wild M, Meyer J and Kasper G 2005 *Aerosol Science and Technology* **39** 782
[2] Kim S C, Harrington M S and Pui D Y H 2007 *Journal of Nanoparticle Research* **9** 117
[3] Huang S-H, Chen C-W, Chang C-P, Lai C-Y and Chen C-C 2007 *Journal of Aerosol Science* **38** 719
[4] Baron P A and Willeke K 2001 *Aerosol Measurement - Principles, Techniques and Applications (2nd edition)*. John Wiley & Sons
[5] Lee K W and Liu B Y H 1982 *Aerosol Science and Technology* **1** 147
[6] Liu B Y H and Rubow K L 1990 *Efficiency, Pressure Drop and Figure of Merit of High Efficiency Fibrous and Membrane Filter Media.* Vth World Filtration Congress, Nice.
[7] Payet S, Boulaud D, Madelaine G and Renoux A 1992 *Journal of Aerosol Science* 7 723
[8] Wang J, Chen D R and Pui D Y H 2007 *Journal of Nanoparticle Research* **9** 109

Risk of dust explosions of combustible nanomaterials

Ritsu Dobashi

Department of Chemical System Engineering, School of Engineering, The University of Tokyo, 7-3-1 Hongo, Bunkyo-ku, Tokyo, 113-8656 Japan

E-mail: dobashi@chemsys.t.u-tokyo.ac.jp

Abstract. Nanomaterials have several valuable properties and are widely used for various practical applications. However, safety matters are suspected such as the influence on health and environment, and fire & explosion hazards. To minimize the risk of nanomaterials, appropriate understanding of these hazards is indispensable. Nanoparticles of combustible materials have potential hazard of dust explosion accidents. However, the explosion risk of nanomaterials has not yet been understood adequately because of the lack of data for nanomaterials. In this presentation, the risk of dust explosions of nanomaterials is discussed.

1. Risk evaluation of dust explosion

Recently, safety management of industrial process is performed by using risk evaluation. It is the most reasonable safety management method to evaluate and minimize the risk. Usually, the risk is expressed as the combination of probability of accident and consequence (loss) by the accident. The potential hazards of nanoparticles are damage on health and fire & explosion hazards. In this study, fire and explosion hazards are taken up. Especially, dust explosion is focused, because the risk will be the most serious in the case of the dust explosion. A typical scenario of dust explosion is shown in Figure 1. In the case of dust explosion, the probability might be governed by ignitability of the dust and the consequence (damage) might be governed by degree of explosion violence. In order to minimize the risk of dust explosion, appropriate risk estimation is needed. Therefore, the evaluations of the ignitability and explosion violence are essential. The well known parameters for the ignitability and explosion violence of combustible dust are as follows [1-3];

- Ignitability: Minimum ignition energy (MIE), Minimum explosible dust concentration (C_{min}), and so on.
- Explosion violence: Maximum rate of pressure rise ($(dp/dt)_{\max}$ or K_{St}), Maximum explosion pressure (P_{max}), and so on.

Dust explosion is a phenomenon that a flame is propagating in combustible particle cloud dispersed in the air. Therefore, above mentioned explosion hazard parameters are governed by the flame propagation mechanism. One of the important processes in the propagation mechanism is gasification of the combustible particle [4, 5]. The particle size has significant effect on the ignitability and explosion violence because the gasification strongly depends on the particle size. Therefore, these parameters of nanoparticles might be much different from those of the particles of micron scale.

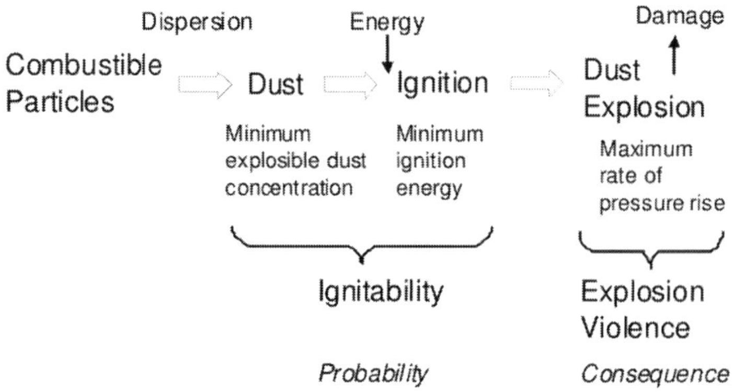

Figure 1. Typical scenario of dust explosion.

2. Measured hazard parameters of dust explosion and their size dependence

There are only little available data on the explosion hazard of nanoparticles. Firstly, the available data on the hazard parameters for the ignitability and explosion violence of combustible dust of micron size particles are presented in this section. Then, the size dependence of these parameters is examined.

2.1. Minimum ignition energy (MIE, *Ignitability*)

Minimum ignition energy (MIE) is the minimum value of the energy that is needed to ignite the dust, which is an index of ignitability. The measured MIE and the size dependence are shown in Figure 2. It is found that values of MIE are order of a few tens mJ. As for the size dependence, MIE tends to decrease as the size of dust particle is decreasing. It means that the dust of smaller particle can be ignited more easily.

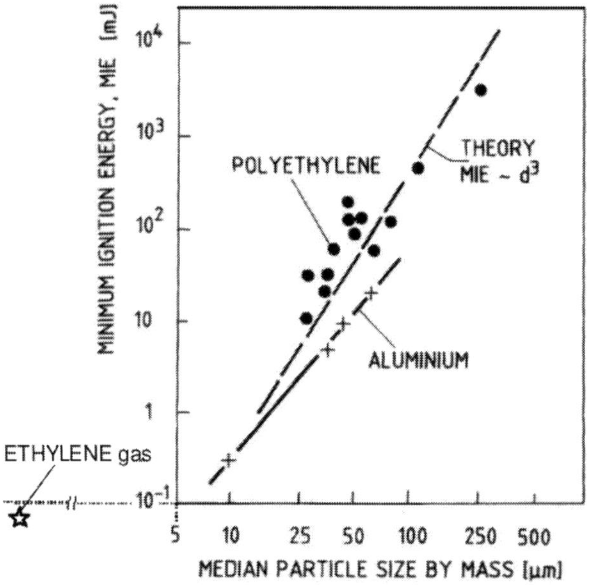

Figure 2. Measured minimum ignition energy (MIE) [1].

2.2. Minimum explosible dust concentration (C_{min}, *Ignitability*)

Minimum explosible dust concentration (C_{min}) is the lower flammable limit of flammable dust concentration, which is also an index of ignitability. The measured C_{min} and the size dependence are shown in Figure 3. It is found that the Minimum explosible dust concentration (C_{min}) tends to decrease as the size of dust particle is decreasing. It means the lower explosion limit becomes smaller for the dust of smaller particle.

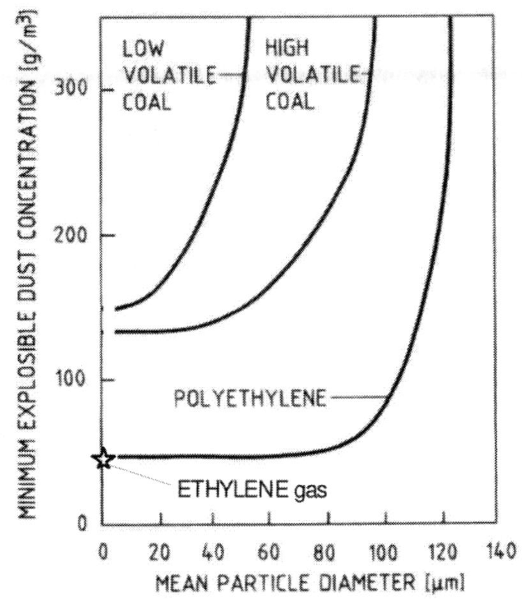

Figure 3. Measured explosible concentration (C_{min}) [1].

2.3. Maximum rate of pressure rise ($(dp/dt)_{\max}$, *Explosion violence*)

Maximum rate of pressure rise is the maximum pressure rising rate, which is an index of the degree of explosion violence. The measured $(dp/dt)_{\max}$ and the size dependence are shown in Figure 4. It is found that the Maximum rate of pressure rise $(dp/dt)_{\max}$ tends to increase as the size of dust particle is decreasing. It means that the dust of smaller particle can cause more violent explosion.

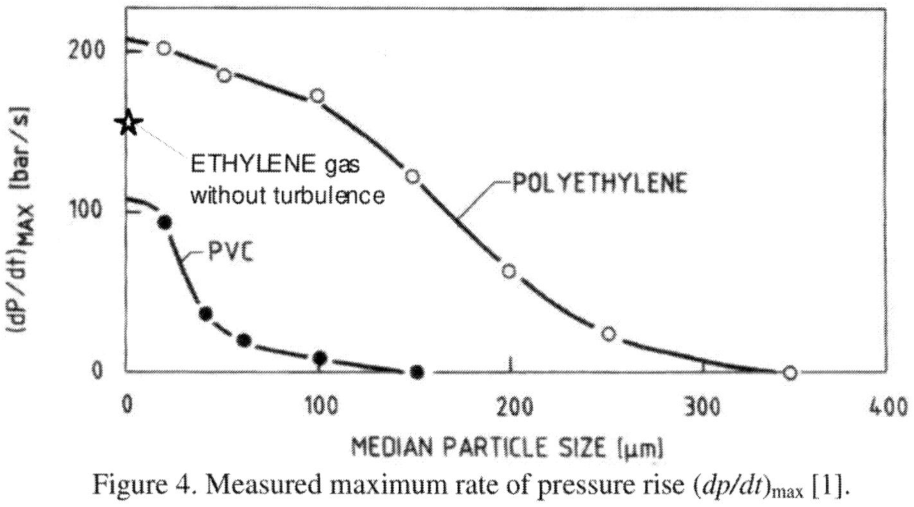

Figure 4. Measured maximum rate of pressure rise $(dp/dt)_{\max}$ [1].

2.4. 3.4 Process of dust explosion and the dependence on particle size

The process of dust explosion is shown in Fig. 5 schematically. Usually, combustion reaction occurs in gas phase. Therefore, in dust explosion, at first, solid particle is gasified (evaporation, pyrolysis, and so on) by heat input from flame to generate combustible gas. A flame is formed by the reaction between the generated combustible gas and oxygen in air. The flame is propagating into the unburdened dust. The particle size strongly affects the explosion behaviour, because the gasification process depends on the particle size. Smaller particles can be gasified quickly because of their large surface area.

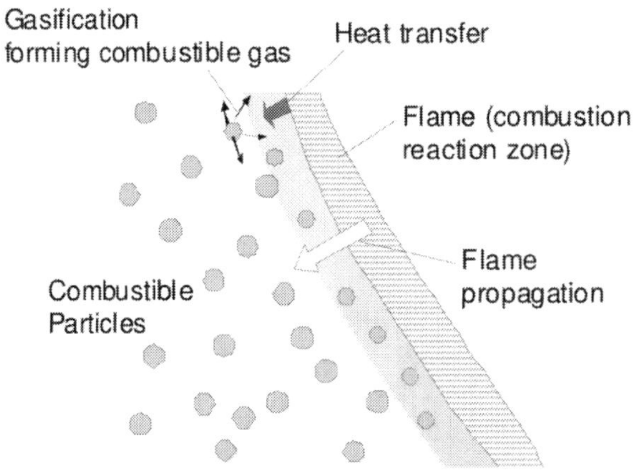

Figure 5. Process of dust explosion.

3. Risk of dust explosion of nanoparticles

3.1. Estimation of the explosion risk of nanoparticles

As shown in the previous section, the ignitability and explosion violence become more significant when the particle size becomes smaller, from the data of MIE, C_{min} , and $(dp/dt)_{\max}$ for the dust of micron size particle. Then, the dust explosion risk of nanoparticles is supposed to become very high. However, it is questionable that this trend could continue endlessly as the particle size is decreasing to nano scale. It is considered that there is a limiting size, below which the risk ceases to increase. However, there are only little available measured data of explosion hazards for nanoparticles.

Considering that the particle size is becoming smaller and smaller, the limit of the smallest particle should be molecule. Then, the comparisons between dust and gas are made on the explosion hazard parameters.

The measured value of MIE of ethylene gas is plotted in Figure 2. The comparison can be made between ethylene (gas) and polyethylene (dust) to expect the trend for smaller particles. Ethylene is monomer of polyethylene, therefore, the combustion characteristic might be similar. Figure 2 shows that MIE of the ethylene gas is far smaller than those of polyethylene dusts. It is supposed that MIE will be decreasing as the size of dust particle is decreasing to nano scale. MIE might not be decreasing endlessly, but it will approach the value of ethylene gas asymptotically.

The measured lower explosion limit (concentration) of ethylene gas is plotted in Figure 3. The figure shows that the lower explosion limit of ethylene (gas) is almost same as the minimum value of C_{min} of polyethylene (dust). It is supposed that the value of C_{min} might not decrease as the particle size is decreasing to nano scale. It is known that the minimum explosion concentration of combustible dust becomes lower than that of combustible gas [6]. There is another report [2, 5], in which the C_{min} of polyethylene (particle size $<$ 10 µm) is 30 g/m^3. This value is lower than that of ethylene gas (34 g/m^3). Table 1 shows the comparison of the minimum explosion concentrations between dust and gas. It is found that the minimum (lean) explosion limits of combustible dusts are much larger than those of combustible gases, compared in the unit of equivalence ratio. The reason of this phenomenon is considered that the combustible dust is non-uniform medium and the local concentration of combustible material can be higher than the combustible gas (uniform medium).

Table 1. Minimum explosible concentration (in Equivalence ratio) [6].

Fuel gas	Lean limit Equivalence ratio *	Dust	Lean limit Equivalence ratio *
Methane	0.50	Adipic acid	0.28
Ethane	0.52	Lactose	0.22
Ethylene	0.40	Poly-ethylene	0.34
Benzene	0.49	Sulfur	0.10
Methanol	0.46	Aluminum	0.29
Ethanol	0.49	Iron	0.35

* Equivalence ratio = (Fuel / Oxygen) / (Fuel / Oxygen)stoichiometric

The measured maximum rate of pressure rise $(dp/dt)_{max}$ of ethylene gas is plotted in Figure 4. The figure shows that $(dp/dt)_{max}$ of ethylene (gas) is smaller than the maximum value of $(dp/dt)_{max}$ of polyethylene (dust). In the measurement for maximum rate of pressure rise, the dust is dispersed by air jet flow in the vessel to form particle cloud (dust). Therefore, $(dp/dt)_{max}$ for combustible dust is measured in turbulent flow condition, beside that for combustible gas is measured in quiescent flow condition without turbulence. If $(dp/dt)_{max}$ for ethylene gas is measured in turbulent flow condition, the value of $(dp/dt)_{max}$ will become a few times larger. Then, as the size of particle is decreasing to nano scale, $(dp/dt)_{max}$ might be increasing, however, it will not be increasing endlessly and approach the value of ethylene gas (with turbulence) asymptotically.

The explosion hazard parameters of nanoparticles are estimated by extrapolating the trend of size dependence in micro scale to nano scale. However, these results are just estimation. The accumulation of measured data of the hazard parameters of the nanoparticles and detailed analysis are needed.

3.2. Special effects of nanoparticles

To examine the explosion hazard of nanoparticles, the effect of nano scale must be understood appropriately. There must be some special effects of nano scale particle other than the increase of surface area. One important effect is aggregation. Nanoparticles can aggregate more easily than the particle of micron scale. When aggregation occurs, the nano scale particles change to become larger aggregated particles with porous structure. The explosion behaviour of such aggregated particles was not understood adequately, but the effective surface area might decrease. In practical condition, most nanoparticles might be aggregated, and then the characteristic scale of the particles will become much larger than the scale of primary nanoparticles. More studies on such special effect of nanoparticles are indispensable.

4. Summary

To minimize the risks of dust explosion of nanomaterials, the potential hazards of dust explosion have to be understood appropriately. However, the potential hazards of dust explosion are not clear because there are only few reported data. In this study, explosion hazard parameters of nanoparticles, such as ignitability and explosion violence, are discussed. There are no available data of these parameters for nanoparticles, therefore, the data of the particles of micron scale and gas are analyzed. Generally the explosion risk will increase as the particle size is decreasing. This trend will continue to nano scale, however, the value of the explosion hazard parameter will be in the range between the value for the particle of micro scale and that of gas. Also aggregation must have important effects on the explosion behaviour of nanoparticles. In any case, it is necessary to accumulate the data and make systematic analysis for achieving the appropriate risk estimation of nanomaterials.

References

[1] Dust Explosions in the Process Industries (Third edition), Rolf K. Eckhoff, Gulf Publishing Co., 2003.
[2] Zalosh, Explosion Protection, SFPE Handbook of Fire Protection Engineering, NFPA, 2-88, 1988.
[3] Hertzberg M, Cashdollar K L, Ng D L and Conti R S 1982 *19th Symp. (Int.) on Combustion* The Combustion Institute 1169
[4] Dobashi R and Senda K 2006 *Journal of Loss Prevention in the Process Industries* **19** 149
[5] Handbook of Safety Engineering (New edition) (in Japanese), Japan Society for Safety Engineering, Corona Publishing Co. 1999.
[6] Sun J-H, Dobashi R. and Hirano T 2003 *Combustion and Flame* **134** 381

Material Scientific Approach to Predict Nano Materials Risk of Adverse Health Effects

Yasuto Matsui, Kenichi Miyaoi, Takeshi Hayashi and Yukio Yamaguchi

Department of Chemical System Engineering, Graduate School of Engineering, The University of Tokyo, Japan

E-mail: matsui@sogo.t.u-tokyo.ac.jp

Abstract. To estimate the potential risk of nano materials, correlations were investigated between material properties and various biomarkers indicating adverse effects on humans. Nano materials have a variety of properties such as solubility, iso-electric point, crystal shape, BET specific surface area and so on. The purpose of our work was to predict relationships between material properties and hazard data by undertaking statistical survey of eleven papers arguing cell viability assays.

The reviewed papers associate cytotoxicity (i) mainly with particle volume and (ii) a certain degree with particle solubility, with relatively large variability of toxicological responses. At present nanomaterials are often very broadly named, defined and categorized based upon only their chief chemical composition or product shape – *e.g.*, "titanium," "carbon black," "nano tubes," *etc.* Such rough, imprecise categorization serves little or no useful purpose when attempting risk assessments for every nano material produced differently, since even materials with the same name can possess different properties and consequently different degrees of hazards.

1. Introduction

More than 300 nanotechnology products such as sunscreens or semiconductors are already on the global market [1]. Supported by government funding, multiple nano-related research projects are being carried out at an unprecedentedly rapid pace. However, many companies have come to withdraw from such research because of the potential adverse effects or uncertain risks [2]. We are trying to develop risk assessment software which can suggest a clear direction of controlling manufacturing processes to support nanotechnology developers.

The concepts are shown in Figure 1. Nano materials possess various properties: specific surface area, zeta potential, solubility, crystal type and so on. It is well known that the diameter or surface area of nano materials correlates highly with inflammation response. In this study, we focused on other parameters and found correlations. We will examine another and attempt to predict whether nano materials could invade into the human body or how they transfer inside the body.

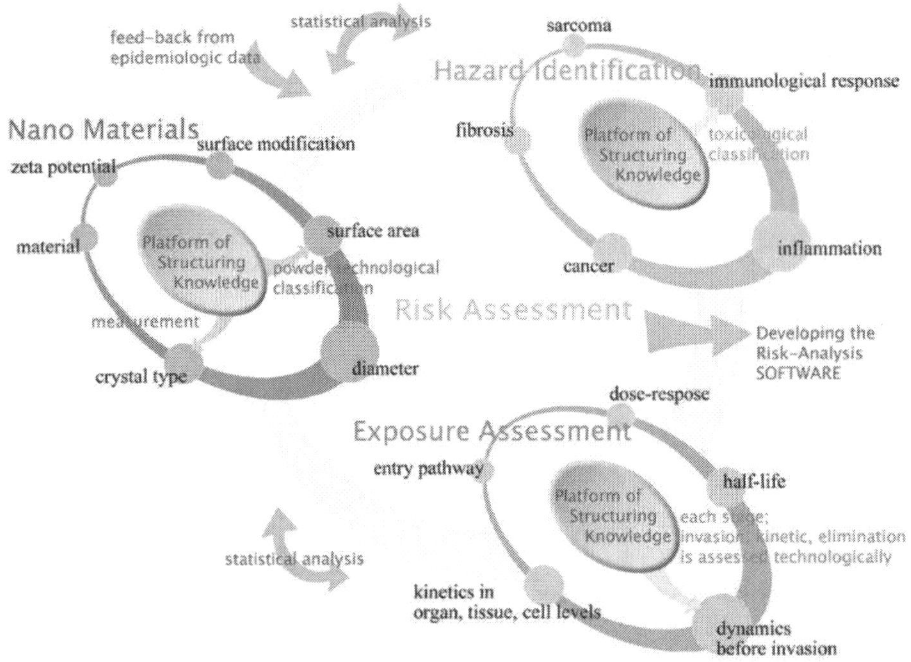

Figure 1. The main concept of nano risk survey.

2. Methods

2.1. Researching papers

For a viability analysis of various types of particles, many papers were searched and selected by using the ISI Web of Knowledge (v4.4). The selected papers consisted of experiments involving cell viability assays performed *in vitro* with nano/sub-micro/micro particles of different chemical species. In the experiments, various types of cells (fibroblasts, osteoblasts, *etc.*) except macrophages were used, and the endpoint was set as the viability of cells after 3 days exposure.

2.2. Dose-mortality analysis *in vitro* (*size effect*)

From the experimental results reported in the selected papers dose and viability data sets were singled out, and scatter diagrams were made with dose as the horizontal axis and "% death" as the vertical axis. "% death" means mortality rate of cells exposed to particles versus unexposed cells calculated by "100 (%) - % of viable cells". Dose was varied by using 4 unit types. "μg/ml" is a unit for particle weight over cell culture media volume. "cm^2/ml" is in one way, a unit for particle surface area computed by BET measurement over media volume, and, in another way, a unit for particle surface area calculated from particle's minor and major axes. When a particle has a spherical shape, calculated surface area (m^2/g) should be $6/\rho D$ where ρ (g/cm^3) is density and D (μm) is diameter. If a particle is fibrous, calculated surface area should be nearly $4/\rho D$ if D(minor axis) << major axis. "No. of particles" is a unit for particle count over media volume. In addition, in the scatter diagrams, each plot had different marks depending on size: one type of key for particles with a primary diameter less than 120 nm and another key for those above 400 nm. Generally "nano-sized" is said to be 1-100 nm, but in this paper the collected data were distributed in 1-120 nm and 400-10 000 nm.

2.3. Dose-mortality analysis *in vitro* (*particle species effect*)

For detecting the chemical species effect or crystal shape effect, the scatter diagram plots above were given different marks depending upon particle species. Here the particle species were identified using chemical composition and particle crystal shape. In addition, regression lines (Y = a + b ln(x)) were drawn for each particle species, and EC50 was evaluated. EC50 was calculated as the concentration of µg/ml needed to reduce cell proliferation to 50% of unexposed controls.

2.4. EC50 comparison between particles (*solubility effect*)
To study the correlation between particle solubility and cell mortality, an x-y scatter diagram was produced with the x-axis showing particle solubility and the y-axis calculated EC50. Solubility data were obtained from "Chemistry Handbook basic edition 2, 4th revised edition". All table calculations or analyses were performed using Microsoft Office Excel 2003.

3. Results

Eleven Papers [3-13] discussing *in vitro* experiments concerning the viability of cells except macrophages were selected. These papers checked the endpoint of cell viability after 3 days exposure. If BET surface area was not described in a paper, calculated surface area was substituted for BET surface area (On spherical particle, calculated surface area (m^2/g) should be $6/\rho D$. On fibrous particle, that should be nearly $4/\rho D$ if D (minor axis) << major axis). The collected cell viability data were used to make scatter diagrams. Figure 2(a) shows a conventional diagram whose horizontal axis is particle dose (µg/ml) and whose vertical axis is %death (100 (%) - viability (%)). Its keys were differentiated in particle diameter between 0-120 nm and 400-10 000 nm. Plots of 0-120 nm and 400-10 000 nm looked well overlapped so that the effect of size on cell viability is likely to be indicated by particle weight or volume (since particle weight and volume show similar values). Figure 2(b) indicates a diagram whose x-axis is the particle dose of BET surface area (cm^2/ml). There seems to be a split between the 0-120 nm group and the 400-10 000 nm group. Figure 2(c) has an x-axis as the particle dose of calculated surface area (cm^2/ml). Here BET surface area of amorphous silica was converted to the smaller value of calculated surface area, but the whole trend of scattering plots was not differed notably. Figure 2(d) has an x-axis as the particle dose of the number of particles (No. of particles/ml). Two groups were split more widely in two and the size effect for cell viability is not likely to be number of particles.

The collected cell viability data are all shown in Figure 3 with each type of mark representing a particle species. Additionally, a regression line Y = a + b ln(x) was drawn on each particle species and EC50 (the concentration of µg/ml that reduced cell proliferation to 50% of unexposed controls) was evaluated.

Seen as Figure 4, an x-y scatter diagram was developed, with the x-axis particle solubility (g/l) and the y-axis EC50 (The solubility of carbon materials was given as 10^{-10} g/l as a matter of practical convenience). There is a possible correlation: the greater solubility, the smaller the EC50 (meaning greater cytotoxicity). The plot for TiO_2 (anatase) looks distant from the correlation and there may be other, more important factors affecting their cytotoxicity (i.e. oxidative stress caused by surrounding visible or ultra violet light), or the data volume may have simply been insufficient.

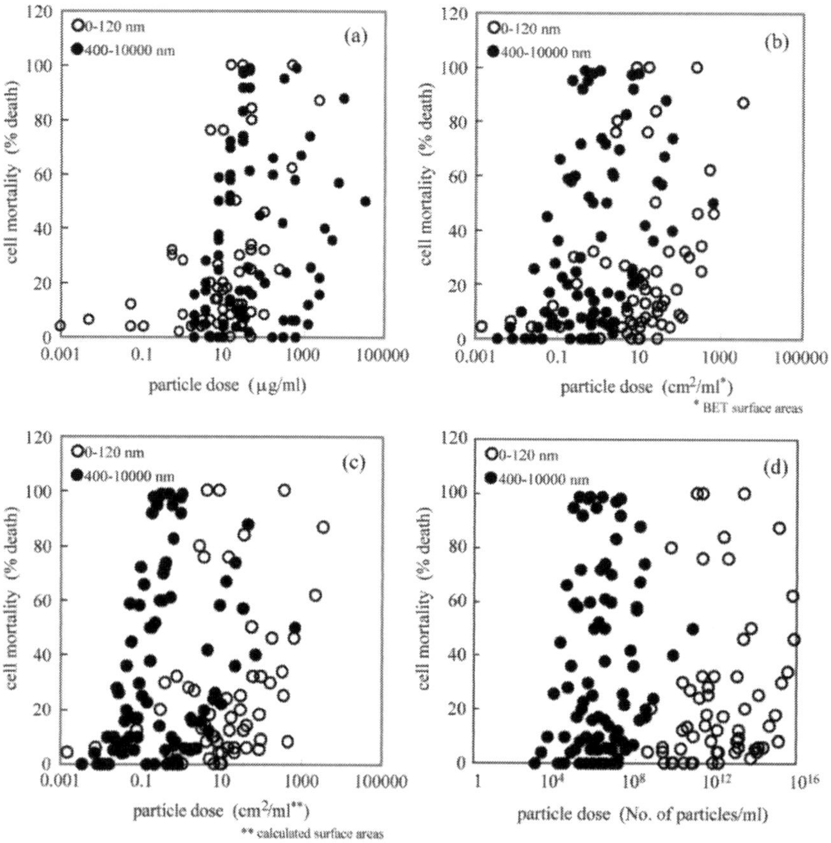

Figure 2. Relationship between the ratio of cell mortality and particle dose in a culture medium.

4. Discussion

To study the effects of particle size and species for hazard assessment, dose-mortality analyses *in vitro* were performed on this paper. Results suggest that the size effect on cytotoxicity is ascribed to particle weight or volume and that another cytotoxic factor is particle solubility.

The effect of particle size on cytotoxicity has been argued in many papers. Nano and micro alumina and titania particles was tested by viability assays, and viable osteoblast density was not always smaller in nano particle cultures than micro particle cultures when particle weight dose (μg/ml) was equivalent [14]. This trend is also seen in Figure 2 of our results showing data of additional particle species and more plots. Another paper argued that cytotoxicity is not dependent on size or chemical species but on total volume [11]. The authors introduced five species of ceramic particles into cell culture and measuring the dose-response curve with varying doses using three criteria: particle number, surface area, and volume. Their results were quite similar to ours in splitting formation of plots on scatter diagrams. On the other hand, the cytotoxic effect of aggregated nano materials was investigated, with little correlation found between BET specific surface area and the cytotoxicity index [15]. Our results also support it and possibly the order of cytotoxicity is explained by solubility.

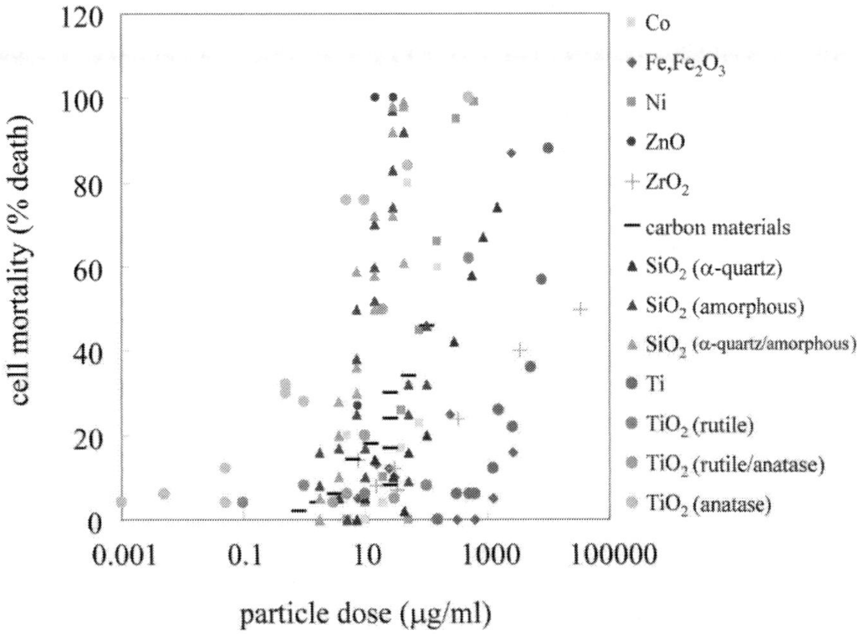

Figure 3. Relationship between the ratio of cell mortality and particle weight in a culture medium.

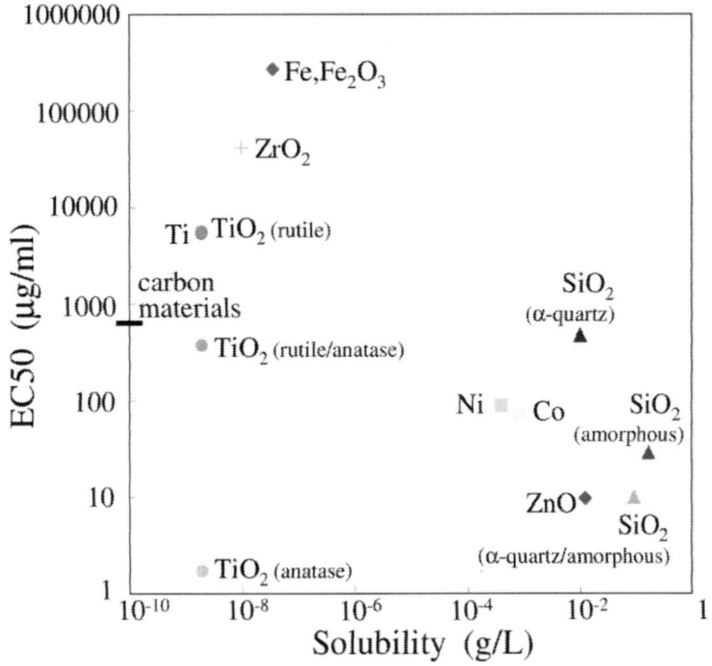

Figure 4. Relationship between solubility and EC50.

Accordingly, the chief factor for cytotoxicity appears to be the particle volume dose and subordinate factor may be particle solubility. A possible mechanism could be that particle volume obstructs intra-cellular adherence or cell-extracellular matrix interaction and slows cell proliferation. Here, equivalent particle surface area dose does not cause equivalent mortality because smaller particles are more susceptible to being pushed away. Additionally, dissolved materials may affect cellular osmotic systems or increase reactive oxygen species.

References

[1] Maynard A D, Aitken R J, Butz T, Colvin V, Donaldson K, Oberdorster G, Philbert M A, Ryan J, Seaton A, Stone V, Tinkle S S, Tran L, Walker N J and Warheit D B 2006 *Nature* **444** 267

[2] Service R F 2006 *Science* **314** 45

[3] Brunner T J, Wick P, Manser P, Spohn P, Grass R N, Limbach L K, Bruinink A and Stark W J 2006 *Environmental Science & Technology* **40**(14) 4374

[4] Tian F R, Cui D X, Schwarz H, Estrada G G and Kobayashi H 2006 *Toxicology in vitro* **20**(7) 1202

[5] Lin W S, Huang Y W, Zhou X D and Ma Y F 2006 *Toxicology and applied pharmacology* **217**(3) 252

[6] Sayes C M, Wahi R, Kurian P A, Liu Y P, West J L, Ausman K D, Warheit D B and Colvin V L 2006 *Toxicological Sciences* **92**(1) 174

[7] Pioletti D P, Takei H, Kwon S Y, Wood D and Sung K L P 1999 *J. Biomed. Mat. Res.* **46**(3) 399

[8] Peters K, Unger R E, Kirkpatrick C J, Gatti A M and Monari E 2004 *J. Mat. Sci.-Mat. in Med.* **15**(4) 321

[9] Sakai T, Takeda S and Nakamura M 2002 *Dental materials Journal* 21(**2**) 133

[10] Elias Z, Poirot O, Fenoglio I, Ghiazza M, Daniere M C, Terzetti F, Darne C, Coulais C, Matekovits I and Fubini B 2006 *Toxicological Sciences* **91**(2) 510

[11] Yamamoto A, Honma R, Sumita M and Hanawa T 2004 *J. Biomed. Mat. Res. Part A* **68A**(2) 244

[12] Pisanic T R, Blackwell J D, Shubayev V I, Finones R R and Jin S 2007 *Biomaterials* **28**(16) 2572

[13] Hart G A and Hesterberg T W 1998 *J. Occup.and Environ. Med.* **40**(1) 29

[14] Gutwein L G and Webster T J 2004 *Biomaterials* **25**(18) 4175

[15] Soto K, Garza K M and Murr L E 2007 *Acta Biomaterialia* **3**(3) 351

Nanosafe 2008: International Conference on Safe production and use of nanomaterials IOP Publishing
Journal of Physics: Conference Series **170** (2009) 012031 doi:10.1088/1742-6596/170/1/012031

Estimation of health risks and safety margins due to inhalation of ultrafine particles and nanoparticles in selected occupational, consumer and environmental settings

Otto Hänninen[1,2], Irene Brüske-Hohlfeld[1], Miranda Loh[2], Tobias Stoeger[3], Wolfgang Kreyling[3], Otmar Schmid[3] and Annette Peters[1]

[1]Helmholtz Zentrum München, German Research Center for Environmental Health (HMGU), Institute of Epidemiology, Munich, Germany

[2]KTL, National Public Health Institute, Finland

[3]Helmholtz Zentrum München, German Research Center for Environmental Health (HMGU), Institute of Lung Biology and Disease, Munich, Germany

E-mail: otto.hanninen@thl.fi

Abstract. Nanoparticles exhibit properties different from those of the same bulk materials leading to unknown toxicological implications that have evoked concern for (1) occupational, (2) consumer and (3) environmental safety.
The current work utilizes epidemiological and toxicological data for screening level assessment of these risks using various suggested health relevant dose metrics (mass, particle number and surface area) to (i) quantify the potential risk levels and to (ii) compare the properties of these alternative risk assessment methods.

1. Materials and methods

Low toxicity low solubility (LTLS) particles (*e.g.* carbon black, polystyrene, and titanium dioxide) have been shown to exhibit similar toxicity, especially per particle surface areas doses [1, 2] and the toxicological data for these particles are used for a baseline risk assessment. The upper limit of particle-related health risks is estimated by using the dose-response function of quartz, a known high toxicity compound with 60 times higher toxicity than the LTLS particles [1].

Besides toxicological studies, epidemiology has shown that especially mass concentrations of urban ambient particles are associated with relative mortality risks for long-term exposures ($PM_{2.5}$; RR_{mass}=10% per 10 µg m^{-3}) and daily exposures (PM_{10}; RR_{mass}=0.6% per 10 µg m^{-3}) [3]. Recently Stölzel *et al.* [4] showed that also particle number concentrations of ultrafine (<100 nm in diameter) particles are associated with mortality (RR_{number}=2.8% per 10^4 cm^{-3}). For application of the epidemiology on ambient particles to other types of LTLS nanoparticles (<100 nm in at least one dimension), we assume that the mortality in the general population is caused by the retained alveolar dose of the non-soluble fraction of ambient particles. Subsequently, the risk factors per ambient particle concentration were converted into risks per accumulated dose of non-soluble particles in the alveolar lung region by using an attenuation model [5- 7], respiratory physiological data, an alveolar particle deposition model [8] and an exponential model for particle clearance from the lungs [9]. The obtained dose-response relationships are reported in more detail in a parallel paper [10].

Occupational concentrations of ultrafine particles (or nanoparticles) at 62 workplaces in various industries were measured by Möhlmann [11], using standard aerosol measurement techniques. In addition to particle number concentrations (size-resolved between 14 nm and 673 nm) and mass concentrations of the respirable fraction, there were mass size distributions from 17 workplaces. The particle number concentration levels at the workplaces ranged from 5.9×10^{3} cm^{-3} to 5.6×10^{6} cm^{-3} (mean 3.5×10^{5} cm^{-3}, sd 3.1×10^{5} cm^{-3}). To provide information on particle surface area, either the particle number or the mass size distribution was used assuming spherical particle shape (and a density of 2 g cm^{-3} for conversion from mass) Particle intake values were calculated using 8-hour working day and inhalation rate of 9.6 m³ per 8 hours (*i.e.* 1.2 m³ h^{-1}) and converted into uptake values using the ICRP model [8] and assuming that no protective equipment was employed.

Exposure levels to printer generated nanoparticles were estimated using the particle emission rates of high emissions printers (27% of the 63 laser printers studied by [12]) and a mass-balance model [13, 14] for occupational settings in an office and for consumers in the case of home use of laser printers assuming 8 hours (office), 4 hours (home office) and 14 hours (family member) exposure times with 10% and 1% (office and home, respectively) printing times. More detailed presentation of the methods regarding the printer particles is given in a parallel full-length paper [15].

Environmental exposures to carbon black from tires were estimated using emission data from a road simulator study (emissions from 300×10^{9} to 3000×10^{9} nanoparticles per vehicle-km [16]) and a traffic volume model for Germany (706×10^{9} km a^{-1} divided into rural, urban and highway fractions, [17]). The average particle intake was estimated from weighted intake fractions for rural, urban and highway emissions [18- 23]. The intake values were converted into alveolar deposition using the ICRP deposition model [8].

Risk characterization: Two different methods, toxicological and epidemiological risk assessment, are used and the results are compared. Toxicological risk is given as the ratio of the alveolar doses in a given exposure scenario to the lowest observed effect level (LOEL) as estimated from the toxicological studies using allometric scaling to convert the cell [24] and animal-based toxicological data into human equivalent values. Toxicological data is available for particle surface and mass doses, but no experimental data was found for particle number doses. Epidemiological data is used to estimate daily and annual excess mortality risks expressed as percentages and cases per million for the printer and tyre carbon black particles. Environmental epidemiology is not applied to the occupational exposures because of the differences in the exposed population characteristics (including age structure and health status). Epidemiological data was not found for particle surface concentration exposures.

2. Results

(1) The occupational exposure risk was expressed as the ratio of the 8-hour working day uptake to the human toxicological LOEL measured as particle mass or as particle surface area. The resulting 8-hour alveolar uptakes were on the average 71.2 times higher than the mass-based observed effect level for LTLS particles (Table 1), indicating a significant risk. Corresponding mean surface area risk measure was 1.8. Potential high toxicity of specific ultrafine particles was estimated using the high known particle toxicity of quartz; the uptake levels were 4.3×10^{3} and 1.1×10^{2} times higher than the lowest observed effect levels for quartz mass and surface metrics, respectively (Table 1).

(2) Consumer and office exposures of laser printer users and risks were calculated using three methods; epidemiological (i) particle number concentration (PNC) and (ii) particle mass, and (iii) toxicological LOEL. The highest risks were modelled for the occupational office and home office. Epidemiological relative mortality risks were 0.43% and 0.08% using the PNC and mass metrics, respectively, corresponding to 34 and 6 annual deaths per million users. Toxicological risk levels were 2.9×10^{-3} and 1×10^{-3} for mass and surface area metrics, respectively (Table 2).

(3) Environmental exposures to carbon black nanoparticles from tyres were estimated for Germany. Additional relative mortality from particle number exposures was estimated to be 0.03% (annual

mortality in Germany 218 deaths) and from daily mass exposures 0.0004% (3 deaths in Germany). Annual mass exposures were estimated to be associated with 7 deaths. Toxicological risk levels were 2×10^5 and 1×10^5 for mass and surface area metrics, respectively.

Table 1. Toxicological assessment of occupational potential risk levels and comparison of mass versus surface based measures on human 8-hour working day uptake (62 industrial working places). Assuming no protective equipment used.

Toxicological measure	Relative toxicity	Statistics on the ratio dose/LOEL Mean	sd	min	max	R<1[b] n
Mass	LTLS[c]	71.2	213	1.6	1010	0
Suface area	LTLS[c]	1.8	4.8	0.001	25.0	50
Mass	High tox[d]	4273	12765	94.3	60577	0
Suface area	High tox[d]	107.1	286	0.08	1499	6

[a] Lowest observed effect level as expected on humans based on allometric scaling
[b] R=dose/LOEL; n=number of workplaces (out of 62) where dose <LOEL (i.e. ratio R< 1.0)
[c] Low toxicity, low solubility nanoparticles [d] High toxicity as estimated using quartz as positive control

Table 2. Population risks associated with laser printers and carbon black particles from tyres.

Risk assessment type		Exposure to laser printer particles in: Offices	Homes or home offices: Singe room	Full residence	Exposure to particles from tyres
1 Daily NC[a] epidemiology					
Daily uptake (number)	10^9 d^{-1}	3.13	3.13	1.09	0.24
Epidemiological RR (number)	%	0.43 %	0.43 %	0.15 %	0.03 %
Annual mortality per million[b]	cases	34	34	12	3
In Germany	cases	c	c	c	218
2 Daily MC[e] epidemiology					
Daily uptake (mass)	µg d^{-1}	0.21	0.21	0.07	0.001
Epidemiological RR (mass)	%	0.08 %	0.08 %	0.03 %	0.0004 %
Annual mortality per million[b]	cases	6	6	2	0.03
In Germany[d]	cases	-	-	-	3
3 Long-term MC[b] epidemiology					
Annual uptake (mass)	µg a^{-1}	e	e	e	0.419
Epidemiological RR (mass)	%	e	e	e	0.0011 %
Annual mortality per million[c]	cases	e	e	e	0.09
In Germany[d]	cases				7
4 Toxicological risk as mass dose					
Daily uptake (mass)	µg d^{-1}	0.21	0.21	0.07	0.001
Ratio to toxicological LOEL[f]	1	0.29 %	0.29 %	0.10 %	0.002 %
5 Toxicological risk as surface dose					
Daily uptake (surface)	cm^2 d^{-1}	0.16	0.16	0.06	0.002
Ratio to toxicological LOEL[f]	1	0.10 %	0.10 %	0.04 %	0.001 %

[a] NC=number concentration; MC=mass concentration
[b] Assuming annual background mortality rate of 0.8%.
[c] Number of printer users in Germany not estimated.
[d] Fraction of population exposed to printers not estimated.
[e] Daily printer use pattern for annual exposures not estimated.
[f] [Uptake]/[Allometrically scaled observed effect level], %

3. Conclusions

Toxicological risks found for occupational settings (R > 1.6 (mass) and 1×10^{-3} (surface area)) were significantly higher than for consumer (printer) and environmental (tyre) scenarios (R < 2.9×10^{-3} (mass) or 1×10^{-3} (surface area)). For consumers and environmental settings the risks are smaller, but based on the epidemiological risk estimates still significant, warranting technological and policy development for exposure reduction especially for the printer particles.

The general linear no-threshold epidemiological risk model projects population level effects below the toxicological thresholds. In the current study the allometric scaling from laboratory experiments to humans was selected specifically for producing highest risks. Nevertheless, all observed toxicological LOEL values correspond to mortality risks, which are substantially higher than $1{:}10^6$, and therefore either risk coefficients from epidemiological studies or toxicological safety factors have to be considered in a risk assessment.

Acknowledgements
This work was conducted as part of the EU FP6 funded project NANOSAFE 2: Safe production and use of nanomaterials (NMP2-CT-2005-515843), Subproject 4, Health, Societal and Environmental aspects. We would like to express our gratitude to Jacques Bouillard, INERIS; over all risk assessment framework; safety issues; physiology based pharmacokinetic modelling of translocation, Carsten Möhlmann (BGIA); occupational exposure and concentration measurements; Peter Hoet, Catholic University of Louvain; general and specific toxicity of nanomaterials, Tony Harker and Sachit Vohra, University College London; life cycle analysis of carbon black used in tires; and Lidia Morawska; nanoparticle emissions from laser printers.

References

[1] Duffin R, Tran L, Brown D, Stone V and Donaldson K 2007 *Inhalation Toxicology* **19(10)** 849
[2] Stoeger T, Reinhard C, Takenaka S, Schroeppel A, Karg E, Ritter B, Heyder J and Schulz H 2006 *Environmental Health Perspectives* **114(3)** 328
[3] Pope C A and Dockery D 2006 *J Air Waste Manag Assoc.* **56** 709
[4] Stölzel M, Breitner S, Cyrys J, Pitz M, Wölke G, Kreyling W, Heinrich J, Wichmann H E and Peters A 2007 *Journal of Exposure Science and Environmental Epidemiology* **17** 458
[5] Wilson W E and Brauer M 2006 *Journal of Exposure Science and Environmental Epidemiology* **16** 264
[6] Hänninen O, Lebret E, Ilacqua V, Katsouyanni K, Künzli N, Srám R J and Jantunen M J 2004 *Atmospheric Environment* **38(37)** 6411
[7] Hussein T, Gltsos T, Ondracek J, Dohanyosova P, Zdimal V, Hämeri K, Lazaridis M, Smolik J and Kulmala M 2006 *Atmospheric Environment* **40** 4285
[8] ICRP 1994. International Commission on Radiological Protection: Human Respiratory Tract Model. *Annals of ICRP* **24** 1 http://www.sciencedirect.com/science/journal/01466453 (accessed January 2009)
[9] Kreyling W 1990 *Journal of Aerosol Medicine* **3(S1)** 93
[10] Hänninen O, Brüske-Hohlfeld I, Cyrys J, Peters A, Wichman E, Schwarze P, Schmid O, Stoeger T and Kreyling W 2009 *Manuscript*
[11] Möhlmann C 2007 Kennzahl 120130. 2. Lfg. IX/2007. In: BGIA-Handbuch Sicherheit und Gesundheitsschutz am Arbeitsplatz. Ed.: BGIA - Institut für Arbeitsschutz der Deutschen Gesetzlichen Unfallversicherung. 2nd edition. Erich Schmidt Verlag, Berlin 2003 – Loose leaflet edition (in German) http://www.bgia-handbuchdigital.de/120130 (January 2009)
[12] He C, Morawska L and Taplin L 2007 *Environmental Science & Technology* **41** 6039
[13] Ilacqua V, Hanninen O, Kuenzli N and Jantunen M F 2007 Indoor Air **17(5)** 372
[14] Nazaroff W W 2008 *Building and Environment* **43(3)** 269
[15] Hänninen O, Brüske-Hohlfeld I, Loh M, Schmid O, Stoeger T, Kreyling W and Peters A 2009 *Submitted*
[16] Dahl A, Gharibi A, Switlicki E, Gudmunsson A, Bohgard M, Ljungman A, Blomqvist G, Gustafsson M 2006 *Atmospheric Environment* **40** 1314
[17] Knörr W, Dünnebeil F, Helms H, Lambrecth U, Höpfner U, Patyk A and Reuter C 2006 Continuation data and computation model: Emissions from transport in Germany 1960-2030. UBA-FB Report, Final Report Summary "TREDMOD: Transport emission model" http://www.ifeu.org/verkehrundumwelt/pdf/IFEU(2005)_TREMOD_Methodology_Report_Summary.pdf (January 2009)
[18] Evans J, Thompson K. and Hattis D 2000 *J. Air & Waste Manage. Assoc.* **50** 1700

[19] Wolff S K 2000 Evaluation of fine particle exposures, health risks, and control options. Department of Environmental Health. Boston, Harvard School of Public Health. Doctoral Thesis.

[20] Marshall J D, Teoh S K and Nazaroff W W 2005 *Atmospheric Environment* **39(7)** 1363

[21] Marshall J D, Riley W J, McKone T E and Nazaroff W W 2003 *Atmospheric Environment* **37(24)** 3455

[22] Greco S L, Wilson A M, Spengler J D and Levy J I 2007 Atmospheric Environment **41(5)** 1011

[23] Greco S L, Wilson A M, Hanna SR and Levy J I 2007 *Environmental Science & Technology* **41(22)** 7675

[24] Brown D M, Donaldson K, Borm P J, Schins R P, Dehnhardt M, Gilmour P, Jimenez L A and Stone V 2004 *Am J Physiol Lung Cell Mol Physiol* **286** L344

Explosion risks from nanomaterials

Jacques Bouillard[1], Alexis Vignes[1], Olivier Dufaud[2], Laurent Perrin[2] and Dominique Thomas[2]

[1]INERIS, Verneuil en Halatte, 60550 France

[2]LSGS-CNRS Rue Grandville, Nancy 54001, France.

E-mail: jacques.bouillard@ineris.fr

Abstract. Emerging nanomanufactured products are being incorporated in a variety of consumer products ranging from closer body contact products (*i.e.* cosmetics, sunscreens, toothpastes, pharmaceuticals, clothing) to more remote body-contact products (electronics, plastics, tires, automotive and aeronautical), hence posing potential health and environmental risks. The new field of nanosafety has emerged and needs to be explored now rather than after problems becomes so ubiquitous and difficult to treat that their trend become irreversible. Such endeavour necessitates a transdisciplinary approach.

A commonly forgotten and/or misunderstood risk is that of explosion/detonation of nanopowders, due to their high specific active surface areas. Such risk is emphasized and illustrated with the present development of an appropriate risk analysis. For this particular risk, a review of characterization methods and their limitations with regard to nanopowders is presented and illustrated for a few organic and metallic nanopowders.

1. Introduction

For the past 10 years, there has been a growing interest in developing nanomaterials in various industrial sectors such as catalysts, in cosmetics, paints and coatings, because of their unique properties. Some of these materials that have traditionally been manufactured are now being developed as new products on the nanoscale. Others show truly new structures, such as carbon nanotubes. Concerning carbon nanotubes (CNTs), their market is positioned to "grow explosively" to reach 540 millions of dollars per year in 2007-2008 and more specifically 340 millions of dollars per year, for multi-walled carbon nanotubes (MWNT) [1]. CNTs production is going to increase in the coming years and consequently potential risks associated to the handling of CNTs will be proportionally more and more important.

Aluminium powders have long been employed as fuel ingredients, explosives, propellants to enhance energy density and stability of propulsion systems. Aluminium is added to improve the burn rate and enhance blast effects in explosives [2, 3].

For any powders set in suspension in air, the ease with which they ignite, burn and propagate varies considerably with important factors that have to be characterized so as to be able to control their explosion. One of them is obviously the particle size, which takes nanopowders now under higher scrutiny for OH&S issues, others are the particle concentration, the degree of de-agglomeration, and the degree of turbulence [4, 5]. So far, literature studies concerning the evaluation of explosion and flammability hazards of powders were essentially carried out on microsized powders, including one on nanomaterials explosivity published in 2004 [6]. This work proposes to review recently measured main explosion safety parameters of multi-walled carbon nanotubes, carbon blacks and aluminium nanoparticles performed in course of the EU Nanosafe2 project.

2. Experimental setups and methods:

2.1. Particle geometrical characteristics

In this study, multi-walled carbon nanotubes were provided by an industrial partner of the Nanosafe2 consortium. Some geometrical characteristics were unknown and were determined during this work. The particle-size distribution, determined by using a laser diffraction analyzer (Mastersizer, Malvern Instrument) and performed in isopropanol, was characterized by the d_{10}, d_{50} and d_{90} of the volumetric distribution. The d_{10}, d_{50} and d_{90} values (*i.e.* particle size at 10%, 50, and 90% of the cumulative particle size distribution) are respectively 20, 200 and 560μm.

It has to be noticed that carbon nanotubes seem to form large agglomerates, which is confirmed by the transmission electronic microscope analysis.

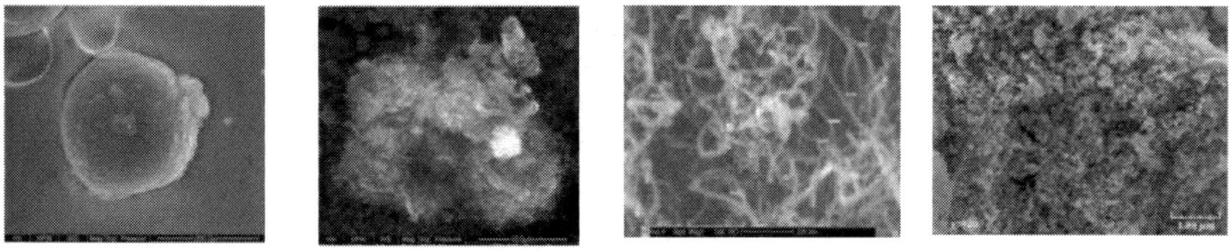

A B C D

Figure 1. TEM images of the MWNTs carbon nanotubes (A-C) and aluminium nanoparticles (D) at different resolutions

TEM observations of CNTs shows particles appearing as large spheres with imbedded nanotube networks having a diameter ranging between 15 and 20 nm.

Similar investigations were carried out with three nano-sized aluminium powders supplied by industrial partners of the consortium. Their surface is covered with an alumina layer representing approximately 20-30% by weight.

BET specific surface areas (S_{BET}) and helium pycnometer density have been measured. By coupling the results obtained by BET and helium pycnometer, an average grain size can be calculated by considering that particles are isolated, cylindrical (or) spherical with a monodispersed size distribution:

$$d_{BET} = \frac{4(or)6}{\rho_{helium} \cdot S_{BET}} \quad (1)$$

2.2. Sample preparation for safety parameter characterisation:

A good reproducibility between the different tests was ensured by systematically drying the samples at 50°C under dynamic vacuum of 1 Pa during 15 minutes and then under a static vacuum of 10 Pa during 4 hours before handling. Particular attention was taken to handle the nanopowders. Nitrile gloves, airline hood, non-woven coverall and HEPA/ULPA vacuum cleaner have been used in addition to the more classical safety barriers.

2.3. Explosion sensitivity and severity

Explosion sensitivity is often referred to as a combination of three parameters characterizing the ability of dust to be ignited and to generate dust explosions. These parameters are the following: the minimum ignition energy (MIE), the minimum ignition temperature (MIT) and the minimum explosive concentration (MEC) [7, 8].

In our study, MIE was determined using a modified Hartmann tube (Kühner AG). It consists in a 1.2 L glass cylindrical vessel in which the powders are dispersed and ignited by an electric spark. Both spark's energy and dust concentration could be modified to characterize the MIE. Turbulence degree is also variable. This apparatus also allows the determination of the MEC [9].

MIT in cloud can be tested by using a Godbert-Greenwald furnace (Chilworth Technology) set in an unconfined environment. This parameter was not systematically measured in our study.

The measurements of dust explosion severity, *i.e.* P_{max} (maximum overpressure), and $(dP/dt)_{max}$, (maximum rate of pressure rise), were performed in a 20 L spherical vessel in accordance with the ASTM Method E 1226 (see Figure 3). Explosion severity is quantified by an explosion index K_{st} which is defined as

$$K_{st} = \left(\frac{dP}{dt}\right)_{max} \cdot V^{1/3} \qquad (2)$$

where V is the explosion volume.

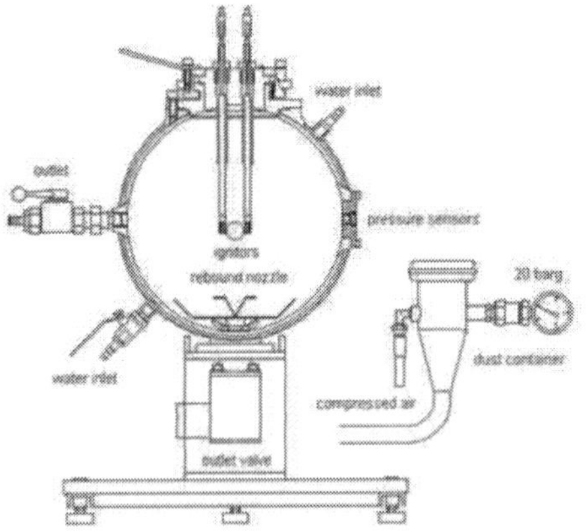

Figure 2. Schematic representation of the 20 L apparatus (Kühner AG).

3. Results and Conclusions

The main results of this study are summarized in the following Table 1.

Table 1. Main safety parameters for CNTs, Carbon blacks and Aluminium

Products	CNT	Carbon blacks				Aluminum		
		Corax N115	Corax N550	Thermal Black N990	Printex XE2	Alex	100 nm	200 nm
MIE (mJ)	> 1J	> 1J	> 1J	> 1J	> 1J	<8mJ	< 1mJ	7 mJ
MEC (g/m3)	45	60	60	60	60	60	-	-
Pmax (bars)	6,6	7,7	7,5	6,7	7,2	8.3	8.2	9.5
(dP/dt)max (bars/s)	227	326	503	240	343	933	1340	2480
Kst	62	88	136	65	93	253	365	675
Explosivity Class	St1	St1	St1	St1	St1	St2	St3	St3
S_{BET} (m^2/g)	195	128	42	9	900	-	23	10.5
d_{BET} (nm)	10	23	70	330	3	-	100	200
d_{agg} (nm)	200 000	90 000	15 000	2000	88000	~80000		

From this table, it is shown that although the primary particles are estimated by the BET method, the resulting average particle diameter (of the order of 10-200 μm) is much larger than the particle primary size indicating that the particles are agglomerated.

The maximum explosion pressure for the studied carbon blacks is 6.6 bars which is slightly lower than the value for MWNTs. The maximum rate of pressure rise is 227 bar/s and the explosion index K_{st} is 62, lower that that of carbon blacks, ranking the explosivity of these powders in a low hazard range. It is seen, however, that for aluminium nanopowders, the explosion sensitivity is high and that the severity can also be very high (K_{st} of 675), greater than that of hydrogen which is considered as a detonable gas. Hence, metallic nanopowders can possibly detonate under certain conditions, thus providing grounds for the need to perform a thorough risk assessment for such types of nanopowders.

References

[1] Frost&Sullivan (2004), An Assessment on the Future of Carbon Nanotubes - Strategic Analysis of the Market & Potential, Market engineering deliverable, www.frost.com

[2] Kwok, Q. S. M., Fouchard, R. C., Turcotte, A., Lightfoot, P. D., Bowes, R., & Jones, D. E. G. (2002). Characterization of aluminium nanopowder compositions. *Propellants, Explosives, Pyrotechnics*, 27, 229 – 240.

[3] Huang Y., Risha G. A., Yang V., and Yetter R. A. (2005). Analysis of Nano-Aluminum Paricle dust Combustion in Different Oxidizer Environments, 43rd Aerospace Sciences Meeting and Exhibit, January 10-13, Reno.

[4] Eckhoff, R. K., 2003, *Dust explosions in the process industries* (3rd edition). Amsterdam – Boston – Paris: Gulf Professional Publishing.

[5] Dahoe, A.E., Zevenbergen, J.F., Lemkowitz, S.M. and Scarlett, B., (1996), Dust explosions in spherical vessels: The role of flame thickness in the validity of the 'cube-root law', *Journal of Loss Prevention in the Process Industry*, 9 (1):33-44.

[6] Pritchard, K ., (2004). Literature Review-explosion hazards associated with nanopowders, HSL/2004/12.

[7] Jacobson, M., Cooper, A. R., & Nagy, J. (1965). *Explosibility of metal powders. Report of investigation 6516*, Bureau of Mines, U.S. Department of the Interior.

[8] Nagy, J. & Verakis, H.C., (1983), *Development and Control of dust explosions*, New York, Marcel Dekker.

[9] IEC standard 1241-2-3 1994

Visualization of nano risk research field to clarify domains year by year

Yasuto Matsui[1], Takeshi Hayashi[1], Kenichi Miyaoi[1], Hironori Tomobe[2], Yuya Kajikawa[2], Katsumori Matsushima[2] and Yukio Yamaguchi[1]

[1]Department of Chemical System Engineering, Graduate School of Engineering, The University of Tokyo, Japan

[2]Innovation Policy Research Center, Institute of Engineering Innovation, The University of Tokyo, Japan

E-mail: matsui@sogo.t.u-tokyo.ac.jp

Abstract. With rising interest of nano technology R&D, nano risk researches have been greatly studied recently. They attract much attention since influence of nano products in the society is not well-known. Now the current state of nano risk research field is not fully investigated, and the object is overviewing this structure until 2008 and predicting the direction of next-coming studies. Nano risk 1611 papers were searched out with certain query and further refinement. And these papers were clustered by bibliometric method. The selected papers were clustered to seven parts and visually seen as aggregated blocks. Each cluster was labeled with proper name by analyzing in detail and the content of each cluster was classified with three terms, *i.e.* "Material", "Hazard" and "Kinetics". The biggest cluster was cluster #0 "atmospheric nanoparticles", and secondly cluster #1 "nanoparticles used in imaging", thirdly cluster #2 "toxicity of manufactured nano materials". Furthermore, historical trend of the number of papers of each cluster was studied year by year. From the all results, short-term future predicting was performed by examining titles of papers or transition of the number of papers in each cluster and by watching the cluster position and gaps between clusters.

1. Introduction

The rapid development of nano science and engineering in the wide range of research fields and industries resulted in various nano-products such as nano-composite, drug delivery system (DDS), and nano-cosmetics. However, the influence of nano-materials in the environment including humans begins to be feared. Therefore, nano risk is recently recognized as a research target and strongly needed now. Furthermore, the analysis of outputs from nano risk researches is desired for the development/design of risk management. To achieve it the current situation survey of nano risk research and forecast for the future is necessary, but it has not yet been done in nano risk.

The purpose of this study was to investigate the current structure of nano risk research field by bibliometric method and to understand what domains are included in the field.

The bibliometric approach is a technique to obtain a whole perspective or state of a large research field. Nano science and engineering is one typical example having been developed rapidly with huge amounts of papers since the 1990s. While one option to assess emerging technologies is expert-based roadmapping [1], a computer-assisted approach is more helpful to see global research trends. Braun focused on the scientific aspects of nanotechnology and described the rapid development since the early 1990s [2]. They could establish an exponential growth of publications in nanoscience and nanotechnology starting in the early 1990s. Meyer characterized the field as more interdisciplinary than other areas of science and explored the contributions by various fields of science and technology to nanoscience [3]. Meyer studied the interrelationships between science and technology [4, 5]. Meyer investigated co-author maps of nanoscience [6], while Calero proposed an author/organization combination and investigated clusters of authors to detect potential partners with similar activities [7]. Chau constructed a web portal about nanotechnology [8]. Huang monitored the research status of nanotechnology based on descriptive statistics and a citation network of countries, institutions, and technology fields [9-11]. Other researches also investigated nanotechnology by bibliometric methods [12-18].

2. Data and Methods

2.1. Data

Defining a research domain via a set of queries is not a rudimentary task [19]. In this paper, we used the following as queries to define nano risk:

$$(A\, and\, B)\, or\, (A\, and\, C)\, or\, D \tag{1}$$

where A denotes "nano* or ultrafine", B denotes "toxic*", C denotes "exposure or dose or injection or administration or inhal", and D denotes "nano* and risk".

We collected citation data on publications from the Science Citation Index (SCI) and the Social Sciences Citation Index (SSCI), compiled by the Institute for Scientific Information (ISI), because SCI and SSCI are two of the best sources for citation data. We also used the Web of Science, which is a Web-based user interface of ISI's citation databases. We searched papers using the above queries as where "*" means wildcard. A total of 17,480 papers were retrieved. The data includes all ISI records, i.e. articles, letters, reviews, editorials, meeting abstracts, and so on. Some of these papers may be not relevant to nano risk because we collected data simply by the query described above. Therefore, we focused on the maximum connected component, which currently comprises 6,475 papers and accounts for 37.0%. We ignored the sidetrack papers of nano risk research field.

2.2. Method

The retrieved data were converted into a non-weighted, non-directed network. Subsequently, the network was divided into clusters using the topological clustering method [20, 21]. Traditionally, co-citation has been used to analyze a citation network. However, because co-citation is accompanied by a time lag to create a link, and analysis of intercitation is more relevant in the similarity of pairs of documents than co-citation [22], we used intercitation as a link. The clustering algorithm is based on modularity Q, which is defined as follows [20, 21]:

$$Q = \sum_{s=1}^{N_m} \left[\frac{l_s}{l} - \left(\frac{d_s}{2l} \right)^2 \right] \tag{2}$$

where N_m is the number of clusters, l_s is the number of links between nodes in cluster s, l is the sum of l_s, and d_s is the sum of the degrees of the nodes in cluster s. In other words, Q is the fraction of links that fall within clusters, minus the expected value of the same quantity if the links fall at random without regard for the clustered structure. Since a high value of Q represents a good division, we stopped joining when ΔQ became minus. A good partition of a network into clusters means there are many within-cluster links and minimal between-cluster links.

After clustering the network, we characterized each cluster by the titles and abstracts of papers that are frequently cited by the other papers in the same cluster. It does not mean that all papers in the cluster study the same topics as covered in these frequently cited papers. In fact, each paper studies its own topics, and each paper has its own unique focus. However, as a first approach, it is reasonable to treat these inter-cited papers as a cluster to investigate the brief structure of a research domain and to consider the frequently cited papers in the cluster as representative of the same. The clustered network is visualized by using a large graph layout (LGL) [23], which is based on a spring layout algorithm where links play the role of spring connecting nodes. Thanks to such layout, papers that cite each other and form a group can be located in closer proximity.

3. Results and Discussion

As a result of clustering the network of nano risk citation, we obtained four main clusters: nano risk, drug delivery system (DDS), dye-sensitized solar cell, and carbon nanotube as a sensing material. The major content of the citation network was study on application and minor is nano risk. The reason clusters of application was mingled with nano risk is that terms "exposure" and "injection" were included in the queries. For example, "electron injection to nano-structured TiO_2 thin film" was used in dye-sensitized solar cells, and "exposure" was usually used like "exposure of NH_3 gas to carbon nanotube sensor". We focused on and analyzed only the cluster *nano risk*. Since the cluster *nano risk* had 1,617papers and its maximum connected component consists of 1,611 papers, we investigated the structure of the citation network of nano risk for the 1,611papers. The result was visualized in Figure 1.

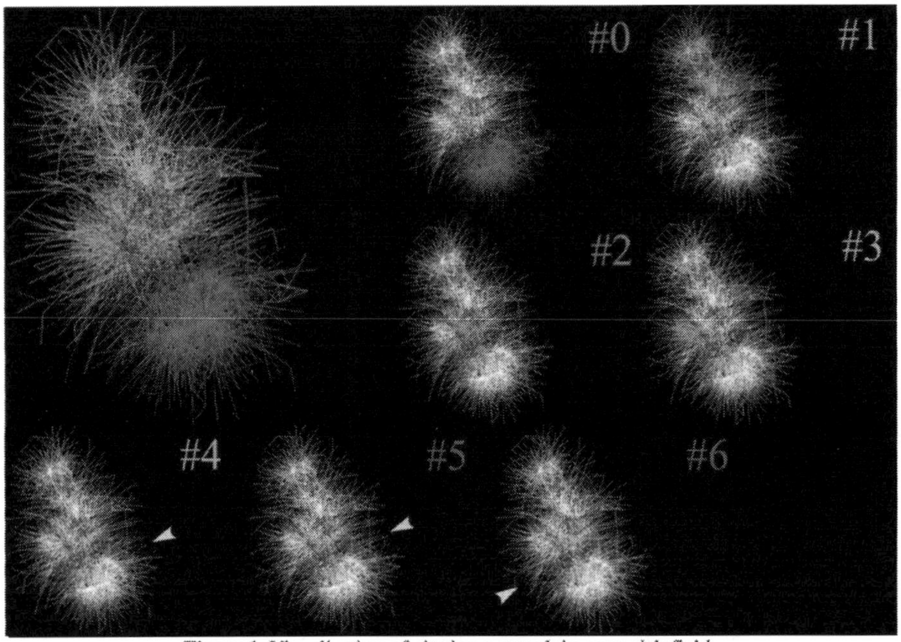

Figure 1. Visualization of citation network in nano risk field.

Nano risk citation network was composed of seven clusters, and each cluster was differently colored in Figure 1. Next, we put the label based on three axes named "Material", "Hazard" and "Kinetics" (Figure 2) to these seven clusters. Table 1 shows this result and each cluster was characterized as follows:

- **Cluster #0**

This cluster has many papers with the title of "ultrafine particle" such as diesel-engine-derived particles, particles generated by friction between car tires and pavement, combustion of woods, or generated in room environment. Accordingly, this cluster in general has interest of particles derived in urban life both outdoors and indoors. Its research theme is inclined to exposure assessment or intra-body kinetics of inhaled particles. There're some epidemiology researches, but researches on hazard or toxicity are rarely seen.

- **Cluster #1**

This cluster has a lot of papers about magnetic nano materials such as Fe_2O_3 and quantum dots such as CdSe, CdTe, ZnSe etc. Quantum dots are studied for imaging in the living organisms by their fluorescence character, and magnetic nano materials are for imaging mainly brain tumors with the technique of magnetic resonance.

- **Cluster #2**

This cluster consists of papers which examine toxicity of industrial nano materials such as C_{60}, CNTs, TiO_2, metals. *In vitro* studies were minor until 2004, and *in vivo* experiments were major using some metals or metal oxides nano particles. However, after 2005 studies *in vitro* explodes with initiating and expanding the synthesis of CNTs or C_{60} or other nano materials. The endpoint of *in vitro* experiments is mainly oxidative stress.

- **Cluster #3**

This cluster has papers on CNTs. Hazard study on CNTs has initiated with 2 *in vivo* experiments [24, 25]. Toxicity of CNTs is studied in many endpoints: inflammation, cytotoxicity, ROS, biocompatibility, *etc*. Intra-body kinetics is hardly studied. The number of *in vitro* studies surpasses *in vivo*.

- **Cluster #4**

This cluster deals with nano particles in air pollution. The aimed point is similar as Cluster #0, but the difference is method: field work for investigating difference between plural cities. Studied particles are diesel particles, sand dusts, fly ashes, etc.

- **Cluster #5**

This cluster has papers written after 2000 and deals with Ag nano particles. Ag nano particles are studied for the existence of anti-bacterial effect.

- **Cluster #6**

This cluster has least paper numbers in the seven clusters, and papers discuss engineering ethic, technology management, and national policy of nano.

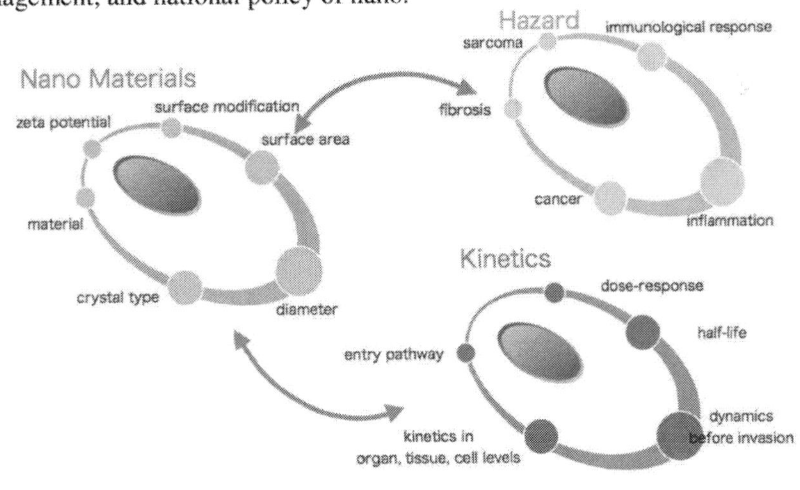

Figure 2. The concept of nano risk assessment.

On the other hand, each number of papers in each cluster year by year was analyzed and shown in Figure 3. From this figure the historical trend of each cluster can be read: Cluster #0 and cluster #1 have conventional themes like atmospheric nanoparticles and nanoparticles used in imaging, with the number of papers in each cluster rising relatively slowly. This is probably because the nano measurement technology has been developed further and quantum dots were emerged around 2002. Clusters #2 and #3 have nearly zero papers until 2001. However, the magnificence of these clusters has been exploding dramatically after 2002, reflecting the increase of R&D for nano materials such as CNTs, C_{60}, ultrafine metals or metal oxides and their toxicity assessment. Cluster #4 seems to be the complement of cluster #0. Cluster #5 gives much attention to Ag nano particles for a good antibiotic property, but the research seems to be in its infancy. Cluster #6 has several papers after 2005, and the reason is papers in this cluster are almost general papers or reviews.

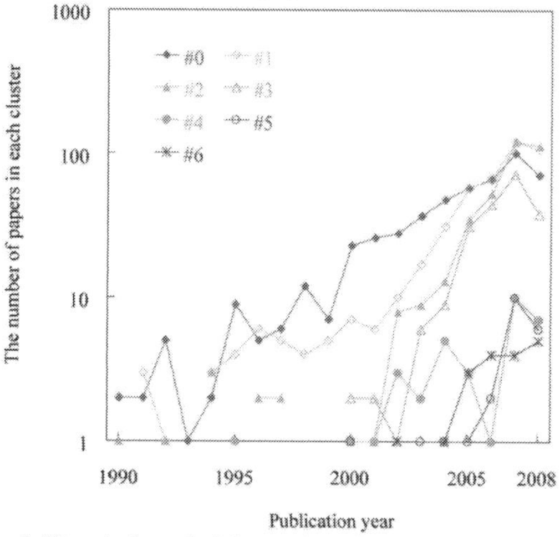

Figure 3. Historical trend of the number of papers in nano risk field.

From the results mentioned above, predicting near future could be partially carried out. By watching Table 1, there're some blank fields and they may be the next research targets. By watching Figure 3, general trend of each cluster is revealed and extrapolating each line in this figure will tell us the more attracting fields in nano risk researches. More specifically, Figure 1 and explanation for each cluster will give us further information: On cluster #1, researches related to quantum dots are largely cytotoxicity assessment so far (until 2008), and after exploring it will come *in vivo* toxicity assessment for practical application in society. On cluster #2, seen from Figure 1, there's a gap between this cluster and clusters #0 or #1. Therefore in recent years, relationship between air pollutant carbon materials and carbon nano materials will probably be clarified, or many kinds of nano materials with quantum dots size will be produced and their toxicity assessment or intra-body kinetics will be investigated. On cluster #3, CNTs *in vitro* test will be yet major in near future but subsequently *in vivo* long-span toxicity tests or exposure experiments will be taken, or CNTs with various modifying chemicals will be investigated in the point of bio compatibility.

Table 1. Labelling of each cluster in nano risk.

No.	Cluster name	Material	Hazard	Kinetics
#0	atmospheric nanoparticles (Feb. 2004, n=511)*	diesel exhaust particles, carbon black	N.D.**	lung deposition
#1	nanoparticles used in imaging (Apr. 2005, n=458)	quantum dot, magnetic nano particle	N.D.	imaging
#2	toxicity of manufactured nanomaterials (Mar. 2006, n=363)	industrial nano material	in vitro, oxidative stress	N.D.
#3	carbon nanotube (Mar. 2006, n=205)	carbon nanotube	cytotoxicity, biocompatibility	N.D.
#4	field work about atmospheric nanoparticles (Jan. 2005, n=34)	diesel exhaust particles, carbon black	N.D.	urban environmental, field work
#5	antibiotic nature of Ag nanoparticles (May 2006, n=22)	Ag nanoparticle	antibacterial	N.D.
#6	engineering ethic and policy about nano (Mar. 2006, n=18)	general about nano	N.D.	N.D.

*:(average publicated date, number of papers within cluster), **: not detected

References

[1] Fleischer T, Decker M and Fiedeler U 2005 *Technological Forecasting & Social Change* **72** 1112
[2] Braun T, Schubert A and Zsindely S 1997 *Scientometrics* **38** 321
[3] Meyer M and Persson O 1998 *Scientometrics* **42**(2) 195
[4] Meyer M 2000 *Scientometrics* **48**(2) 151
[5] Meyer M S 2001 *Scientometrics* **51**(1) 163
[6] Meyer M 2006 *Scientometrics* **68**(3) 545
[7] Calero C, Buter R, Valds C C and Noyons E 2006 *Scientometrics* **66**(2) 365
[8] Chau M, Huang Z, Qin J, Zhou Y and Chen H 2006 *Decision Support Systems* **42** 1216
[9] Huang Z, Chen H, Yip A, Ng G, Guo F, Chen Z-K, Roco M C 2003 *Journal of Nanoparticle Research* **5** 333
[10] Huang Z, Chen H, Chen Z-K, Roco M C 2004 *Journal of Nanoparticle Research* **6** 325
[11] Huang Z, Chen H, Yan L, Roco M C 2005 *Journal of Nanoparticle Research* **7** 343
[12] Hullmann A and Meyer M 2003 *Scientometrics* **58**(3) 507
[13] Marinova D and Mc Aleer M 2003 *Nanotechnology* **14** R1
[14] Eto H 2003 *Scientometrics* **58**(1) 5
[15] Kostoff R N, Murday J S, Lau C G Y and Tolles W M 2006 *Journal of Nanoparticle Research* **8** 193
[16] Kostoff R N, Stump J A, Johnson D, Murday J S, Lau C G Y and Tolles W M 2006 *Journal of Nanoparticle Research* **8** 301
[17] Zhou P and Leydesdorff L 2006 *Research Policy* **35** 83
[18] de Miranda Santo M, Coelho G M, dos Santos D M and Filho L F 2006 *Technological Forecasting and Social Change* **73** 1013
[19] Zitt M and Bassecoulard E 2007 *Information Processing and Management* **42** 1513
[20] Newman M E J 2004 *European Physical Journal B* **38**(2) 321
[21] Newman M E J and Girvan M 2004 *Physical Review E* **69**(2) 026113 Part 2
[22] Klavans R and Boyack K W 2006 *Journal of the American Society for Information Science and Technology* **57** 251
[23] Adai A T, Date S V, Wieland S and Marcotte E M *2004 Journal of Molecular Biology* **340**(1) 179
[24] Lam C W, James J T, Mc Cluskey R and Hunter R L 2004 *Toxicological Sciences* **77**(1) 126
[25] Warheit D B, Laurence B R, Reed K L, Roach D H, Reynolds G A M and Webb T R 2004 *Toxicological Sciences* **77**(1) 117

The NANOFEU project: objectives and tools

Didier Calogine[1], Stephane Duplantier[1], Carine Chivas[2], Eric Guillaume[2], José-Marie Lopez-Cuesta[3], Claire Longuet[3], Benoit Minisini[4], Francis Gensous[5] and Loïc Chesné[5]

[1]INERIS, Parc Technologique ALATA, BP 2, 60550 Verneuil-en-Halatte, France

[2]LNE, CEMATE – Fire Behaviour Division, 78197 Trappe, France

[3]EMA, CMGD, 6 Avenue de Clavières, 30319 Ales, France

[4]ISMANS, 44 Avenue Bartholdi, 72000 Le Mans, France

[5]PlasticsEurope France, 14 rue de la République, 92800 Puteaux, France

E-mail: didier calogine@ineris.fr

Abstract. The NANOFEU project supported by the French Research Agency (ANR) aims to characterise the fire behaviour of nanoparticles filled polymer materials. Day after day, new applications of nanoparticles appear in industry. Among multiple applications in various domains, the role played by nanostructures particularly in combustion and flame retardancy phenomena needs to be quantified by modelling, experimental and numerical validations, since the use of these new components is increasing to upgrade polymer performances. Their employment could represent nowadays a valuable alternative or could be used in synergy with the conventional flame retardants systems. Nanocomposites based on several categories of polymers will be designed, incorporating suitable nanoparticles. Multiscale modelling of material, investigation of thermal degradation processes, and influence of interfacial modifications of nanoparticles, characterisation of effluents, particles released (size and morphology) and smoke toxicity will be made, developing original experimental and numerical means. We will particularly focus on fire performance, smoke toxicity and morphological modification of particles in the effluents, to weight the various impacts of the introduction of these nanofillers. A traceability of nanoparticles affected by combustion will be done through the analysis of effluents. In a first part, this paper includes a detailed presentation of the project by clarifying the partners' roles and objectives expected at the end of the project. In a second part, the foreseen experimental and numerical tools will be developed in order to improve the knowledge of mechanisms involved in combustion.

1. Introduction

Nanoparticles have changed the perception of the concept of polymer composite. Nanocomposites are the emerging material of the 21st century. The importance of these products is growing from industrial and research point of view. These materials represent with their complexity new economical, scientific, technical and social challenges. Estimated at 40 billion Euros in 2001, the world market of nanotechnologies would exceed the 1000 billion Euros in the next ten years.

Nanocomposites have been observed to exhibit enhanced mechanical properties [1]. They represent a combination of unique properties, such as increased heat resistance temperature, reduced permeability, reduced flammability and improved mechanical properties. As for any new development, advantages and drawbacks of these new products must be precisely evaluated.

It is useful to have resistant materials in domains pertaining to buildings and to transport in case of fire. To reduce the flame propagation, some fire-retardants systems usually employed cannot satisfy the environmental requirements.

The non-halogenated flame resistant systems represent consequently a way of substitution but to reach the aimed performances, a strong mass percentage is necessary. Consequently another way lies in using nanoparticles with appropriate properties. It consists in developing systems containing particles allowing to modify the thermal degradation or the inflammation of polymer materials. The association of organo-modified nanoparticles or those treated on surface with polymers modified by the incorporation of specific monomers, as well as with additive flame retardants non-halogenated, aimed at obtaining high levels of protection against fire.

In this context, the NANOFEU project aims to get performances in terms of fireproofing of nanocomposites while analyzing the released products in terms of gaseous effluent.

2. Description of the project

The primal reason for incorporating fire retardants into these materials is to reduce hazard due to fire. Fire retardants strategies aim to reduce the ignitability or the rate of heat release of burning material but the results produce toxic gas and smoke responsible to several deaths [2].

Polymer nanocomposites offer a wide range of promising applications because of their enhanced properties arising from the reinforcement using nanoparticles. They are often reinforced by stiff fillers to improve mechanical properties. The efficiency of this reinforcement depends on the filler aspect ratio, the filler mechanical properties and the adhesion between the matrix and the filler. It is expected that their presence within the fireproofed composite improved effects barriers in developing a synergy between the various actors within the matrix. Further development depends on the fundamental understanding of the hierarchical structures and behaviours which requires multiscale modelling and simulation strategies to provide the coupling of various time scales and length scales [3]. According to the physical scale studied, a hierarchy of models can be applied using various numerical tools from molecular dynamics simulation to macroscale models. At mesoscopic scale studies initiated by Nielsen [4] allows to set up laws taking into account physical properties of both matrix and inclusions.

The NANOFEU project, started in January, 2008 for 3 years, is a research program carried out in collaboration between five complementary partners: "Institut national de l'environnement industriel et des risques" (INERIS), "Laboratoire national de métrologie et d'essais" (LNE), "Ecole des Mines d'Alès" (EMA), "Institut Supérieur des Matériaux et Mécaniques Avancés" (ISMANS) and Plastics Europe France. All the partners will share their knowledge in combustion, polymers and nanoparticles fields in this joint project.

2.1. Presentation of partners

Under the supervision of the French Ministry for Environment, Energy, Sustainable Development and Land settlement, INERIS coordinator of this project, assesses the processes, the techniques of prevention and potential effects of accidents or pollutions. This institute evaluates the risks pertaining to the development of new technologies.

The LNE, connected with the Ministry-delegate for Industry, has for missions the certification of products and the global piloting of the French metrology. This laboratory has specific devices for the characterization of the materials behaviour in fire.

Placed under the technical supervision of the Ministry-delegate for Industry, EMA developed its skills in the field of the Sciences of materials and in particular in the study of high diffusion materials (polymers, asphalts, concretes) and the interfaces from microscopic to macroscopic scales.

ISMANS brings to this project his assessment in numerical design of materials and functional devices and also in virtual engineering and non linear mechanics. This partner is charged of the modelling and the molecular simulation of nanocomposites behaviour in fire.

Plastics Europe France represents the French analogue of the European syndicate of the plastics producers. It includes the various companies working in the domain such as ARKEMA, SOLVAY, LVM.

2.2. Objectives

The aim of this project is to acquire knowledge on the efficiency obtained by the addition of nanostructure within polymer and to define the toxicity inherent to the development of nanocomposites [5].

For that, comparisons are elaborated with regard to properties and behaviours of polymer with various additives such as nanoparticles, modified nanoparticles, flame retardants and combinations of nanoparticles and IFR.

Objective consists in studying new concepts of fireproofing such as effects barriers and in developing effects of synergy particularly by combining aspects of nanoparticles and intumescences.

Flammability and fire toxicity depend on scenario and type of sample. That is why several shapes will be treated. Bench-scale determination of toxic product yields in fire effluents will be produced associated to particular methodologies.

The flame-retardants efficiency of samples will be improved by simultaneous test analysis to define the thermal degradation characteristics of products. Indeed nanoparticles have an influence on the thermal stability of polymers according to two mechanisms: by barrier effect in volatilization gases or combustive and by local modification of the polymer morphology (creation of an interphasis between matrix and nanofiller).

Relations between structures and properties at various scales will be studied by leaning on methodological steps and experimental approaches [6]. For that, the dynamic molecular tool at nanoscopic scale using specific potential of correlation will be investigate to obtain thermal properties of samples such thermal conductivity and resistivity of microscopic structure. Methods of validation such as simulation by flash method can complete this procedure for the determination of sample physical properties.

This project proposed to improve one's knowledge of the role played by nanostructure in the polymer matrix. Procedure is summed up in two stages:

1. Use of experimental methods for the measure of thermo physical parameters
2. Study, modelling and simulation of nanocomposites combustion.

3. Materials and methods

A number of experimental and numerical procedures will be used in the framework of this project:

3.1. Preparation of the nanocomposites

Nanocomposites can be synthesized by different routes. Most commonly used processes are realized using melt-intercalation, in-situ polymerization, solution and emulsion polymerization.

Samples used in this project will be made by EMA by melt mixing of the components.

Figure 1. Kind of elaborated samples

3.2. Experimental techniques used to characterize the sample
Studies show that, concerning the toxicity of nanoparticles, effects at medium and long term depend on size and on form of particles. That is why analysis of samples is primordial. Studies of structural and morphological characteristics were carried out.

Various techniques will be used, such as Transmission Electron Microscopy (TEM) to observe the uniformity of dispersion in the polymer matrix, Thermogravimetry Analysis (TGA) to determine the influence of nanofillers on the thermal comportment of composite, Scanning Mobility Particle Sizer for the determination of size distribution of particle …

Figure 2. ELPI impactor and AFM spectrometer

3.3. Combustibility characteristics
On one hand, the flammability properties of samples with different percentages of fillers will be measured by fire calorimeter. The fire calorimeter is a commonly used device to measure the mass loss rate per unit area and the heat release per unit area for a given constant external radiative heat flux. Chemical energy release date, mass loss rate and time of ignition will be evaluated. Ignitability and flammability tests will be performed according to standard procedures.

On the other hand identification and quantification of combustion gases and particularly their toxic components from different fire scenarios were undertaken by continuous Fourier Transform Infrared spectroscopy (FTIR). The spectra of the decomposition and combustion products of various materials will be collected.

Figure 3. Teawarson calorimeter and TFIR metrology

3.4. Numerical tools

In the last decade molecular modelling tools were extensively used to study the properties of nanocomposites systems at different scales. For example, Density Functional Theory (DFT) was used to evaluate the mechanical or electrical properties of nanotubes whereas Molecular Dynamics (MD) was used to have a molecular insight into the interactions between polymers and nanoparticles.

On other hand, it appears from experimental studies that the chemical reactivity of polymer in presence of nanocharges is different too in the bulk. Consequently, some questions arise on radical stability, new reaction pathways catalysed by nanoparticles and so on. Moreover, the presence of combustion products can play a role in the aggregation of the nanoparticles and consequently affect the size and the morphology of particles released.

In this project, a methodology will be developed to assess the role of nanoparticles in the thermal degradation of polymer resin. DFT, MD and hybrid methods associated with atomistic models will be used for this purpose.

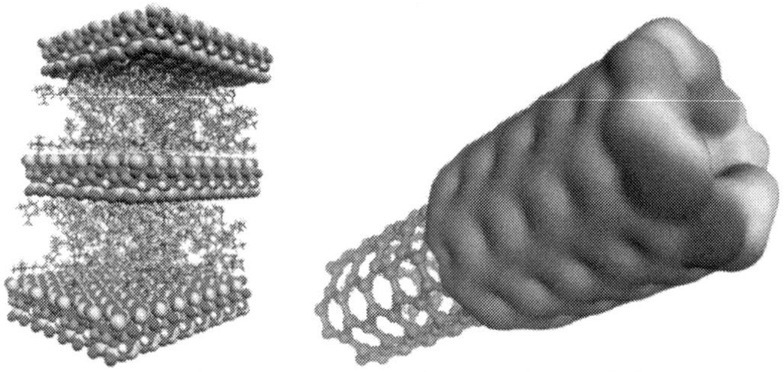

Figures 4. Examples of dynamic molecular simulations

References

[1] Valavala P K and Odegard G M 2005 *Rev. Adv. Mater. Sci.* **9** 34
[2] Stec A A, Hull T R, Purser J A and Purser D A 2008 *Fire Safety Journal* **44(1)** 62
[3] Zeng Q H, Yu A B and Lu G Q 2008 *Prog. Polym. Sci.* **33**, 191
[4] Nielsen L E 1967 *J. Macromol. Sci. Chem.* **5(1)** 929
[5] Chivas C, Guillaume E, Saragoza L, Ducourtieux S, Sainrat A and Macé T 2008 Biannual Conference on modification, degradation and stabilisation of polymers, Belgium
[6] Picken S J, Vlasveld D P N, Bersee H E N, Ozdilek C and Mendes E 2008 "Structure and mechanical properties of nanocomposites with Rod- and Plate-Shape nanoparticles" in Nanocomposites: Ionic conducting materials and structural spectroscopies, P. Knauth and J. Schoonman (eds.)

Nanotechnologies: Risk assessment model

F. Giacobbe, L. Monica and D. Geraci

ISPESL – National Institute for Occupational Safety and Prevention, Italy

E-mail: francesco.giacobbe@ispesl.it

Abstract. The development and use of nanomaterials has grown widely in the last years. Hence, it is necessary to carry out a careful and aimed risk assessment for the safety of the workers.
The objective of this research is a specific assessment model finalized to the workplaces where the personnel work manipulating nanoparticles. This model mainly takes into account the number of exposed workers, the dimensions of particles, the information found in the safety data sheets and the uncertainties about the danger level coming from the exposition to nanomaterials. The evaluation algorithm considers the normal work conditions, the abnormal (*e.g.* breakdown air filter) and emergency situations (*e.g.* package cracking). It has been necessary to define several risk conditions in order to quantify the risk by increasing levels ("low", "middle" and "high" level). Each level includes appropriate behavioural procedures.
In particular for the high level, it is advisable that the user carries out urgent interventions finalized to reduce the risk level (e.g. the utilization of vacuum box for the manipulation, high efficiency protection PPE, etc). The model has been implemented in a research laboratory where titanium dioxide and carbon nanotubes are used. The outcomes taken out from such specific evaluation gave a risk level equal to middle.

1. Methods

The method has been structured keeping in mind the European strategy defined by the European Commission [1, 2] through responsible approach. Waiting for further information around the connected toxicological effects to the exposure to the nanomaterials it is opportune to operate with the maximum caution.

The risk assessment is implemented for each workplace and work activity typology; it is based on the following ten factors: A) numerousness of the exposed workers, B) frequency of exposure, C) frequency of direct manipulation, D) dimensions of the nanoparticles, E) nanoparticles behaviour (*e.g.* dispersion or agglomeration), F) effectiveness of Personal Protection Devices (PPD) used, G) work organization/procedures, H) toxicological characteristics of the substances, I) risk of fire and explosion and J) suitability of workspaces and installations. The aforesaid factors are denominated "factors level risk".

Since the utilization of nanomaterials presents nowadays unknowns about the effective level of danger, the risk assessment takes into consideration these important aspects through the help of an appropriate index denominated "corrective factor". Such index assumes a value within the range 0.5 and 2.0 in accordance to the established level of scientific knowledge. And particularly it assumes the following values: 0.5 - good scientific knowledge; 1.0 - sufficient scientific knowledge; 2.0 - insufficient scientific knowledge.

The evaluation algorithm is:

$$EvaluationRisk = \sum_{i=A}^{J} (FactorLevelRisk)_i \cdot CorrectiveFactor$$

The evaluation algorithm considers both the normal work conditions, the abnormal (*e.g.* breakdown air filter) and emergency situations (*e.g.* package cracking).

The evaluation outcome shall consist of several risk levels subdivided in an increasing way ("low", "middle" and "high" level).

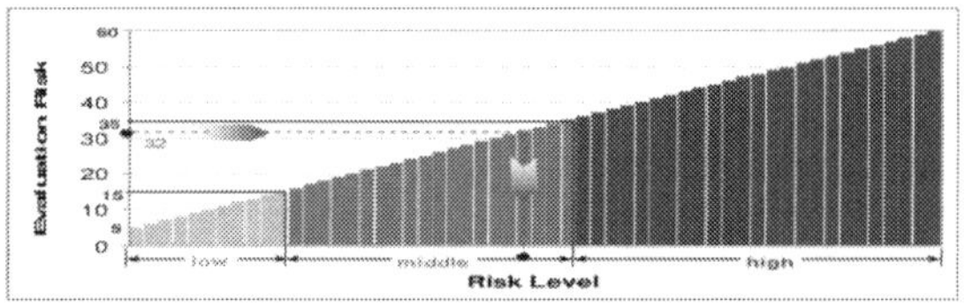

Figure 1. Risk level are subdivided in: "low" (5 ÷ 15); "middle" (16 ÷ 35); "high" (36 ÷ 60). *E.g.* evalutation risk=32

The "high" level risk needs the activation of immediate interventions to reduce the final result of the evaluation at least to the "middle" level risk.

Table 1. Factors level risk for factors specific (A ÷ J).

FACTORS LEVEL RISK / FACTORS	LOW 1	MIDDLE 2	HIGH 3
A numerousness of the exposed workers	1 ÷ 2 unit	3 ÷ 5 unit	greater 6 unit
B frequency of daily exposure	< 2 h/day	> 2 h/day and < 6 h/day	> 6 h/day
C dimensions of the nanoparticles	> 70 nm	> 10 nm and < 70 nm	< 10 nm
D nanoparticles behaviour	high tendency agglomeration	middle tendency agglomeration	high tendency dispersion
E frequency of direct manipulation	< 2 h/day	> 2 h/day and < 4 h/day	> 4 h/day
F effectiveness of Personal Protection Devices (PPD) used	PPD used: • rubber gloves (hands) • safety glasses or goggles (eyes) • laboratory coats (skin) • musk with filter HEPA (respiratory)	partial PPD used	no PPD used
G work organization/procedures	good work practices examples: • training for laboratory personnel must be informed of the risks associated with workplace hazards • equipment cleaning/maintenance • use and maintenance of Personal Protection Devices (PPD)	simple and limited procedures	any procedure or free access to the workspaces
H toxicological characteristics of the substances (material safety data sheet)	safety advice concerning dangerous substances and preparations: • S28 (after contact with skin, wash immediately with water (to be specified by the manufacturer) • S38 (in case of insufficient ventilation, wear suitable respiratory equipment)	safety advice concerning dangerous substances and preparations: • S22 (do not breathe dust) • S26 (in case of contact with eyes, rinse immediately with plenty of water and seek medical advice) • S36 (wear suitable protective clothing) • S37 (wear suitable gloves) • S39 (wear eye/face protection)	risk phrases: • R36 (irritating to eyes) • R37 (irritating to respiratory system) • R40 (limited evidence of a carcinogenic effect)
I risk of fire and explosion	no considered	improbable	probable
J suitability of workspaces and installations	• ≤ class 100 clean room • use of chemical fume hood (if nanomaterials are in gas phase or aerosol) • use of dry box (if nanomaterials are in powder)	1000 ≤ class clean room ≤ 10000	≥ class 100000 clean room

2. Results

The evaluation model has been experimented within research laboratories which are using nanomaterials. In the specific the research studies are finalized to the realization of photovoltaic cells using electric properties of carbon nanotubes health effects of particles produced for nanotechnologies and metal oxide nanoparticles (TiO_2). The above nanomaterials have been purchased in single stocks of reduced dimension in weight (amount less than 10 g). The single products employed present themselves under dust features and are equipped of Material Safety Data Sheet (MSDS) that report the information referring to the risks connected with the use. The realization of photovoltaic cells doesn't involve the direct handling of the nanomaterials in solid state. These are initially scattered in water [3] in the way to reduce and control the density (SWNTs have a density equal to about 1 g/cm^3 [4]). The nano dusts in TiO_2 are worked with solvents so that to obtain a cream paste. Every single working phase is developed avoiding the contact with skin and the breathing of aerial dispersion dusts. Researchers work by wearing the protection devices (gloves and masks) and carry out the manual operations using a dry box. Nowadays specific procedures concerning the way how to establish rules for both the premises entrances and possible emergency situations (*e.g.* example the SWNTs dust outlet without any control from own confections because an accidental overturning) haven't been drawn up yet.

The first results have been initially published in scientific journal *Human and Experimental Toxicology*, that is the primary source of the published data [5].

Below the results of the application of the evaluation model are shown, shared in four possible conditions (Table 2 and 3).

Table 2.

Event	Nanomaterial	Particle Formation	Condition (*)	Exposure Route	Exposure Source or Worker Activity
a	SWNT	dry powed and liquid suspension	N	• inhalation • dermal	dispersing in water
b	SWNT	liquid suspension	N	• dermal	depositing and fixing on plate
c	TiO_2	dry powed	A	• inhalation • dermal	dry box partial failure of the air treatment system
d	TiO_2	in air	E	• inhalation • dermal	casually spilling powder

(*) N - normal A - abnormal E – emergency

Table 3.

Event	A	B	C	D	E	F	G	H	I	J	Σ (A ÷ B)	Corrective Factor (**)	Evaluation Risk	Level Risk (***)
a	1	2	2	3	1	1	2	3	1	1	17	2	34	middle
b	1	2	2	3	1	1	2	3	1	1	17	2	34	middle
c	1	1	2	2	1	1	2	2	1	3	16	2	32	middle
d	1	1	3	2	1	1	3	2	1	3	18	2	36	high

(**) 0.5 - good scientific knowledge 1 - sufficient scientific knowledge 2 - insufficient scientific knowledge
(***) 5 ÷ 15 – "low" level risk 16 ÷ 35 – "middle" level risk 36 ÷ 60 – "high" level risk

3. Discussion

Examining the results it is possible to make different important considerations. First of all it is positive, for the activities of search, the use of safety operational precautions, already in use at the laboratories of chemistry (use of dry box and of personal protection devices). The personnel during the activities that need to manipulate nanomaterials don't underestimate the possible consequences connected to a direct contact and maintains high the level of attention and concentration in the respect risk phrases and safety advice present on the materials safety data sheet of the SWNTs and nano TiO_2 used.

Journal of Physics: Conference Series **170** (2009) 012035 doi:10.1088/1742-6596/170/1/012035

Best practices for nanomaterials:

- Written procedures must document worker exposure potential, required engineering controls and PPE, methods for safe storage and handling, procedures to minimize exposure during production or analysis, waste disposal protocols, and spill containment practices.
- High-efficiency particulate air filtered vacuum systems should be used. Laminar flow filtered laboratory hoods and negative pressure around equipment can reduce worker exposure potential. Also, filtration on vacuum pumps for equipment can minimize air emissions and contamination.
- Personal respirators are not recommended until ongoing studies on seal effectiveness are completed.
- Personal protective clothing should be worn at all times to minimize skin contact.

Nanoparticles in liquid or dry particulate form should be treated as hazardous or special waste. In the absence of consensus standards, best practices for asbestos disposal should be used for insoluble nanoparticles.

In relation to the simulation we underline the following principal factors that have characterized the evaluation of the level of risk for safety and health:

- low numerousness of the exposed workers
- low frequency of direct manipulation
- limited frequency of exposure
- use of Personal Protection Devices
- medium toxicological characteristics of the substances

References

[1] Commission of the European Communities. A code of conduct for responsible nanosciences and nanotechnologies research. Commission Recommendation of 07/02/2008, 2008.

[2] Commission of the European Communities. Nanosciences and nanotechnologies: An Action Plan for Europe 2005-2009. Communication from the Commission to the Council, the European Parliament and the Economic and Social Committee, 2005.

[3] Borm P, Klaessig F C, Landry T D, Moudgil B, Pauluhn J, Thomas K, Trottier R and Wood S 2006 *Toxicological Sciences* 90(1) 23.

[4] Oberdorster G, Oberdorster E and Oberdorster J 2005 *Environmental Health Perspectives* 113 823.

[5] Giacobbe F, Monica L, Geraci D. *Risk Assessment Model of Occupational Exposure to Nanomaterials*. Human and Experimental Toxicology. In press.

Current Standardisation for Nanotechnology

Delphine Bard[1], David Mark[1] and Carsten Möhlmann[2]

[1]Health and Safety Laboratory, Harpur Hill, Buxton, UK

[2]BGIA-Institute for Occupational Safety and Health, Sankt Augustin, Germany

E-mail: carsten.moehlmann@dguv.de

Abstract. Standardisation and standards provide an important mechanism to support both innovation and the application of regulations. There is currently no specific regulation for any nanomaterials. Health, safety and environmental protection aspects associated with nanomaterials are however in principle covered to different levels by current EU regulatory framework. There are a number of national, European and international organisations developing standards associated with the development, description and use of nanomaterials as well as the protection of human health and the environment from the production and use of chemicals and consumer products, including nanomaterials. These organisations have also established specific committees on nanotechnology. This paper outlines the different relevant regulations and standards. This paper will mainly be focused on a European health and safety perspective.

1. European Union regulations

There is no specific regulation for any nanomaterials. However, national and European regulations are applicable to nanomaterials. The main European Union health and safety regulations for the protection of workers include:

- General Directive 89/391/EEC from 12 June 1989 on the introduction of measures to encourage improvements in the safety and health of workers at work.
- Directive 98/24/EC from 7 April 1998 on the protection of the health and safety of workers from the risks related to chemical agents at work (fourteenth Directive within the meaning of Article 16(1) of Directive 89/391/EEC).
- Directive 2004/37/EC of the European Parliament and of the Council of 29 April 2004 on the protection of workers from the risks related to exposure to carcinogens or mutagens at work (Sixth individual Directive within the meaning of Article 16(1) of Council Directive 89/391/EEC) (codified version) (Text with EEA relevance)
- Council Directive 89/655/EEC of 30 November 1989 concerning the minimum safety and health requirements for the use of work equipment by workers at work (second individual Directive within the meaning of Article 16 (1) of Directive 89/391/EEC)
- Directive 1999/92/EC of the European parliament and of the council of 16 December 1999 on minimum requirements for improving the safety and health protection of workers potentially at risk from explosive atmospheres (15th individual Directive within the meaning of Article 16(1) of Directive 89/391/EEC)

- Directive 67/548/EEC (substances) – 1999/45/EC (preparations): Classification, packaging and labelling of Dangerous substances/preparations. 7th amendment as Directive 92/32/EEC
- Under this directive, an evaluation must be made for any substance or preparation manufactured within or imported into the EU and placed on the EU market and results in classification of the substance/preparation as dangerous for one or several end-points concerning physical-chemical properties, health or environmental effects. Classification and labelling is therefore a useful tool for risk management of chemicals. All marketed substances and preparations must be classified and labelled, irrespective of the quantity placed on the market. The labelling is the first and often the only information on the hazards of a chemical that reaches the user, which could be a consumer or a worker. In addition the classification has a large number of downstream consequences within EU legislation.
- Annex V of the 67/548/EEC directive is divided in three parts that contain testing methods for chemicals that address all areas of concern.
- Council Regulation (EEC) No 793/93 on the evaluation and control of the risks of existing substances (Existing Substances Regulation (ESR)).

Health, safety and environmental protection aspects associated with nanomaterials are in principle covered to different levels by current EU regulatory framework.

The risks to be assessed under the Directive 98/24/EC due to the presence of hazardous chemical agents may be one or more of the following:

- Risk of fire and/or explosion.
- Risk generated due to hazardous chemical reactions, which may affect the health and safety of workers.
- Risk due to inhalation.
- Risk due to absorption through the skin.
- Risk due to contact with the skin or eyes.
- Risk due to ingestion.
- Risk due to penetration through the parenteral route.

The Directive 98/24/EC lays down minimum requirements for the protection of workers from risks to their safety and health arising, or likely to arise, from the effects of chemical agents that are present at the workplace or as a result of any work activity involving chemical agents.

In carrying out the obligations concerning risk assessment laid down in Articles 6(3) and 9(1) of Directive 89/391/EEC, the employer shall first determine whether any hazardous chemical agents are present at the workplace. If so, then the employer shall assess any risk to the safety and health of workers arising from the presence of those chemical agents, taking into consideration the following:

- their hazardous properties,
- information on safety and health that shall be provided by the supplier, (e.g. the relevant safety data sheet in accordance with the provisions of Directive 67/548/EEC or Directive 88/379/EEC),
- the level, type and duration of exposure,
- the circumstances of work involving such agents, including their amount,
- any occupational exposure limit values or biological limit values established on the territory of the Member State in question,
- the effect of preventive measures taken or to be taken,
- where available, the conclusions to be drawn from any health surveillance already undertaken.

The Directive 98/24/EC requires that the risk to the health and safety of workers using or handling hazardous chemical agents shall be eliminated or reduced to a minimum. The hierarchy of control measures, which should be applied to all hazards, consists of:

- Eliminate the use of the hazardous substance.

- Substitute the use of the hazardous substance a less hazardous substance.
- Enclose the process to reduce the potential for exposure to the hazardous substance.
- Control the exposure of the substance at source using engineering means.
- Provide adequate protective equipment (such as Respiratory Protective Equipment - RPE).

Elements of the precautionary principle exist in different regulation approaches and REACH (Registration, Evaluation, Authorization and Restriction of Chemicals), which came into effect on the 1st of July 2007, explicitly mentions the precautionary approach. Chemicals notified under REACH should include hazard data for nano-sized materials where these are supplied. The threshold value for the registration of a new chemical substance is 1 ton.

In 2008, the Commission and EU agencies will start to review existing documents that support implementation of the various directives with regard to their applicability and appropriateness to nanomaterials.

2. Voluntary codes and voluntary schemes

To address the growing concerns about nanomaterials a number of organisations or industries have released and or adopted voluntary measures in the management of engineered nanomaterials such as codes of conduct or voluntary schemes.

These voluntary codes of conduct include:

- Code of conduct for responsible nanotechnology (European Commission)
- Responsible nanocode for businesses (Royal Society, UK)
- Code of conduct for nanotechnology in consumer products (Switzerland)
- Code of conduct for the protection of employees, customers and the environment (established by several manufacturers of nanomaterials)
- Guideline for operations with nanomaterials at workplaces (VCI, BAuA, Germany)

There are also voluntary schemes set up at national levels, where organisations and industries are encouraged to submit data on engineered nanomaterials including data on toxicity or exposure levels. These schemes include:

- Voluntary reporting scheme for engineered nanomaterials (DEFRA, UK)
- Nanoinventory (Switzerland)
- Voluntary Nanoscale Materials Stewardship Program (EPA, USA)

3. Standards

Pre-normative and co-normative research may also be required before certain projects may commence or be completed. It is foreseen that standards may be required to support legislation, particularly in the areas of health, safety and the environment, and so the work programme will be influenced by what these requirements are, and when they will be needed.

There are many national, European and international organisations developing standards associated with the development, description and use of nanomaterials as well as the protection of human health and the environment from the production and use of chemicals and consumer products, including nanomaterials. These organisations have also established specific committees on nanotechnology. The European Committee for Standardization CEN established CEN/TC 352 "Nanotechnologies"; the ISO established ISO/TC 229 "Nanotechnologies"; the IEC founded TC 113 "Nanotechnology standardization for electrical and electronic products and systems". According to the Vienna agreement most of the new normative projects are carried out in conjunction with and under lead of ISO/TC 229 "Nanotechnologies". The Organisation for Economic Co-operation and Development (OECD) also established a Working Party on Manufactured Nanomaterials and a Working Party on Nanotechnology. In parallel different national "mirror" groups are working meanwhile on this subject.

These organisations are working to develop and standardise instruments and test methods for the support of appropriate health, safety and environment legislation and regulations of nanomaterials. It includes work on the development and standardisation of:

- Nomenclature and definitions of terms in nanotechnology;
- Test methods to characterise nanomaterials;
- Instruments and test methods for measurement of airborne nanoparticles in the workplace;
- Test methods to detect and identify nanoparticles in the workplace and in the environment;
- Protocols for toxicity and eco toxicity testing;
- Protocols for whole life cycle assessment of nanomaterials, devices and products;
- Risk assessment tools relevant to the field of nanotechnologies;
- Test methods to assess the performance efficiency of engineered and personal control;
- Occupational health protocols relevant to nanotechnologies;
- Materials performance.

In the context of the NANOSAFE 2 project (EU FP6 project; contract number NMP2-CT-2005-515843) the following areas are of interest:

- exposure measurement and characterisation;
- exposure control;
- toxicity;
- fire and Explosions;
- life cycle analysis;
- risk assessment.

Experience gained in these fields through the NANOSAFE2 project will lead to interactions with CEN/TC 352 for the development of future standards.

4. European standards

The first CEN standard for nanotechnology terminology was published in 2008: CEN ISO/TS 27687:2008 Nanotechnologies - Terminology and definitions for nano-objects - Nanoparticle, nanofibre and nanoplate.

A number of work items are also under development including:

- prCEN ISO/TR 11808 Guide to nanoparticle measurement methods and their limitations.
- Guide to labelling of manufactured nanoparticles and products containing manufactured nanoparticles.

Furthermore, there are many other European standards associated with the protection of human health and the environment from the production and use of chemicals and consumer products, which may be of relevance to nanomaterials. Nevertheless these standards must be assessed whether they will apply to nanomaterials or amendments are required. For example in the workplace, the standard method of assessing worker exposure to airborne particles currently involves measuring the mass concentration and chemical composition of health-related aerosol fractions (inhalable or respirable) in the breathing zone. The exceptions are for fibres (*e.g.* asbestos), where the number concentration is determined by microscopy, and for micro-organisms, that are collected on a growth medium, where the number of colony forming units is counted. For the ambient atmosphere, particulate pollution is measured currently according to the PM 10 fraction and the finer PM 2.5 fraction to be introduced soon. A few protocols for measurement of submicrometer particles in environmental air exist and research is currently being carried out in a number of EU states to investigate particle number concentrations and particle number size distributions, prior to the possible introduction of number-based limits for ambient particles.

These other standards include:

EN 14530: 2004, „Workplace atmospheres - Determination of diesel particulate matter - General requirements" specifies a method for the determination of diesel particulate matter in the workplace atmosphere by collecting the particles on a filter and determining the elemental and organic carbon component by successive heating of the sample and measuring the carbon dioxide emitted at two different temperatures.

EN 14175-3:2003, "Fume cupboards – Part 3: Type test methods", is part of a series of standards on assessing the performance of fume cupboards.

EN 12469: 2000, "Biotechnology - Performance criteria for Microbiological Safety Cabinets", gives test methods for assessing the performance of microbiological safety cabinets in preventing the release of micro-organisms into laboratory air.

EN 60335-2-69:1999, "Safety of household and similar electrical appliances – Part 2: Particular requirements for wet and dry vacuum cleaners, including power brush, for industrial and commercial use", provides performance standards for industrial vacuum cleaners. Part 2 of the standard specifies performance of vacuum cleaners in 3 hazard classes according to particle penetration at the most penetrating particle size (MPPS).

In EN 1882: 1993, "Standard for testing and categorising HEPA and ULPA filtration systems", test aerosols of solid sodium chloride particles or liquid di-octyl phthalate (DOP) are used to test the filtration efficiency of filters used in ventilation and air cleaning systems.

For personal protective equipment, EN 143: 2000, "Filters – particles", and EN 13274-7: 2002, "Respiratory protective devices. Methods of test. Determination of particle filter penetration", two protocols are given for testing the particle penetration efficiency of the filter material used in the respirator or dust mask. These do not include particle penetration through face seals or valve seals.

EN 943-1:2002, "Protective clothing against liquid and gaseous chemicals, including liquid aerosols and solid particles – Part 1: Performance requirements for ventilated and non-ventilated "gas tight" (Type 1) and "non gas tight" (Type 2) chemical protective suits".

EN 13982-2:2005, "Protective clothing for use against solid particulate chemicals. Part 2: Test methods for chemical protective clothing providing protection to the full body against solid particulate chemicals (type 5 clothing)".

These two standards use a similar methodology to that used for assessing inward leakage of respirators – solid sodium chloride test aerosol with monitoring inside and outside the suit worn by live subjects following a prescribed exercise routine.

The following four standards describe different test methods for the determination of explosion and combustion characteristics of dusts. Special adaptations for the testing of nanomaterials have to be taken into account and were tested in the NANOSAFE 2 project (for example, a system that employs powder confinement is recommended for these tests). This will facilitate control over the atmosphere within the test chamber such that, for example, the powder can be protected from oxidation until the point of ignition.

EN 1127-1: 1998, "Explosive atmospheres – Explosion prevention and protection Part 1. Basic concepts and methodology".

The occurrence of an explosive atmosphere depends on the presence of a flammable material, the degree of dispersion of the material and the concentration of the material in air within the explosibility range.

EN 13821: 2002, "Potentially explosive atmospheres — Explosion prevention and protection — Determination of minimum ignition energy of dust/air mixtures".

EN 14034-1:2004, "Determination of explosion characteristics of dust clouds – Part 1: determination of the maximum explosion pressure Pmax of dust clouds".

EN 14034-2:2006, "Determination of explosion characteristics of dust clouds – Part 2: determination of the maximum rate of pressure rise (dp/dt) max of dust clouds".

5. ISO standards

The first ISO standard for nanotechnology terminology was published in 2008 as an outcome of ISO/TC 229 Nanotechnologies: ISO/TS 27687 "Nanotechnologies - Terminology and definitions for nano-objects - Nanoparticle, nanofibre and nanoplate", also adopted by CEN. Nano-objects are considered to have at least one dimension below 100 nm. In advance of this, ASTM International had published a similar terminology standard: ASTM E2456 - 06 "Standard terminology relating to Nanotechnology".

Concerning the health and safety aspects of nanotechnologies, ISO/TR 12885 "Health and safety practices in occupational settings relevant to nanotechnologies", published in September 2008, describes current procedures applied in manufacture and processing of nanomaterials.

Further project groups in ISO TC 229 are busy working on following topics:

- ISO/CD 29701 – "Nanotechnologies – Endotoxin test on nanomaterial samples for in-vitro systems".
- ISO/CD 10801 "Nanotechnologies – Generation of silver nanoparticles for inhalation toxicity testing".
- ISO/CD 10808 "Nanotechnologies – Monitoring silver nanoparticles in inhalation exposure chambers for inhalation toxicity testing".
- ISO/TR 13014 "Nanotechnologies – Guidance on physico-chemical characterisation of engineered nanoscale materials for toxicological assessment".

Other existing international standards to be considered when working with nanomaterials, are the following:

ISO/TR 27628: 2007, "Workplace Atmospheres – Ultrafine, nanoparticle and nano-structured aerosols – Exposure characterization and assessment", gives guidance on the current methods of assessing human exposure to ultrafine, nanoparticle and nano-structured aerosols in workplace atmospheres.

ISO 10882-1:2001, "Health and safety in welding and allied processes - Sampling of airborne particles and gases in the operator's breathing zone - Part 1: Sampling of airborne particles", gives guidance of how to measure exposure to welding fume in the workplace.

ISO 8672: 1993, "Air quality - Determination of the number concentration of airborne inorganic fibres by phase contrast optical microscopy - Membrane filter method".

ISO 10312: 1995, "Ambient air - Determination of asbestos fibres - Direct transfer transmission electron microscopy method"

ISO 13794: 1999, "Ambient air - Determination of asbestos fibres - indirect transfer transmission electron microscopy method"

ISO 14966:2002, "Ambient air - Determination of numerical concentration of inorganic fibrous particles - Scanning electron microscopy method"

ISO 14644, "Clean rooms and associated controlled environments", is a suite of 8 standards with the main aim of protecting components (especially semiconductors, computer chips, etc.) from damage by airborne particles. There are 9 classes of air cleanliness assessed by monitoring particle number size distributions using Optical Particle Counters (OPCs).

6. Conclusions:

There is currently no regulation specific to nanomaterials. Health, safety and environmental protection aspects associated with nanomaterials are however in principle covered to different levels by current EU regulatory framework. In 2008, the EU Commission and Agencies will start to review existing documents that support implementation of the various directives with regard to their applicability and appropriateness to nanomaterials.

A number of national, European and international organisations are developing standards that either focus on or have sections dealing with the protection of human health and the environment against the production and use of chemicals and consumer products containing nanomaterials. ISO/TC229 has recently published ISO/TR 12885 "Health and safety practices in occupational settings relevant to nanotechnologies" (September 2008), which describes current procedures applied in manufacture and processing of nanomaterials. Much work is still to be carried out and multiple co-operations are required. Furthermore, there are knowledge gaps in toxicological and exposure data. Such additional knowledge is necessary for a full risk assessment on the manufacture, handling or use of nanomaterials. In the meantime the precautionary principle should be followed to minimise possible risks.

Guidelines for Safe Handling, Use and Disposal of Nanoparticles

Gh. Amoabediny[1,5], A. Naderi[2], J. Malakootikhah[3], MK. Koohi[4], SA. Mortazavi[5], M. Naderi[6] and H. Rashedi[1]

[1]Department of Biotechnology, Faculty of Chemical Engineering, University of Tehran, Iran

[2]Department of Occupational Health, Faculty of Medical Science, University of Tarbiat Modares, Iran

[3]Nanotechnology Standardisation Committee, Iran Nanotechnology Initiative, Tehran, Iran

[4]Faculty of veterinary medicine, University of Tehran, Iran

[5]Department of Nanosafety, Research Centre for New Technology in life science Engineering, University of Tehran, Iran

[6]University of Amirkabir, Iran

E-mail: amoabedini@ut.ac.ir

Abstract. Health, safety and environmental (HSE) risks of a technology is an inseparable part of it which threatens all exposed employees. It has been proved for many years that exposure to particles, in an occupational setting, could be linked with the onset of lung diseases, such as pneumoconiosis, chronic obstructive pulmonary disease (COPD), and mesotelioma and lung cancer. Nanoparticles, due to their unique characteristics including; small size, shape, high surface area, charge, chemical properties, solubility, and degree of agglomeration can cross cell boundaries or pass directly from the lungs into the blood stream and ultimately reach to all of the organs in the body. This is the reason why they may pose higher risk than the same mass and material of larger particles. Moreover, biodegradation of nanoparticles by some kinds of fungi (like wood decay fungi) may result in metabolites which may be toxic to microorganisms under aerobic and anaerobic conditions. Bacteria and living cells can take up nanoparticles, providing the basis for potential bioaccumulation in the food chain. Considering Iran's prominent position in nanotechnologies and fast-growing in research and industrial activities, controlling nanoparticles related HSE risks should be highly considered. In general, there are three main approaches to risk and exposure control: engineering techniques, administrative means and personal protective equipments. These complementary approaches especially engineering techniques should be considered starting with the design stage of an industrial process. Administrative means of control constitute an additional approach when the other methods have not achieved the expected control levels. Administrative means of control must never substitute for engineering techniques, which always be performed according to standard practices. In some situations, due to insufficiently advanced technology and prohibitive costs,

Journal of Physics: Conference Series **170** (2009) 012037 doi:10.1088/1742-6596/170/1/012037

engineering measures can not be implemented. In these situations, performing administrative means of control constitute other ways of limiting the occupational exposure risks. Accordingly, to minimize the risks from known and unknown health, safety and environment hazards in research and occupational settings of the country, guideline for safe handling, use and disposal of nanoparticles has provided.

1. Introduction

Nanotechnology is defined as: research and technology development at the atomic, molecular, or macromolecular levels using a length scale of approximately one to one hundred nanometers in any dimension; the creation and use of structures, devices and systems that have novel properties and functions because of their small size; and the ability to control or manipulate matter on an atomic scale [1]. For biomedical application, this definition has been expanded to include particles greater than 100 nm, such as liposomes, in order to encompass particle sizes that take advantage of anatomical considerations, such as vascular gaps surrounding tumours [4]. Nanotechnologies, allow us to build new materials, atom by atom. This often endows the materials with properties that are very different from ordinary materials. At the *nano* level, the behaviour of particles is dominated by quantal effects. The particles may be confined to a small structure, distributed over large surfaces or demonstrate an entire series of unique phenomena and properties not encountered in larger materials [2].

The field of nanotechnology is advancing rapidly and will likely revolutionize a broad range of consumer, medical, and industrial sectors. As with any new technology, we are faced with many unknowns; all of which raise questions concerning occupational safety and health. Nanotechnology also has the potential to improve the environment, both through direct applications of nanomaterials to detect, prevent, and remove pollutants, as well as indirectly by using nanotechnology to design cleaner industrial processes and create environmentally responsible products. However, there are unanswered questions about the impacts of nanomaterials and nanoproducts on human health and the environment [1].

Currently, a broad range of processes have been influenced by nanotechnology which will pose likely higher exposure potential to workforces in the nanotechnology occupational settings than consumers of final products. Considering inadequate information, until the results from research studies can fully elucidate the characteristics of nanoparticles that may potentially pose a health risk, precautionary measures are warranted [7, 8]. Concerning Iran's prominent position in nanotechnologies and fast-growing in research and industrial activities, minimizing the risks from known and unknown health, safety and environment hazards of handling, use and disposal of nanoparticles in nanotechnology workplaces (research laboratories and industrial firms) is considered as a priority. To fulfil the main objective, a document titled *"Guideline for Safe Handling, Use and Disposal of Nanoparticles"* has been issued to address the potential health, safety and environment hazards of nanomaterials and available good practices for mitigating the risks. This guideline will help the decision makers to:

- develop site-specific controls that will protect workers and the environment,
- offer reasonable guidance for managing the uncertainty associated with nanomaterials whose hazards have not been determined and reducing to an acceptable level the risk of worker injury, worker ill-health and negative environmental impacts and
- promote consistency in policy and procedures between the nanotechnology workplaces.

During reviewing literatures, it was identified, among the 30 industrialized countries of the organization for economic co-operation and development (OECD), that United States, England, Germany, European commission and Australia have developed good practices documents in the safety of manufactured nanomaterials [13]. Following the review of the documents, this paper attempts to address briefly the good practices which could be suitable and effective in reducing the risks.

2. Nanomaterials potential health, safety and environmental hazards

Potential routes of nanoparticle exposure include inhalation, dermal, oral, and in the case of biomedical applications, parenteral (Figure 1) [4]. Uptakes of nanoparticles by inhalation or ingestion are likely to be the major route in terrestrial organisms [3, 5, 10]. Discrete nanoparticles are deposited in the lungs to a greater extent than larger respirable particles and deposition increases with exercise due to increase in breathing rate and change from nasal to mouth breathing and among persons with existing lung diseases or conditions. These nanoparticles may also enter the bloodstream from the lungs and translocate to other organs like brain [3, 6, 8, 9, 10]. This is the reason why they may pose higher risk than the same mass and material of larger particles.

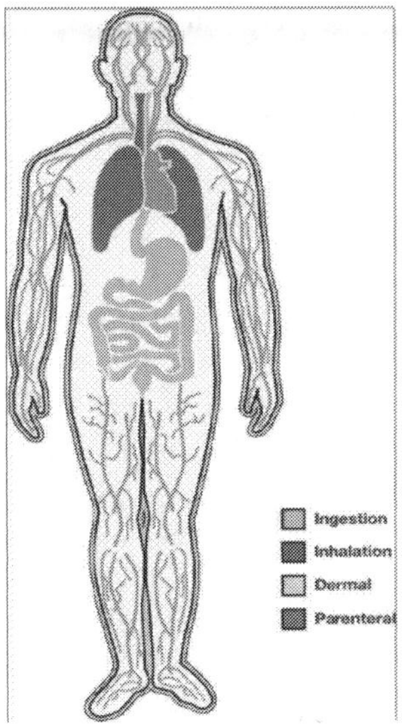

Figure 1. Potential routes of nanoparticle exposure [4].

According to the researches no particle with an aerodynamic diameter of 1 nm, or 0.001 micrometre, reaches the alveoli, while 80% are deposited in the nose and pharynx. The other 20% are deposited in the tracheobronchial region. At this size, retention of inhaled nanoparticles is nearly 100% [2, 6]. For particles larger than 5 nm, deposition is predominantly in the alveolar region of the lungs. The deposition fraction of inhaled nanoparticles is greater in the alveolar and tracheo-bronchial regions of human lungs, compared to the larger-diameter inhaled particles. Once deposited, nanoparticles may also remain in the lungs longer than larger particles, due to decreased clearance and increased retention of nanoparticles. Particle deposition in these regions may be important in the development of airways diseases, such as chronic obstructive pulmonary disease (COPD) or asthma [6]. Studies also support a direct role for inhaled nanoparticles in systemic disease, such as cardiovascular disease. For example, CNTs have been shown to induce platelet aggregation in vitro and enhance thrombosis in vivo [4].

Since the late 1980s, toxicological evidence has been emerging indicating that the health effects associated with inhaling nanoaerosols may not be closely associated with particle mass. However, current research indicates that particle size, surface area, and surface chemistry (or activity) may be more important metrics than mass and bulk chemistry [6, 7, 9]. The toxicity and health risk of nanoparticles may also be a factor of the properties like agglomeration state, size distribution, shape, biopersistence/durability/solubility, porosity, physical properties, crystal structure/crystallinity and trace impurities/contaminants [7].

There is some evidence that dermal exposure to nanoparticles may lead to direct penetration of nanoparticles into the epidermis and possibly beyond into the blood stream [3, 9, 11, 25]. However, at this time, it is not known if skin penetration of nanoparticles would result in adverse effects as these studies have not been reported in animal models [3]. The limited *in vivo* studies that have been conducted to address the issue of cutaneous toxicity have identified only mild irritation as an adverse response to topical nanomaterial application. Nanoscale metal oxides, for example, are currently used in commercially available sunscreens, and have undergone extensive animal and clinical testing to fulfil regulatory requirements. These studies found minimal irritancy potential, and no evidence of photo-irritation, sensitization, or photo-sensitization [4]. In contrast, some recent studies have revealed some cutaneous penetration by ultrafine beryllium particles and the formation of cutaneous nodules [2].

Ingestion is another route whereby nanoparticles may enter the body. Ingested particles smaller than 20 μm (20 000 nm) can pass through the intestinal barrier and enter the bloodstream. Ingestion can occur from unintentional hand to mouth transfer of materials and direct ingestion of contaminated drinking water or particles absorbed on vegetables or other foodstuffs [2, 3, 5]. Ingestion may also accompany inhalation exposure because particles that are cleared from the respiratory tract via the mucociliary escalator may be swallowed [3, 8]. Little is known about possible adverse effects from the ingestion of nanoparticles.

Although insufficient information exists to predict the fire and explosion risk associated with nanoparticles, the general trend is for the violence of the dust explosion and the ease of ignition to increase as the particle size decreases or specific area increases. Decreasing the particle size of combustible materials can reduce minimum ignition energy and increase combustion potential and combustion rate, leading to the possibility of relatively inert materials becoming highly combustible. Depending on their composition and structure, some nanomaterials may initiate catalytic reactions and increase their fire and explosion potential that would not otherwise be anticipated from their chemical composition alone [2, 3, 8, 12].

Moreover, nanomaterials released to soil can be strongly sorbed to soil due to their high surface areas and therefore be immobile. Bacteria and living cells can take up nanosized particles, providing the basis for potential bioaccumulation in the food chain. Certain nanomaterials are being designed for release as reactants in the environment, and therefore are expected to undergo chemical transformation. One example of this is iron (Fe0). Biodegradation of nanoparticles by some kinds of fungi (like wood decay fungi) may result in metabolites which may be toxic to microorganisms under aerobic and anaerobic conditions [1].

3. Exposure assessment

Particle sampling and measurement is extremely helpful in understanding exposure and risk in workplace scenarios [8]. In general, personal sampling is preferred to ensure an accurate representation of the worker's exposure, whereas area samples (*e.g*, size-fractionated aerosol samples) and real-time (direct-reading) exposure measurements may be more useful for evaluating the need for improvement of engineering controls and work practices. Several of the instruments and techniques like condensation particle counters (CPC), Electron microscopic analysis (SEM and TEM), Scanning or Stepped Mobility Particle Sizer (SMPS), Electrical Low Pressure Impactor (ELPI), Diffusion Charger and Tapered Element Oscillating Microbalance (TEOM) are readily available to measure or estimate directly or indirectly different metrics as mass, number and surface area of nanoparticles [2, 3, 6, 7, 8, 11, 28]. Regardless of the metric and method selected for exposure monitoring, it is critical that measurements must be conducted before production or processing of a nanomaterial to obtain background exposure data [3, 7, 11]. Currently, there is not one sampling method that can be used to characterize exposure to nanosized aerosols. Therefore, any attempt to characterize workplace exposure to nanoparticles must involve a multifaceted approach incorporating many of the sampling techniques mentioned above, all relevant characteristics of nanoparticle exposure being measured.

By using a combination of these techniques, an assessment of worker exposure to nanoparticles can be conducted. This approach will allow a determination of the presence and identification of nanoparticles and the characterization of the important aerosol metrics. However, since this approach relies primarily on static or area sampling some uncertainty will exist in estimating worker exposures [3, 6, 7, 8]. Of the three primary physical exposure metrics (mass, surface area and number), there is strong evidence to suggest that occupational nanoaerosols should be monitored with respect to surface area [2, 7].

4. Occupational Exposures

Few workplace measurement data exist on airborne exposure to nanoparticles that are purposely produced and not incidental to an industrial process. In general, it is likely that processes generating nanomaterials in the gas phase, or using or producing nanomaterials as powders or slurries/suspensions/solutions (*i.e.* in liquid media) pose the greatest risk for releasing nanoparticles. In addition, maintenance on production systems (including cleaning and disposal of materials from dust collection systems) is likely to result in exposure to nanoparticles if it involves disturbing deposited nanomaterial. Exposures associated with waste streams containing nanomaterials may also occur [3]. Table 1 summarises the most common routes and potential sources of occupational exposure to a nanomaterial [5, 7, 11]. Clean In Place (CIP) technologies may be used to eliminate the opening of process vessels and reduce the potential for unbound engineered nanoparticle (UNP) releases during cleaning operations. Minimization of maintenance activities by task planning (identification of required tools, replacement parts, etc) may help reduce exposure time by shortening maintenance times [7].

In processes involving high pressure (*e.g.* supercritical fluid techniques), or with high energy mechanical forces, exposures could occur in the case of failure of sealing of the reactor or the mills [10]. The potential for exposure to nanoparticles also exists when transferring nanomaterials within or outside the work area [7]. Factors affecting exposure to UNP include (a) the amount of material being used; (b) whether the material can be easily dispersed (in the case of a powder) or from airborne sprays or droplets (in the case of suspensions), or is fixed on or within a matrix (and generally do not present an exposure risk); (c) the degree of containment; (d) duration of use or presence in exposure areas; and (e) state of agglomeration or aggregation [7,14].

Table 1. Potential sources of occupational exposure to nanomaterials for various synthesis methods [5, 7, 11].

Process Synthesis	Particle Formation	Exposure Source or Worker Activity	Primary Exposure Route
Gas Phase	In air	Direct leakage from reactor, especially if the reactor is operated at positive pressure.	Inhalation
		Product recovery from bag filters in reactors.	Inhalation / Dermal
		Processing and packaging of dry powder.	Inhalation / Dermal
		Equipment cleaning/maintenance (including reactor evacuation and spent filters).	Dermal (and inhalation during reactor evacuation)
Vapour Deposition	On substrate	Product recovery from reactor / dry contamination of workplace.	Inhalation
		Processing and packaging of dry powder.	Inhalation / Dermal
		Equipment cleaning/maintenance (including reactor evacuation).	Dermal (and inhalation During reactor evacuation)
Colloidal	Liquid suspension	If liquid suspension is processed into a powder, potential exposure during spray drying to create a powder, and the processing and packaging of the dry powder.	Inhalation / Dermal
		Equipment cleaning/maintenance.	Dermal
Attrition	Liquid suspension	If liquid suspension is processed into a powder, potential exposure during spray drying to create a powder, and the processing and packaging of the dry powder.	Dermal
		Equipment cleaning/maintenance.	Dermal

5. Transportation

All shipments of nanomaterials, regardless of whether they meet the definition for hazardous materials or not, should be consistently packaged using the equivalent of a DOT-certified Packing Group I (PG I) container. Any nanomaterial being shipped by air that meets the definition of dangerous goods according to the International Civil Aviation Organization (ICAO) must be packaged, marked, labelled, and shipped, with an accompanying properly prepared dangerous goods declaration, in accordance with the ICAO technical instructions. Therefore, nanomaterials must be packaged, marked, labelled, shipping papers prepared and shipped in accordance with 49 CFR 100 to 185. It should have a secondary seal, such as tape seal, or a wire tie to prevent a removable closure from inadvertently opening during transport. The outer package should be filled with shock absorbing material that can (a) Protect the inner sample container(s) from damage (b) Absorb liquids that might leak from the inner container(s) during normal events in transport [15, 16].

6. Nanomaterials spills

No specific guidance is currently available on cleaning up nanomaterial spills or contaminated surfaces. Available standard approaches to cleaning up powder and liquid spills include the use of HEPA-filtered vacuum cleaners, wetting powders down, using dampened cloths to wipe up powders and applying absorbent materials/liquid traps. Damp cleaning methods with soaps or cleaning oils is preferred [3, 7, 17]. To deal with spills and contaminated surfaces current good practices should be modified or new standard operation procedures (SOPs) provided. These should involve small (typically involving less than 5 mg/ml of material) of nanoparticle-containing powder or solutions and larger spills [18]. Using an absorbent walk-off mat where the clean-up personnel will exit the access controlled area, barriers to minimize air currents across the surface affected by the spill and providing safety equipments such as eyewash fountains, first aid kits, safety showers, multi-purpose fire extinguisher (ABC) and spill kits are also recommended [8, 15, 19].

Currently, portable peristaltic pumps are used by few organisations to transfer liquid to waste containers in order to prevent potential spills and reduce aerosolisation of the material. These pumps, because they work on positive displacement, are less prone to producing aerosols as opposed to conventional high pressure pumps [20].

7. Labelling

Currently, there is no generic labelling requirement for either manufactured nanoparticles (MNPs) and products containing manufactured nanoparticles (PCMNPs) used by both suppliers and users. Labelling and specifications are important tools for addressing possible adverse effects of nanomaterials on health and the environment. It is recommended that the term *nano* should only be used on a product label if the product does in fact contain manufactured nanoscale entities or produces a nano-enabled effect. Labelling is recommended for:

- MNPs,
- PCMNPs, except where the nanoparticle component of the product is intimately bound and could not be released under reasonable and foreseeable conditions of use or disposal,
- PCMNPs which are components of complex systems (*e.g.* a vehicle, mobile phone or game console), which could be expected to release MNPs under reasonable and foreseeable conditions of use or disposal,
- by-products, for which MNPs are generated as by-products.

Additional to that required by any legal obligation or known risks of the product, it is recommended that content of labels for PCMNPs or MNPs provide required information for consumer, professional and business-to-business uses. These information including a list of ingredients, technical specification, claims of intellectual property or indications of compliance with specific standards, instructions for handling, maintenance, cleaning, storage or disposal of the product, first aid or other medical treatment, name, address and contact details of the producer, identification data that can be used to trace the product and minimum professional competence. This should be advised in instructions and, wherever possible, on labels permanently attached to the product itself, or where that is impractical, on packaging in which the product is intended to be kept by the consumer. In the event that this is not possible, such information should be combined with accompanying instructions [21]. Figures 2 and 3 show minimum information appeared on inner packaging label for dry particulates and non-particulates.

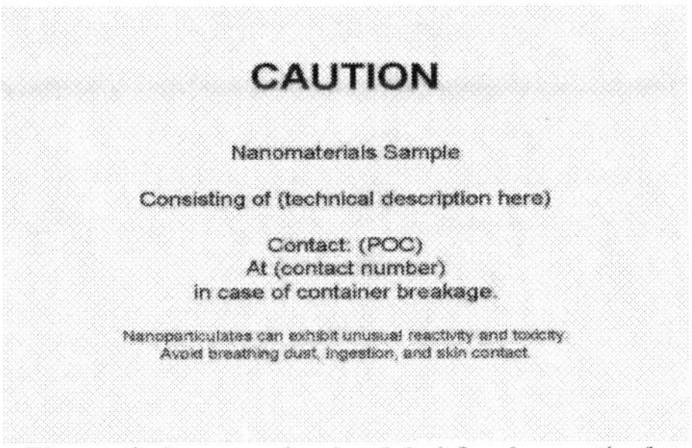

Figure 2. Inner packaging label for dry particulates [16].

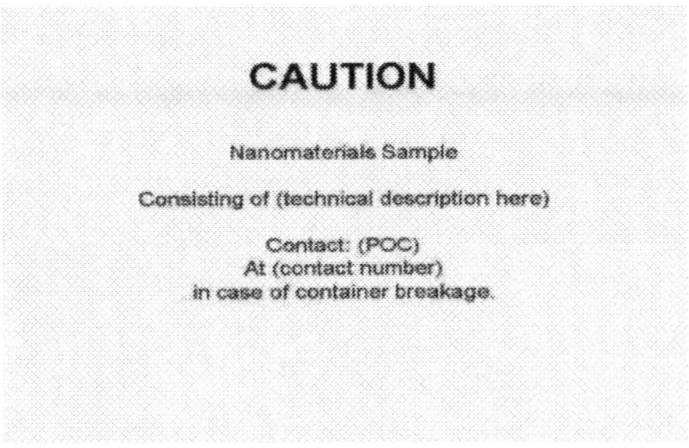

Figure 3. Inner packaging label for non-particulates [16].

8. Exposure Control Strategies

8.1. Engineering controls

Exposure to nanoparticles should be prevented, preferably by avoiding so far as is reasonably practicable the use of a hazardous substance by substituting a substance or process which eliminates or reduces the risks to health. If, however, this is not possible then exposure should be controlled by applying protection measures appropriate to the activity and consistent with the hierarchy of control [8]. Many harmful chemicals and processes are currently used as a matter of tradition although less harmful alternatives exist. Substitution of less harmful chemicals and processes for particularly hazardous ones has the advantage of completely removing a hazard from a workplace so that the potential for exposure is gone. Substitution is a complex process that requires research and experimentation. A secondary benefit of substitution can be reduction or elimination of waste and the costs associated with its disposal, regulatory compliance, liability, and environmental impact. It is important to compare the chemicals being considered for substitution not only for toxicity but also routes of entry, vapour pressure, flammability, particle size, safe disposal, etc. A slurry/brick/waxy material or a material encapsulated in a dissolvable plastic bag or gelatine capsule instead of a dry powder, flow coating or dipping instead of spraying, water blasting instead of abrasive blasting are some examples of possible substitutions [22]. Until more information become available, cautionary measures should be acted.

Consideration should be given to the following prudential limits when selecting analogous precautionary measures:

- measures should be consistent in scope and nature with comparable measures from comparable areas;
- measures should be proportional to the chosen level of protection and the scope of the harm (*e.g.*, severity, irreversibility, uniqueness, numbers affected, temporal and spatial extent);
- measures should be chosen with due consideration of costs and benefits (cost effective);
- and measures (and underlying assumptions) should be continuously reviewed in light of new information and understanding [7].

For most processes and job tasks, the control of airborne exposure to nanoaerosols can be accomplished using a wide variety of engineering control techniques similar to those used in reducing exposure to general aerosols. Engineering control techniques such as source enclosure (*i.e.*, isolating the generation source from the worker) and local exhaust ventilation systems should be effective for capturing airborne nanoparticles. Current knowledge indicates that a well-designed exhaust ventilation system with a high-efficiency particulate air (HEPA) filter should effectively remove nanoparticles [3, 7].

Employees may be isolated from hazardous operations, processes, equipment, or environments by distance, by physical separation, barriers, control rooms, isolation booths, closed systems or containers (*e.g*, sealed reactor vessels, closed storage containers or vessels, pumps enclosures, valve isolation, glove boxes) and by capture ventilation. As a further precaution, regulated areas can be established around enclosed operations with access only to a limited number of essential employees. Air locks with interlocked doors and computerized card readers and doors with alarms add an extra measure of isolation and prevent unauthorized entry [2, 7, 22]. According to an international survey, respondents were asked whether "nano-specific" facility design and engineering controls were used to safely manage worker exposure. Furthermore, respondents were asked whether the organization utilized cleanrooms, fume hoods, biological safety cabinets, laminar flow clean benches, glove boxes, glove bags, a closed piping system, pressure differentials (negative or positive), isolated heating, ventilation, and air conditioning (HVAC) systems, or other controls specifically for handling nanomaterials. Overall reports of engineering controls are reported in Figure 4.

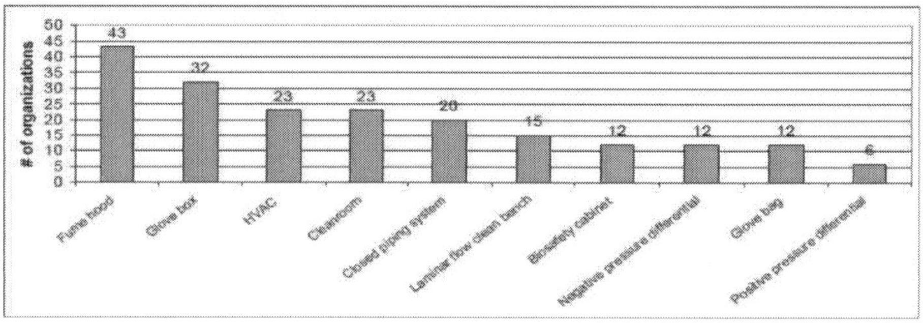

Figure 4. Reports of "nano-specific" facility design and engineering controls [23].

Fume hoods were the most widely reported engineering control. The results of a similar survey, performed by Conti *et al.* indicated that again fume hoods had highest use compared to other engineering controls [27]. Fume hoods were used with a variety of materials and phases but the highest usage was among organizations that worked with solutions, which could be an indication that fume hoods were used more as a barrier of protection against harmful vapours than nanomaterials. Fume hoods were less likely to be used when the nanomaterial was in a dry powder form. This may have been due to the potential loss of dry powder form material and the risk of inhalation stemming from air turbulence generated by the fume hood exhaust system. Most reports of fume hood use were associated with the handling of nanopowders, carbon nanotubes, dispersions, and fullerenes. European and North American organizations reported the highest (82%, 72% respectively) and organizations from Asia reported the lowest (52%) use of fume hoods.

The survey results indicated that again North American and European organizations had the highest use of glove boxes in their nanomaterial operations (64% and 45% respectively) especially in working with nanopowders and carbon nanotubes. This result is in contrast to Asian countries, where only 36% reported using glove boxes. Those organizations working with colloidal dispersions were the least likely to report using a glove box. Nearly all organizations working with nanowires, nanocrystals and carbon black reported using glove boxes [23]. General ventilation is often ineffective and, in some situations, recirculation of a fraction of the air may be prohibited by the regulations in force [2, 22].

Nanoparticle behaviour varies greatly depending on the dimensions and degree of agglomeration. Airborne, non-agglomerated nanoparticles behave much like gases, thus allowing rapid diffusion over long distances. For this reason, the engineering control systems, such as enclosure and ventilation, must be designed according to the gaseous and particulate properties of the nanoparticles [2, 11, 24, 25]. Wet scrubbers or electrostatic precipitators in the final stage of filtration are recommended. This capture principle, involving electrostatic attraction, is particularly effective for very fine particles [2].

Biological safety cabinets (BSCs) are designed to protect the operator, the laboratory environment and work materials from exposure to infectious aerosols and splashes that may be generated when manipulating materials containing infectious agents, such as primary cultures, stocks and diagnostic specimens [26]. Based on the results of the previously mentioned survey higher instances of reported use of biological safety cabinets came from organizations working with a range of smaller amounts of nanomaterials at any given time. One organization indicated that their biological safety cabinet, type

2b2, did not recirculate air like conventional biological safety cabinets. The air was HEPA-filtered before being exhausted, thus preventing the emission of nanomaterials into the environment. While this type of cabinet was available commercially, it did not appear to be widely utilized for nanomaterial applications.

The cabinets were used by organizations working with a range of smaller amounts, particularly nanopowders in powder or suspended form or colloidal dispersions. North American organizations reported marginally higher use of this control compared to Asian or European organizations. In general, compared to North American organizations, Asian organizations used fewer "high–end" engineering controls. Some respondents stated that, when handling dry powders, fume hood exhaust fans would be "off" to prevent nanomaterials loss. That was an important finding because several other organizations reported using certain engineering controls less to protect workers from exposure than to prevent the loss of the nanomaterial or to protect the material from the ambient environment [23, 27].

8.2. Administrative controls

Administrative means of control constitute an additional approach when the other methods have not achieved the expected control levels. In these situations, reduction of work periods, modification of work practices, personal hygiene measures, housekeeping and preventive maintenance constitute other ways of limiting the occupational exposure risks [2, 3, 7]. As examples of good practices, hand-washing , showering, changing and cleaning clothes facilities should be provided to prevent the inadvertent contamination of other areas (including take-home) caused by the transfer of nanoparticles on clothing and skin. The storage and consumption of food or beverages and smoking in workplaces should be prevented where nanomaterials are handled. Dry mopping, sweeping, dusting, cleaning using compressed air or portable blowers or fans are prohibited. Work areas should be cleaned at the end of each work shift (at a minimum) using either a HEPA-filtered vacuum cleaner or wet wiping methods [2, 3, 7, 15].

It is also recommended to prepare written operating procedures, sufficient operational training, to conduct regular and timely inspection of process, manufacturing, operational and exposure control equipment and ancillary systems (including ventilation and filtration equipment), and regular and timely preventative and corrective maintenance and repair of such equipment [7]. It is generally emphasized to keep the number of potentially exposed worker as small as possible. Lastly, routine monitoring and health and medical surveillance should be carried out as needed [7, 8]. Currently, no specific measurable health effects have been uniquely associated with exposure to nanomaterials (other than those already associated with larger variants of the same materials) [8].

The survey results generally revealed that organizations working with nanomaterials use conventional chemical safety methods through the life-cycle of nanomaterials [20, 27]. Concerning this, developing and implementing a chemical hygiene plan satisfying the criteria in 29 CFR 1910.1450 has been required in some nanomaterial laboratories [15, 18, 29]. At this time, improving existing related laws and regulations (particularly laws and regulations for chemicals) or developing really needed new ones should be highly considered. Regulations such as Registration, Evaluation and Authorisation of Chemicals (REACH) program in the European Union and the Toxic Substances Control Act (TSCA) and RCRA in the U.S have good capabilities to regulate nanosubstances. The existing regulatory network will be modified if necessary. Here are a few examples of regulations applicable to nanoparticles and nanostructures that could be modified [2].

8.3. Personal protective equipments (PPE)

Nanoparticle exposure may often be attributable to the wearing of inadequate PPE [18]. Protective clothing that would typically be required for a wet-chemistry laboratory would be appropriate and could include but not limited to:

- closed-toed shoes made of a low permeability material,
- long pants without cuffs, a long-sleeved shirt,

- gauntlet-type gloves or nitrile gloves with extended sleeves,
- chemical splash goggle and Laboratory coats [15, 18].

Standard tests showed that non-woven fabrics, like high density polyethylene textile (Tyvek type), seems to be much more efficient against nanoparticles penetrations. Thus, it is advised to avoid the use of protective clothing made with cotton fabrics [2, 30]. Diffusion tests performed with nanoparticles showed that nanoparticles may penetrate through commercially available gloves. The tests revealed that gloves material is not the only criteria, elaboration process and thickness are major issues as well. As a result of this, use of at least two layers of gloves is advised [15, 30].

The use of respirators is often required when engineering and administrative controls do not adequately keep worker exposures to an airborne contaminant below a regulatory limit or an internal control target [2, 3]. Based on the previous survey results, just a little over half, indicated that employees used respiratory protection when working with nanomaterials. It was also revealed that: respirators were used at 100% of organizations that stated they worked in the chemicals sector and 93% of those in nanomaterials manufacturing; respirators were used by only 50% of the other business categories [20]. Intermediate protection is assured by using a PAPR (powered air-purifying respirator) with adequate APF (Assigned Protection Factors), which includes high-efficiency filtration and a pump supplying a full-face mask. In cases where high-efficiency dust filters are insufficient, airline respirators or self-contained breathing apparatus are necessary [2, 7, 14, 15].

Currently, there are no specific exposure limits for airborne exposures to engineered nanoparticles (except 0.1 mg/m^3 for ultrafine particles of TiO_2) although occupational exposure limits and guidelines (*e.g.*, OSHA, NIOSH, ACGIH) exist for larger particles of similar chemical composition. In determining the effectiveness of controls or the need for respirators, it would therefore be prudent to consider both the current exposure limits and guidelines (*e.g.*, PELs, RELs, TLVs) and the increase in surface area of the nanoparticles relative to that of particles for which the exposure limits or guides were developed [3]. It is believed that P-100, FFP3 and P3 cartridge-type respirators or respirators provide higher level of protection than others [8, 15, 16, 24, 27, 28].

8.4. Waste disposal

As mentioned above, many organizations working with nanomaterials use conventional chemical safety methods through the life-cycle of nanomaterials. It is also obvious that regulation in Europe regarding nanoparticles is based on existing laws and regulations for chemicals [2]. Due to these, at this time many organizations characterize all nanomaterial wastes as potentially hazardous wastes and use related regulations to manage their wastes [8, 20, 31]. In the UK, the List Of Waste Regulations 2005 (LOWR) provides a list of hazardous properties, including examples, such as H5 "Harmful substances" and H6 "Toxic substances". H5 would be a reasonable worst case assumption for many insoluble and soluble nanoparticle types but H6 would describe CMAR types (carcinogenetic, mutagenic, asthmagenic or a reproductive toxin) and would be a reasonable worst case assumption for carbon nanotubes or other nanomaterials having a fibrous nature [8].

In the U.S, subtitle C of RCRA covers the storage, transportation, treatment, disposal, and cleanup of hazardous wastes. Nanomaterials that meet one or more of the definitions of a hazardous waste (i.e., a waste that is specifically listed in the regulations and/or that exhibits a defining characteristic) potentially would be subject to subtitle C regulations. RCRA regulations set out several requirements for generators of hazardous wastes. Most notably, RCRA requirements for generators vary based on the amount of hazardous waste that they generate in a calendar year. Because nanoscale materials may present novel properties at comparatively small quantities, the current 100 kg annual threshold to qualify as a CESQG (conditionally exempt small quantity generators) may allow the on-site storage and management of nanomaterials for extensive periods of time. EPA may review whether to vary storage and management quantity thresholds based on the actual hazard posed by the nanomaterials rather than their quantity [32]. Moreover, nanoscale materials that meet the definition of "chemical substances" under the Toxic Substances Control Act (TSCA), but which are not on the TSCA Inventory, must be reported to EPA according to section 5(a) of the Act, which provides for pre-manufacture review [1]. It is concluded that despite EPA's sweeping powers to regulate hazardous waste management and its comprehensive regulatory framework, there are several areas of potential interest where EPA may wish to determine whether its current regulations will have unintended consequences when applied to nanoscale waste materials [32].

9. Conclusions

It has been currently known that some engineered nanomaterials will present new and unusual risks, but there is very little information on how these risks can be identified, assessed and controlled. In contrast, good occupational hygiene practices and existing knowledge on working with hazardous substances provide a useful basis for working safely with nanomaterials. But where existing knowledge fails, new research is needed to fill the gaps. Until more information become available nanomaterials should be considered as hazardous materials. Lastly, in the future, many types of nanoparticles may turn out to be of limited toxicity but precaution should be used until more is known.

References

[1] Morris J and Willis J 2007 "Nanotechnology White Paper", U.S.Environmental Protection Agency
[2] Ostiguy Claude *et al.* 2006 "Nanoparticles – Actual Knowledge about Occupational Health and Safety Risks and Prevention Measures" IRSST, report- 470, Montréal (Québec)
[3] U.S. National Institute for Occupational Safety and Health, Centers for Disease Control and Prevention, "Approaches to Safe Nanotechnology—An Information Exchange with NIOSH," July 2006.
[4] Stern S T and McNeil S E 2008 *Toxicological Science* **101(1)** 4
[5] University of New Hampshire, "Nanomaterials Safety Program," May 21, 2007.
[6] ISO/TR27628, "Workplace Atmospheres—Ultrafine, nanoparticle and nano-structured aerosols – Exposure characterization and assessment," First edition, Switzerland, 2007.
[7] E 2535 – "Standard Guide for Handling Unbound Engineered Nanoscale Particles in Occupational Settings," ASTM, USA, 2007.
[8] British Standard Institute (BSI), "Nanotechnologies – Part 2: Guide to safe handling and disposal of manufactured nanomaterials," PD 6699-2:2007, UK, 2007.
[9] Ostiguy Claude *et al.* 2006 "Health Effects of Nanoparticles" IRSST, report- 469, Montréal (Québec)
[10] European Commision, Health & Consumer Protection Diretorate, Scientific Committee on Emerging and Newly Identified Health Risks (SCENIHR), "The appropriateness of existing methodologies to assess the potential risks associated with engineered and adventitious products of nanotechnologies," 10 March 2006.
[11] Aitken R J, Creely K S and Tran C L 2004 "Nanoparticles: An Occupational Hygiene Review" HSE, UK
[12] Pritchard D K 2004 "Literature review – explosion hazards associated with nanopowders" Health & Safety Laboratory (HSL), UK
[13] Organisation for Economic Co-operation and Development (OECD), "Current Developments/Activities on the Safety of Manufactured Nanomaterials," OECD Environment, Health and Safety Publications Series on the Safety of Manufactured Nanomaterials, Berlin, Germany, 25-27 April 2007.
[14] Health and Safety Executive, "Nanotechnology – Horizons Scanning Information," Note No HSIN 1, UK, 2004.
[15] Department of Energy (DOE), Nanoscale Science Research Centers, "Approach to Nanomaterial ES&H," USA, Rev. 2 – June 2007.
[16] Stanford Linear Accelerator Center, "Nanomaterial Safety Plan," 4 August 2008.
[17] Gerritzen Gina *et al.* 2006 "Review of Safety Practices in the Nanotechnology Industry," Phase One Report: Current Knowledge and Practices Regarding Environmental Health and Safety in

the Nanotechnology Workplace, the International Council on Nanotechnology (ICON)
[18] Virginia Commonwealth University, Office of Environmental Health and safety, “Nanotechnology and Nanoparticles- Safe Working Practices Information,” 2007.
[19] Iowa State University, “Laboratory Safety Manual,” USA.
[20] Gerritzen Gina *et al.* 2006 "A Review of Current Practices in the Nanotechnology Industry" Phase two report: survey of current practices in the nanotechnology workplace, the International Council on Nanotechnology
[21] British Standard Institute (BSI), “Guidance on the Labeling of Manufactured Nanoparticles and Products Containing Manufactured Nanoparticles,” PAS 130, First published, UK, 2007.
[22] Senn Eileen *et al.* 2000 "Controlling Chemical Exposure" Concise guidance on 16 components of industrial hygiene controls, New Jersey Department of Health and Senior Services, Division of Epidemiology, Environmental and Occupational Health Service
[23] Gerritzen Gina *et al.* 2006 "A Review of Current Practices in the Nanotechnology Industry" Phase two report: survey of current practices in the nanotechnology workplace, the International Council on Nanotechnology
[24] Mark David, “Nanomaterials: a risk to health at work?” First International Symposium on Occupational Health Implications of Nanomaterials, UK, 12-14 October 2004.
[25] Texas A&M University, “Interim Guideline for Working Safely with Nanotechnology,” USA, 2005.
[26] World Health Organization, “Laboratory Biosafety Manual,” Third edition, Geneva, 2004.
[27] Conti Joseph A. *et al.* “Health and Safety Practices in the Nanomaterials Workplace: Results from an International Survey,” American Chemical Society, Environmental Science & Technology, 2008.
[28] VCI & BauA, “Guidance for handling and use of nanomaterials in the workplace,” Germany, 2007.
[29] U of MN Nanofabrication Center, “Laboratory Safety Plan,” USA, 2004.
[30] Golanski Luana, Guiot Arnaud and Tardif Francois "Efficiency of Fibrous Filters and Personal Protective Equipments against Nanoaerosols" Nanosafe, Dissemination report, January 2008.
[31] Massachusetts Institute of Technology (MIT), “Chemical Hygiene Plan,” Department of Aeronautics & Astronautics, USA, January 2008.
[32] American Bar Association, “RCRA Regulation of Wastes from the Production, Use and Disposal of Nanomaterials,” Section of Environment, Energy and Resources, USA, June 2006.

Nanosafe 2008: International Conference on Safe production and use of nanomaterials IOP Publishing
Journal of Physics: Conference Series **170** (2009) 012038 doi:10.1088/1742-6596/170/1/012038

The NanoCare project: a German initiative on health aspects of synthetic nanoparticles

Katja Nau[1] and Harald F. Krug[1, 2]

[1]Forschungszentrum Karlsruhe GmbH, Institute of Toxicology and Genetics, 76021 Karlsruhe, Germany

[2]Empa, Materials-Biology Interactions Laboratory, 9014 St. Gallen, Switzerland.

E-mail: katja.nau@itg.fzk.de

Abstract. Nanotechnology is increasingly considered to be the future technology. It will enable science and industry to provide new and better product solutions for the society. NanoCare is a German project, funded by the German Federal Ministry of Education and Research (BMBF), which aims to broaden knowledge about synthetic nanomaterials with regard to the potential impacts of nanomaterials on human health. 13 partners from industry, universities and research institutes are contributing their expertise to this partnership.
The work plan of the NanoCare project is composed of three different parts: (1) the generation, (2) the management, and (3) the transfer of knowledge. The production of synthetic nanoparticles, the subsequent analysis of primary particles, aggregates and agglomerates, as well as the behaviour in biological media and effects on biological systems are focused in the generation of knowledge. In addition to the production and characterization of new synthetic nanoparticles (metal oxides like zirconium dioxide or zinc oxide), titanium dioxide and Carbon Black will be established as reference materials. This enables the comparison of the results of all partners of our project. Various analytical methods for characterization will be applied, for example: transmission and scanning electron microscopy, inductive coupled plasma mass spectroscopy (ICP-MS), atomic absorption spectroscopy (AAS) and the Brunner-Edward-Teller method (BET). *In vitro* studies will systematically investigate biological mechanisms of action of nanoparticles and the dependency on their size, shape, zeta potential and other important properties. *In vitro* data will be complemented by *in vivo* studies. Another work package deals with the measurement of working place exposure and agglomerate stabilities. Established measurement devices and methods will be developed further in order to determine aerosols and nanoparticles directly at the workplace during ongoing work processes. The stabilities of the agglomerated nanoparticle powders are additionally investigated with three different methods to assess deagglomeration probabilities which also influence the possible exposure. Data created within the NanoCare project consortium will be interpreted together with information from literature and then published for the public in a data base on the World Wide Web (www.nanopartikel.info). Furthermore, the results will be presented and discussed with the interested public, politicians and non-governmental organizations (NGOs) at dialogue events. Together with two other BMBF-funded projects (INOS, TRACER) NanoCare will help to standardize analytical procedures and will substantially increase knowledge about the biological activities of nanomaterials.

Organization of the NanoCare project

NanoCare is organized in different work packages (WP, Figure 1). An internal and an external council act as advisors. The objectives of the WP 1-5 intend to produce (primary) data for work place based health effects of nanoparticles. These data will be interpreted together with information from literature and then published for the public in a data base on the World Wide Web by WP 6. Furthermore, the results will be presented and discussed with the interested public, politicians and non-governmental organizations (NGOs) at dialogue events (WP 7).

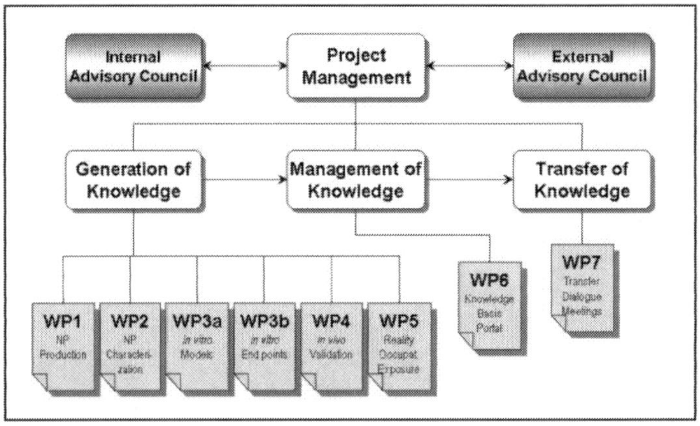

Figure 1. Organization chart of the NanoCare project.

WPs 1 & 2 deal with the identification, preparation, standardization and characterization of nanoparticles. The production of synthetic nanoparticles is not new and numerous particles are already used in different applications. Nano-scaled Titanium dioxide and Carbon Black for example are well characterized. For that reason these particles serve as reference materials within the NanoCare project and will be compared with other novel and/or modified synthetic particles. The industrial partners of the project chose and produced inorganic and hardly soluble materials (Table 1). The control and adjustability of surface chemistry during the production process of these novel nanoparticles results in specific characteristics and opens new applications.

Table 1. Particles used within the NanoCare project in alphabetical order.

Material
AlOOH (Boehmite)
$BaSO_4$
CeO_2
$SrCO_3$
ZnO
ZrO_2 mixed oxides (Ti-Al-Zr, Ti-Zr)
TiO_2 (reference material)
Carbon Black (reference material)

Methods of synthesis, source materials and modifications are equally defined to ensure that all project partners work with the same material, hence the comparability of the results is guaranteed. Therefore, methods and measuring equipments will be standardized. Importantly, agglomeration- and aggregation-behaviour of the primary particles as well as the stability of the agglomerates and aggregates will be considered, both in aerosols and in solution. To assess the risk potential of nanoparticles the physico-chemical characteristics as well as the behaviour in physiological media and biological test systems will be analyzed.

WP 3 and WP 4 deal with the biological effects. In WP3 the toxicity of synthetic nanoparticles is investigated in various cell-based *in vitro* systems. *In vitro* experiments are carried out with cell cultures of different types and origins, with different methods of analysis and with various particles to enable a broad evaluation of the adequate test system. For this purpose, fifteen cell lines representing the organs lung, skin and gastrointestinal tract as well as cells of the immune system were selected. Cell biological and toxicological methods are applied and, if necessary, adapted to the needs of particle testing. The aim is the development of standard operation procedures (SOPs) for toxicological *in vitro* particle testing. An SOP dealing with the preparation of suspensions of nanoscaled metal oxides for biological testing has already been published [1]. A second SOP concerning phagocytes is published on our web page [www.nanopartikel.info/methoden.html].

Other main aspects of this work package are the investigations of the cellular uptake and the crossing of the pulmonary air-blood-barrier by nanoparticles. An additional aim is the evaluation and pre-validation of *in vitro* cell systems as an alternative for animal experiments. Thereby, the localization of particles at or in cells is investigated with different nano-biophysical methods like AFM/FS. Furthermore, the *in vitro vector model*, which is based on primary rat macrophages and mimics some of the elemental cell effects related to fine dust toxicity, is adapted to its use with nanomaterials [2]. With this model system effects on the cellular metabolism, stimulatory effects and particle-induced ROS secretion, can be analyzed simultaneously and a biological screening and monitoring of the potential hazard of nanomaterials is possible. This represents a helpful tool to correlate *in vitro* data with results from *in vivo* animal experiments.

Animal studies were carried out within the work package 4. In this WP *in vivo* inhalation studies (5 days of inhalation and 3 weeks observation) as well as 28-day subacute inhalation toxicity studies (inhalation: 6 h/d, 5 day/week on 4 consecutive weeks) with selected particles are performed in rats to analyze pulmonary toxicity. To date, titanium dioxide, zinc oxide, zirconium oxide, cerium oxide and carbon black were used in the short term animal studies, while AlOOH was investigated in the subacute inhalation study. Technical procedures of atmosphere generation and determination of particle size were established and standardized. There is evidence that nanomaterials show a tendency to form agglomerates or aggregates in the test atmosphere and that only a small part of the particles is nanoscaled [3]. An SOP concerning this short-term inhalation study for testing of nanomaterials has already been published [4].

WP5 "Reality – Exposure measurements at workplaces" is divided into three parts: (1) stabilities of agglomerates under mechanical stress and modelling of physical particle dynamics; (2) development and standardization of measurement devices and methods; (3) measurement and characterization of possible exposure to aerosols and nanoparticles directly at the workplace during ongoing work processes.

First results of the dusting experiments (with low mechanical stress) show that there is a release of nanoscaled particles subjected to the particle powder [unpublished data]. Several SOPs concerning the activities in this work packages are in preparation.

WP6 & 7: Knowledge transfer

The third pillar of the project encompasses the research, compilation and interpretation of published data from the literature and internal data of the toxicity of nanomaterials. This information will be addressed to the public and published on the web site www.nanopartikel.info (in German language). We established this "knowledge base" (Wissensbasis) and focus on the materials analyzed within the NanoCare project. We also aim to answer to frequently asked questions and give the opportunity to ask an expert.

The main concern of the NanoCare project focuses on the allocation of a competent and coherent knowledge base about chances and security aspects of nanomaterials for stakeholder, politicians, NGOs and interested layman. The organization and realization of two stakeholder dialogs (2007 and

2009) and three citizen dialogs (2008 in Hamburg, Munich and Dresden) is a further important part of this work package.

We prepare high quality results due to the manifold competences of the NanoCare partners. All together we work on the basis of an objective and targeting discussion about potential chances and risks of nanomaterials. More than 70 NanoCare members contribute to this aim coordinated by a project leader and his management assistant at the research center Karlsruhe.

Taken together, the efforts of the NanoCare project are to ensure that new scientific insights into the effects of synthetic nanoparticles on human health are efficiently and adequately communicated to the public.

Acknowledgements:
We would like to thank all the members of the NanoCare project for support and discussion. The project is funded by the German Federal Ministry of Education and Research BMBF (NanoCare; Förderkennzeichen 03X0021A).

References

[1] Schulze C, Kroll A, Lehr C M, Schäfer U, Becker K, Schnekenburger J, Schulze Isfort C, Landsiedel R and Wohlleben W 2008 *Nanotoxicology* **2(2)** 51
[2] Bruch J, Rehn S, Rehn B, Borm P J and Fubini B 2004 *Int. J. Hyg. Environ. Health* **207(3)** 203
[3] Ma-Hock L, Gamer A O, Landsiedel R, Leibold E, Frechen T, Sens B, Linsenbuehler M and van Ravenzwaay B 2007 *Inhal. Toxicol.* **19(10)** 833
[4] Ma-Hock L, Burkhardt S, Strauss V, Gamer A O, Wiench K and Landsiedel R 2009 *Inhal Toxicol.* **21(2)** 102

Nanosafe 2008: International Conference on Safe production and use of nanomaterials IOP Publishing
Journal of Physics: Conference Series **170** (2009) 012039 doi:10.1088/1742-6596/170/1/012039

Development of methodology for alternative testing strategies for the assessment of the toxicological profile of nanoparticles used in medical diagnostics. NanoTEST – EC FP7 project

Maria Dusinska[1], Lise Maria Fjellsbo[1], Eldbjorg Heimstad[1], Mikael Harju[1], Alena Bartonova[1], Lang Tran[2], Lucienne Juillerat-Jeanneret[3], Blanka Halamoda[3], Francelyne Marano[4], Sonja Boland[4], Margaret Saunders[5], Laura Cartwright[5], Sara Carreira[5], Susan Thawley[5], Maurice Whelan[6], Christoph Klein[6], Christos Housiadas[7], Katarina Volkovova[8], Jana Tulinska[8], Milan Beno[8], Katarina Sebekova[8], Lisbeth E. Knudsen[9], Tina Mose[9], José V. Castell[10,11], Maya R. Vilà[10], Lourdes Gombau[10], Mark Jepson[12], Giulio Pojana[13] and Antonio Marcomini[13]

[1]Norwegian Institute for Air Research, Norway

[2]Institute of Occupational Medicine, United Kingdom

[3]Centre Hospitalier Universitaire Vaudois, Switzerland

[4]University Paris Diderot Paris7, France

[5]University Hospitals Bristol NHS Foundation Trust, United Kingdom

[6]EC – Directorate General Joint Research Centre, Italy

[7] National Centre for Scientific Research Demokritos, Greece

[8]Slovak Medical University, Slovakia

[9]University of Copenhagen, Denmark

[10]Advanced In Vitro Cell Technolgoies SL, Spain

[11] Hospital Universitario La Fe. Valencia, Spain

[12]Department of Biochemistry, University of Bristol

[13]University Ca' Foscari of Venice, Italy

Journal of Physics: Conference Series **170** (2009) 012039 doi:10.1088/1742-6596/170/1/012039

E-mail: dusinska@szu.sk

Abstract. Nanoparticles (NPs) have unique, potentially beneficial properties, but their possible impact on human health is still not known. The area of nanomedicine brings humans into direct contact with NPs and it is essential for both public confidence and the nanotech companies that appropriate risk assessments are undertaken in relation to health and safety. There is a pressing need to understand how engineered NPs can interact with the human body following exposure. The FP7 project NanoTEST (www.nanotest-fp7.eu) addresses these requirements in relation to the toxicological profile of NPs used in medical diagnostics.

1. Introduction

The overall aim of this project is to develop alternative testing strategies and high-throughput toxicity-testing protocols using *in vitro* and *in silico* methods which are essential for the risk assessment of these NPs. To be able to achieve this ambitious goal, the specific aims of NanoTEST are: a) to carry out a detailed characterization of selected NPs in order to define the major physico-chemical properties b) to study specific and nonspecific interactions of NP with molecules, cells and organs and to develop *in vitro* methods which can identify the toxicological potential of NPs; c) to validate *in vitro* findings in short-term *in vivo* models, to study manifestation of particle effects in animals and humans, and to assess individual susceptibility in the response to NPs; d) to perform both Structure-Activity modeling and physiologically-based pharmacokinetic (PBPK) modeling of NP; e) to adapt the most advanced and promising assays for high-throughput automated systems and to prepare them for validation by the European Centre for the Validation of Alternative Methods (ECVAM).

2. Selection of Nanoparticles and their characterization

The NanoTEST project will focus on the NPs used either in medical applications or as control or benchmark NPs: a) TiO2,as benchmark; b) paramagnetic metal oxides: Fe_3O_4, and rare earth oxides, which are applied for contrast enhancement as well as for cancer treatment in various biomedical applications; c) Metal fullerenes, which contain paramagnetic metals in the fullerene cage, proposed for use in MRI, d) polymeric materials utilised for the delivery of macromolecules (polylactic glycolic acid, PLGA, chitosan, hyaluronic acid), to be used as controls in order to ascertain the behaviour of nanoparticles at the biological level; e) Quantum dots (QD), used as alternatives for organic dyes for cellular labelling, imaging and tracking. The detailed description of main physico-chemical properties of selected NPs is an integral part of this project and selected reference NPs are being exhaustively characterised prior to use as exposure material for toxicological testings. All pertinent physico-chemical properties of concern, such as size distribution, shape, specific surface area, porosity, chemical composition, purity, impurities of concern, surface chemistry, surface charge, crystal structure, are being determined by applying various analytical techniques in order to ensure high quality of results. In addition, dispersion and stability of selected NPs in aqueous and biological media, as well as their interactions with culture media components, are being investigated. A broad range of different techniques, such as SEM, TEM, AFM, FFF, isotherm N_2 absorption/desorption measurements (BET) for, ESR, NMR, UV/fluorescence measurements ICP-MS, HPLC-MS and GC-MS analysis are being employed to produce a well-characterised bank of NPs. At present two nanomagnetite NPs dispersions in water are being examined. Eventually, specific analytical protocols will be developed for the quantitative determination of the uptake and distribution of selected NPs in biological tissues after *in vitro* and *in vivo* exposure experiments.

3. Cellular models and biomarkers for toxicity testing *in vitro* and *ex vivo*

The mechanism of action of particles is likely to involve oxidative stress and inflammation. We will therefore focus on the cross-cutting areas of cellular toxicity, oxidative stress, inflammation, immunotoxicity, genotoxicity and related endpoints. We will define representative organs which can be used as predictors to evaluate in cellular models the biological response of exposure to selected medical NPs. We aim to propose relevant *in vitro* toxicological models based on reference cell lines of these organs in order to identify reference biological markers and endpoints which can be used to test the possible toxicity of medical NPs. To identify these biomarkers, we will utilize different cell cultures (*in vitro* models), organotypic cell culture and small fragments of organs (*ex vivo* models), exposing them to NPs in clinical use. This will allow us to determine the mechanism of action and to predict the biological and physiological response of exposure to medical NPs and finally to evaluate the potential health hazards of NPs. Whilst it is not possible to consider all organs which may be potentially affected by exposure to medical NP, we intend to use a range of representative organs and to define representative cell lines of these organs. We will focus on cells derived from eight different biological target systems; blood, vascular system, liver, kidney, lung, placenta, digestive, renal and central nervous systems. All *in vitro* studies will be co-ordinated and harmonized as much as possible by using the same batch of NPs, the same handling and safety procedures of NPs, the same exposure time, concentration of NPs, same protocols with similar endpoints, in order to be able to compare results and to evaluate the most relevant biomarkers and target cell types. The common draft safety protocols for handling nanoparticles are being prepared to be used by all partners. Development of *in vitro* models using the cell lines and cells from different organs is in progress together with optimizing protocols for biomarkers of oxidative stress, inflammation, immunotoxicity, cellular toxicity and genotoxicity (Figure 1). Experimental protocols (Standard operating procedures) for biomarker detection to be used across the cell systems are under development.

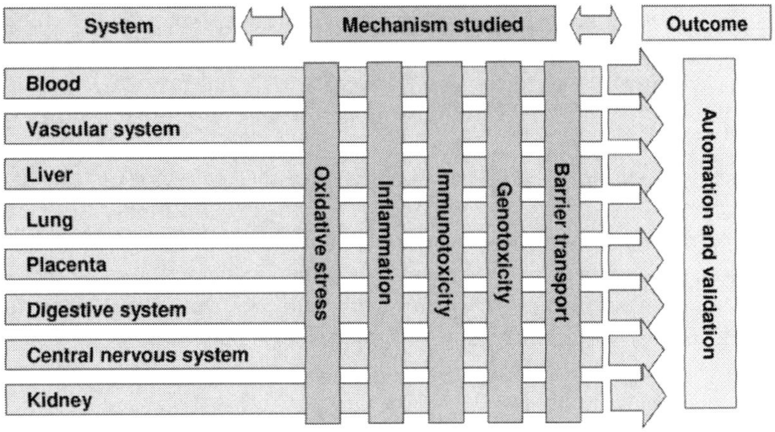

Figure 1

An important cross-cutting area that NanoTEST will focus on is barrier transport of NPs. We will evaluate transport of NP across different barriers such as brain, lung, vascular, digestive system, kidney, placenta (Fig 1). Preliminary results from the placenta will be presented. The placenta is the interface between the maternal and foetal circulation during gestation and plays a very important role in maintenance of a healthy pregnancy and a successful delivery. Anything that interferes with placental function has the potential to have an adverse effect upon foetal development and may impact upon health in later life. Women may come into contact with nanoparticles through occupational, environmental or medical exposure and this can happen at any stage of pregnancy including prior to them even being aware that they are pregnant. Therefore it is essential that we determine whether nanoparticles that are manufactured for use in medical procedures and which could be released accidentally pose any risk to the foetus. To determine this, we need to evaluate both transport characteristics and toxicity since whether a particle is able to transfer across the placental barrier will have a significant impact on where it accumulates and what cell types it may be in contact with. Ethically, this cannot be investigated in pregnant women and therefore NanoTEST will utilise two models to investigate these aspects: the *ex vivo* isolated human perfused placenta model and the *in vitro* human cell barrier model developed using the BeWo placental cell line. Transport experiments have commenced with a benchmark model particle (Fluoresbrite®) which show that, within the BeWo model, particles of 40-60 nm in diameter are capable of passing across an intact cell monolayer. Investigations are under way to clarify the mechanisms which are likely to involve endocytic pathways. Assays will be carried out in both systems to evaluate any toxicity resulting from exposure to the medical nanoparticles under investigation in this project.

4. *In vivo* animal –validation study

The toxicological profile of NPs *in vitro* will be validated in an experimental *in vivo* model. Rats will be used as experimental animals for sensitivity considerations. The route of exposure will be chosen according to the medical usage of the studied NP but most likely animals will be exposed by NP through intravenous injections. There will be a single exposure at the beginning of the study with subsequent sacrifice after 1, 2, 3 and 4 weeks. The effect of exposure to NP will be tested in the following organs: heart/aorta, lung, brain, blood, spleen, bone-marrow, liver, and kidney addressing the same biomarkers and endpoints as the *in vitro* study. Gross pathological and histopathological examinations will be undertaken according to OECD guidelines (425).

5. PBPK modelling and QSARS

One focus of NanoTEST is to identify descriptors relevant for the experimentally determined toxicity endpoints and suitable for modelling from the available data. The selected descriptors in addition to the experimentally determined physicochemical properties (size, surface area, solubility etc.) will be the starting point to develop Structure-Activity models using various statistical multivariate data analysis techniques. The computational algorithms which will be used for the physicochemical characterisation and derivation of descriptors important for the reactivity of the NPs will be chosen based on the particle type and composition, i.e. inorganic and organic structures. PBPK modelling will explore the possibility of formulating a simple compartment-based model to describe the biodistribution of NPs. Due to the limited amount of information available, only the salient features of partition between blood and different organ/tissues will be considered (i.e. small number of compartments). The work will use an existing compartment model as a starting point. In a second stage we will employ computational fluid-particle dynamics to analyze the behaviour of NPs in (i) the respiratory system and (ii) the cardiovascular system.

6. Database and development of standard operating procedures

The Consortium will generate an enormous amount of data to assess from all investigated systems. For this purpose a common NanoTEST (NAPIRAhub) database is under development and will be available for all partners. All the information obtained will be statistically evaluated and collected into a common database. Common approaches for handling, analyzing data and for comparison of data from different systems will be followed. Standard operating procedures for biomarker detection to be used across the cell system will be included in the databank. The data from the experiments will be collated, organised in a database and appropriate statistical analyses will be carried out to determine the nature and extent of NP cell uptake, transport and toxicity over time using the various outcome measurements.

7. Automation assay

NanoTEST aims to develop automated assays to generate high quality data for input into modelling activities, and for future robust testing applications. A short-list of candidate assays for automation will be established. Priority will be given to protocols involving the study of principal mechanisms of action such as oxidative stress and inflammation. An evaluation study of the suitability for automation will be performed in context of the process flow of the selected SOPs and the capabilities of the automated testing platform. Detailed verification studies will be carried out to assess the performance of the assay using high quality reference substances and NPs. Once the candidate assays for automation are selected, an automated SOP, based on the established manual procedure, will be created. The complete automated assay will be verified with reference substances (e.g. positive controls).

8. Conclusion

A better understanding of how properties of NPs define their interactions with cells, tissues and organs in exposed humans is a considerable scientific challenge but one that must be addressed if there is to be safe and responsible use of biomedical NPs. This project will address many key elements in evaluation of NP uptake and exposure and toxicological effect using different biological systems and Structure-Activity models and PBFK modelling.

NanoTEST will integrate the investigation of toxicological properties and effects of NPs in several target systems by developing a battery of *in vitro* assays using cell cultures, organotypic cell culture and small organ fragments (*ex vivo*) derived from different biological systems; blood, vascular system, liver, kidney, lung, placenta, digestive, renal and central nervous systems. As NP action is likely to involve oxidative stress we will focus on the cross-cutting areas of inflammation, cellular toxicity, immunotoxicity, genotoxicity and related endpoints. Following development of Standard Operating Procedures and generation of a common database and in parallel with *in silico* assays (QSAR, PBPK modelling), NanoTEST will evaluate toxic effects and interactions of NPs used in nanomedicine. There are a number of different NP characteristics which will influence transport and toxicity including size, surface area, coating and charge. With the use of a suitable panel of NPs of the highest purity we will determine how these characteristics relate to possible adverse health effects. We aim to develop a battery of *in vitro* and *in silico* assays to evaluate toxic effect of NPs used in nanomedicine. Results will be validated in an experimental ethically approved experimental *in vivo* model. The most advanced and standardised techniques will be adapted for automation and prepared for validation by ECVAM. Finally, we will propose recommendations for evaluating the potential risks associated with new medical NP, which will be communicated to the scientific and industrial community.

Acknowledgement

Supported by EC FP7 [Health-2007-1.3-4], Contract no: 201335.

Nanosafe 2008: International Conference on Safe production and use of nanomaterials IOP Publishing
Journal of Physics: Conference Series **170** (2009) 012040 doi:10.1088/1742-6596/170/1/012040

NANOINTERACT: A rational approach to the interaction between nanoscale materials and living matter?

Iseult Lynch[1], Sara Linse[2], C. Vyvyan Howard[3], Maciej Stępnik[4], Konrad Rydzyński[4], John Hanrahan[5], Wim de Jong[6], Dominique Langevin[7], Joachim Rädler[8], Wolfgang Parak[8], Yuri Volkov[9], Marek Radomski[9], Robert Thomas[10], Jacob Klein[10], Andrew A. Barron[11], Colin Janssen[12], Fiona M. Lyons[13], Francis Quinn[14], Bert Swennen[15], Peter Cuypers[16], Angela Duffy[17] and Kenneth A. Dawson[1]

[1]Centre for BioNano Interactions, University College Dublin, Belfield, Dublin 4, Ireland.

[2]Biophysical Chemistry, Lund University, PO 124, 22100 Lund, Sweden

[3]University of Ulster, Coleraine, United Kingdom

[4]Nofer Institute of Occupational Medicine, Lodz, Poland

[5]Glantreo Ltd, The Rubicon Centre, CIT campus, Bishopstown, Cork.

[6]National Institute for Public Health and the Environment, Bilthoven, The Netherlands,

[7]Université Paris-Sud Bât. 510, 91405 Orsay cedex, France

[8]Center for NanoScience, Ludwig-Maximilians-Universität München, Munich, Germany.

[9]Department of Clinical Medicine and School of Chemistry, Trinity College Dublin, Ireland

[10]Physical and Theoretical Chemistry Laboratory, University of Oxford, OX1 3QZ, UK

[11]Rice University, 6100 Main Street, Houston, TX 77005

[12]Ghent University, Jozef Plateaustraat 22, B-9000 Gent, Belgium

[13]Intel Performance Solutions Ltd., Lexlip, Ireland

[14]L'Oreal SA, 25-29, quai Aulagnier, 92600 Asnières sur Seine, France

[15]Umicore, Broekstraat, 31, rue du Marais, 1000 Brussels, Belgium

[16]DSM Innovation Center B.V. Heerlen, The Netherlands

[17]Medtronic Vascular Galway Ltd., Parkmore Business Park West, Ballybrit, Co. Galway.

E-mail: kenneth@fiachra.ucd.ie

Abstract. The importance of understanding the interactions between nanoscale materials and living matter has now begun to be appreciated by an extraordinaryly large range of stakeholders, including researchers, industry, governments and society, all of whom appreciate both the opportunities presented by and challenges raised by this arena of research. Not only does it open up new directions in nanomedicine and nanodiagnostics, but it also offers the chance to implement nanotechnology across all industry in a safe and responsible manner.
The underlying reasons for this arena as a new scientific paradigm are real and durable. Less than 100 nm nanoparticles can enter cells, less that 40 nm they can enter cell nucleus, and less that 35 nm they can pass through the blood brain barrier. These are fundamental length scales of biological relevance that will ensure that engineered nanoscience will impinge on biology and medicine for many decades to come.
One important issue is the current lack of reproducibility of the outcomes of many experiments in this arena. Differences are likely a consequence of such things as uncontrolled nanoparticle aggregation leading to unpredictable doses being presented to cells, interference of the nanoparticles themselves with many of the tests being applied, differences in the degree of confluency of the cells used, and a host of other factors.
NanoInteract has shown how careful control of all aspects of the test system, combined with round robin type approaches, can help resolve these issues and begin to ensure that the field can become a quantitative science. The basic principle of NanoInteract is that given identical nanomaterials, cells and biological materials, and using a common protocol, experiments must yield identical answers. Thus, any deviations result from errors in (applying) the protocol which can be tracked and eliminated, until quantitatively reproducible results are obtained by any researcher in any location. This paper outlines the NanoInteract programme, illustrates key advances, and highlights early successes. (www.nanointeract.net)

1. Introduction

It is clear that as applications and uses of nanotechnology increase, so too will our exposure to nanoparticles and nanomaterials. Our ancestors were exposed to only very limited types of particles throughout human evolution [1] and there were few particles smaller than 70 nanometers in diameter in the air throughout prehistory, until we learned to harness fire for use. This may mean that we have not built up any defences to nanoparticles [1]. Certainly where certain sizes and types of particle have been released into the environment, profound and unpredictable consequences for human health emerged, for example in the case of asbestos-induced mesothelioma. In this instance, the impact on many sectors of the developed worlds' economy, from manufacturing to insurance, has been highly damaging.

The problem facing nanotechnology is that there exists no clear understanding of the nature and origin of the risks involved in the manufacture, processing, and use of nanoparticles, both research and regulatory frameworks being at early stages of development.[1] Juxtaposed with this concern is the growing opinion that it will be increasingly difficult to commercialise, indeed even to seek venture funding for commercialisation of nano-systems, without having addressed the issues associated with the (perceived) risk. The potential socio-economic benefits in addressing these issues in a deep, structural (and timely) fashion can hardly be overstated.

Thus it is becoming increasingly clear that unless the issues regarding the safety of nanoparticles and nanotechnology are resolved, the technology may never reach it's full economic potential, and fulfil its promise to revolutionize medicine, computers and other arenas.

The NanoInteract programme has been designed to address the issues related to uncertainty and irreproducibility in the assessment of nanotoxicity, and to develop a complete understanding of the mechanisms of nanoparticle interaction with living systems. It brings together approaches from physical chemistry, biophysics and protein chemistry, classical toxicology and ecotoxicology and modern biological technologies to develop a new fundamental framework, and to begin to validate the new approaches which offer potential for higher throughput screening of the estimated 30 000 nanoparticles that are currently being developed in laboratories and industries around the world.

A key concept that has emerged in NanoInteract is the idea that nanoparticles in a biologically relevant environment (cell media, plasma, *etc*) draw to themselves a number of proteins and lipids that form a sort of dynamic 'corona' in slow exchange with the environment [2, 3]. It is this nanoparticle-protein complex that represents the "biological identity" of the nanoparticles, and is this that interacts with the cellular machinery. Interestingly, natural organic matter also appears to function in a similar manner [4]. Such interactions with proteins can have dramatic impacts on the dispersion properties of the nanoparticles, as well as on the protein aggregation behaviour [5, 6].

An overview of the NanoInteract work-programme and the conceptual framework is shown in Figure 1. The project combines the classical toxicology and ecotoxicology approaches in order to assess their applicability to the questions, within a programme where attention to systems parameters is applied at all stages of the experimental design. Thus, as illustrated in Figure 1, key stages in the research process including: (a) ensuring that the nanoparticles used are of the highest purity and monodispersity possible, with any eluting species removed where possible, and characterised where they cannot be removed; (b) dispersion in the relevant biological media for the experiment, such as cell culture medium containing 10% foetal calf serum from a common source, with the dispersion properties characterised over timescales relevant to the experiments using a range of complementary techniques and taking into account measurement artefacts such as the weight-bias which can result in a high number of small particles being hidden by a small number of large particles; (c) characterisation of the protein and biomolecule corona that associates to the nanoparticles in the test media, specifically identification of the proteins and their residence times, as only those proteins that are associated with the nanoparticles for a timescale of several minutes can be considered biologically relevant; (d) connection of the protein corona with the mechanism of uptake utilised by the nanoparticles, including identification of the receptors and transport proteins involved in the uptake and transport processes, and elucidation of the final sub-cellular localisation and the kinetics of nanoparticle uptake; and (e) assessment of the biological impact of nanoparticles using a range of advanced biological techniques such as proteomics and transcriptomics, and connection of the impacts to the nature of the nanoparticle corona, and the final sub-cellular localisation.

As so much is unknown, unproven and irreproducible in the field at present, one of the most important issues to resolve is to establish a rational, quantitative basis for nanotoxicology, and within NanoInteract the approach taken to date has been the application of round-robin type approaches, where results are confirmed independently in up to three different locations (partner or collaborator laboratories). The basic principle of a round-robin (and of the NanoInteract project) is that given identical nanomaterials, cells and biological materials, and using a common protocol, experiments must yield quantitatively identical answers. Thus, any deviations result from errors in (applying) the protocol which can be tracked and eliminated, until quantitatively reproducible results are obtained by any researcher in any location. This ensures that the results are unbiased, scientifically sound, and not open to alternative interpretation. Within NanoInteract we have successfully applied this approach to identification of the protein corona surrounding polymeric nanoparticles [3], nanoparticle-induced fibrillation studies [5] and to genotoxicity studies using silica nanoparticles [7]. Currently, the round-robin approach is being applied to the development of a standard nanoparticle protein corona, in conjunction with NIST (National Institute of Standards and Technology in the United States).

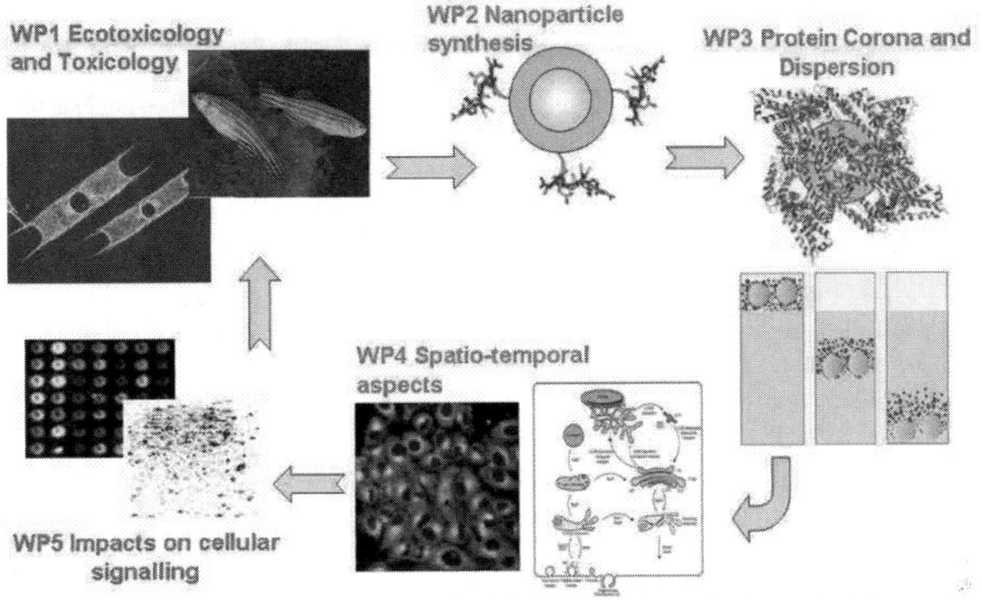

Figure 1. Overview of the NanoInteract programme, and the logic flow which aims to develop a rational framework in which to build our understanding of the mechanisms of nanoparticle interactions with living systems. The programme is seen as being a continuous feedback loop of information.

2. Materials and Methods

2.1. Round-robin

A round-robin defines a useful protocol in challenging cases where the science is not yet understood. Everybody gets same answer using the same protocol proving that the test is robust. Anything leaving room for interpretation in a protocol is identified by the round robin process via a total process of evaluation and comparison of results obtained by the different partners. Given identical materials, and a well developed protocol carefully applied, experimental results must converge. Widespread differences that cannot be resolved by a round robin processes suggest that either the test is unsuitable for nanomaterials or that the protocol is incomplete. Options to proceed include finding a new test or evolving the protocol.

2.2. Nanoparticle protein corona

Nanoparticles that enter the blood stream become coated with proteins. Recovery of the adsorbed proteins by centrifugation with controlled washing steps and times, in conjunction with consideration of the affinities and association / dissociation rates of the proteins, ensures that only truly particle-associated proteins are recovered and identified [3]. Copolymer nanoparticles, 0.5 or 2 mg in 0.05 or 0.2 ml of 10 mM Tris-HCl, pH 7.5, 0.15 M NaCl, 1 ml EDTA were incubated with increasing amounts of plasma on ice. The final volume was constant within each experiment. After 1h, the samples were moved to 23°C to promote aggregation. The particles were pelleted by centrifugation (13 kRPM, 2 minutes), washed three times with 0.5 ml 10 mM Tris-HCl, pH 7.5, 0.15 M NaCl, 1 ml EDTA and the vials changed after each washing step. Bound proteins were removed from the particles by adding SDS-PAGE loading buffer and separated by 12% or 15% SDS-PAGE. Bands were cut out from SDS PAGE gels, digested by trypsin, and studied by mass spectroscopy (nanoscale liquid chromatography quadrupole time-of-flight MS/MS). Spectra were analyzed by MASCOT software to identify tryptic peptide sequences matched to the International Protein Index (IPI) database (www.ebi.ac.uk/IPI). A second independent identification experiment was performed using nano-electrospray liquid chromatography mass spectrometry (Nano-LC MS/MS). Spectra were searched using the SEQUEST algorithm [8] against the International Protein Index (IPI) database.

2.3. Nanoparticle-induced Fibrillation

Dialysis-related amyloidosis involves the fibrillation of protein β2-microglobulin (β2m) which occurs at increased serum concentration in patients with renal failure. The molecular events leading from the native folded monomer to the fibre are not resolved, but nucleation is the key step, after which fibrillation proceeds rapidly. The effect of nanoparticles on β2m fibrillation was studied at 37°C in the absence and presence of nanoparticles under a set of solution conditions. The protein concentration was 40, 80 or 125 μM (as determined by absorbance at 277 nm), in 6, 10 or 20 mM sodium phosphate buffer, pH 2.5, with 0.02% NaN_3, and 50 mM NaCl. Buffer stock was added to the monomer solution, pH checked on a thoroughly rinsed pH electrode, and the solution was then filtered (0.2 μM filter) before it was incubated at 37°C with or without 0.01 mg/ml nanoparticles (copolymeric nanoparticles, cerium dioxide nanoparticles or quantum dots). Experiments were carried out in three different locations, using an identical protocol. To improve the statistics and eliminate irreproducibility due to small differences in solution conditions, experiments were set up in groups from the same solution, with 48 to 120 samples in each group. Each sample of 0.5 ml was shaken in a 1.5 ml Eppendorff tube at 250 rpm close to horizontal orientation. To monitor the appearance and growth of fibrils, aliquots from the tubes were pipetted at different time points to a black 96 well plate and the thioflavin T fluorescence (20 μM thioflavin T added) was measured at 475 nm with excitation at 435 nm in a plate reader [5].

2.4. Silica genotoxicity (Comet assay)

The aim of this study was to perform the most complete, thorough and well-controlled investigation of the genotoxicity of commercially available and laboratory synthesized SiO_2 nanoparticles performed to date using the Comet assay test at the same doses used by Pacheo *et al.* [9], and validating the findings from the study at two independent laboratories. Its purpose is primarily to show that careful application of protocols, standards and methodologies renders the Comet test of these materials quantitatively reproducible.

Silica nanoparticles from Sigma (Ludox, alumina and sodium counter-ion stabilised, both 30nm) and non-stabilised amorphous silica nanoparticles from Glantreo (nominally 30, 80 and 400 nm) were assessed at physiologically relevant concentrations of 4 μg/ml and 40 μg/ml with 3, 6 and 24 hours incubation times using cultured 3T3-L1 fibroblasts supplied to all partners from a central stock, and using a common protocol carefully applied. Nanoparticles were sonicated in an ultrasonic bath for 15 minutes prior to dilution in ultrapure water and subsequent preparation of working solutions in cell culture media. Cells were dosed immediately and incubated under normal cell culture conditions (37°C, CO_2, *etc.*).

A detailed protocol for the Comet assay was described and rigorously applied at the two test locations. Each particle type was tested in detail with multiple repetitions and multiple time points at the first site (UU) and selected particles were independently chosen at a separate laboratory at the 6h time point (NIOM) and the results compared. Additionally, slides from each location were scored at the other location, for complete validation of the results. Comets were analysed using a Nikon Eclipse E400 fluorescence microscope using a x 40 dry lens (NA 0.75). Comet analysis was carried out using Komet 5.0 (Andor™ Technology), % tail DNA and olive tail moment were selected for statistical analysis. For each individual slide 50 cells were analysed at random locations across the slide.

3. Results and Discussion

3.1. Protein Corona

The exchange times (of the 'hard corona') can be so slow that many early biological responses are already defined by these associated biomolecules [10]. It is therefore these that define the biological identity of the nanoparticle, and it is important to learn their identity and more broadly to develop methods to assess them. Significant progress is being made in this arena.

Following incubation of nanoparticles with plasma, the proteins bound to the nanoparticles are recovered by centrifugation and washing. The washing and centrifugation protocol was investigated in detail, to ensure that only proteins genuinely bound to the nanoparticles are recovered, rather than those that are present simply due to their abundance in plasma. After one wash, the amount of albumin is still significant but not after three washes. A total washing time of 20 minutes was implemented to retain proteins with relatively high affinity and slow exchange. Five proteins were consistently associated with the particles.

In two independent experiments this lead to identification of human serum albumin (69 kDa; $p<10^{-6}$), apolipoprotein AIV (43 kDa; $p<10^{-6}$), apolipoprotein E (34 kDa; $p<10^{-6}$), apolipoprotein AI (28 kDa; $p<10^{-6}$), and apolipoprotein AII (8.7 kDa; $p<5\cdot10^{-5}$). The molecular weights of these proteins match those estimated from SDS-PAGE in the present experiments (shown in Figure 2). Judging from the band identities, the most abundant protein on the particles is apolipoprotein AI, and the amount of HSA is low in comparison. The concentration of apolipoprotein A1 in human plasma is only about 1-2 mg/ml, compared to 30mg/mL for HSA.

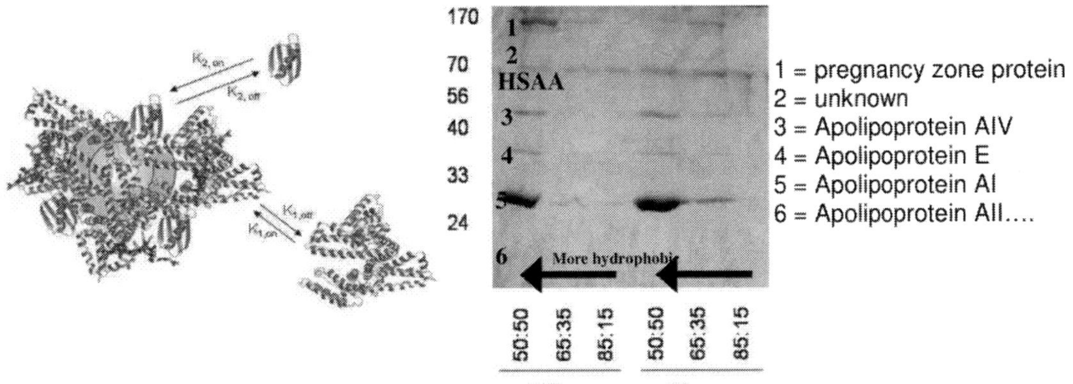

Figure 2. Left: Illustration of the dynamic nanoparticle protein corona where rapid exchange occurs with the surrounding proteins, and proteins with high affinity replace lower affinity higher abundance proteins.
Right: 1D-gel showing the major proteins bound to copolymer particles of increasing surface hydrophobicity. Note the very low binding of HSA (human serum albumin) to the particles despite it being the major component of human plasma [3]. The fact that identical materials with different size and surface charge bind different proteins into their coronas has deep implications for regulation of nanomaterials, and suggests that in the future nanoparticles may be able to be classified according to the proteins that they bind, and potential impacts from nanoparticles may be predicted based on protein coronas.

Table 1. Selective binding of different apolipoproteins to polystyrene nanoparticles of different sizes and different surface charges.

Lipoproteins	100nm NH_3	50nm NH_3	100nm plain	50nm plain	100nm COOH	50nm COOH
Apolipoprotein A-I	X	X	X	X	X	X
Apolipoprotein A-II		X				X
Apolipoprotein A-IV		X	X	X	X	X
Apolipoprotein B-100	X	X	X		X	
Apolipoprotein C-I	X	X			X	X
Apolipoprotein C-III		X		X		X
Apolipoprotein D		X	X	X		
Apolipoprotein E	X	X	X			X
Apolipoprotein F		X				
Apolipoprotein L1		X			X	
Beta-2-glycoprotein 1 (apolipoprotein H)	X		X	X	X	X

Interestingly, apolipoprotein A1 is the major component of HDL (high density lipoproteins), and is found on chylomicrons, large lipoprotein particles created by the absorptive cells of the small intestine, which are used in cholesterol transport *in vivo* [11, 12]. This has potentially significant implications for the uptake and transport of nanoparticles, as, for example, apolipoprotein E has been shown to enhance transport to the brain [13]. More recent results from the NanoInteract project have shown that changing the size and / or surface charge of nanoparticles influences the proteins that bind to the nanoparticles in a biological media, including resulting in specific binding of Apolipoprotein E or Apolipoprotein F which is known to promote transport across the placenta, as shown in Table 1 [10].

3.2. Fibrillation Assays

β2m-fibril growth appears to follow a (seed-independent) homogeneous nucleation-and-growth mechanism, requiring high concentration of the monomer only, at pH values less than 4.8, where unfolding of the protein is rather extensive. Nucleation is a stochastic event, so fibrillation by this mechanism is characterised by a lag phase that increases as we approach the heterogeneous nucleation regime. However, once the process is started it quickly goes to completion.

Experiments (in total 432 tubes) were set up at 50 mM NaCl to enable the collection of statistical amounts of data in a reasonable time. Under these conditions the presence of 0.01 mg/ml nanoparticles was observed to enhance in the rate of fibrillation significantly. The 'lag phase' (likely a reflection of the time for a critical nucleus to form) for two groups of experiments in tubes are presented in the curves shown in Figure 3 (left), and in the form of histograms in Figure 3 (right). From the time dependence of fibril formation, it appears that the enhanced rate of β2m fibrillation in the presence of the nanoparticles is due to a significantly shortened lag phase, or time to form a critical nucleus. Indeed, in the absence of nanoparticles, the lag phase is so long that fibrillation has not even begun during the timeframe in which the process has gone to completion in the solutions with nanoparticles. More recent experiments using amyloid-β (one of the proteins associated with Alzheimer's disease) have shown the opposite trend, with selected nanoparticles slowing down or even reversing the fibrillation process.[6]

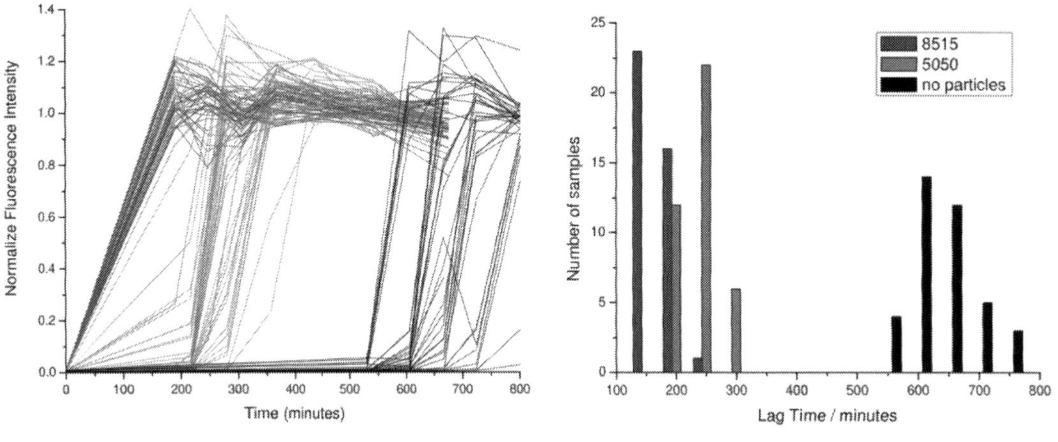

Figure 3. Left: ThT fibrillation assays in the absence (black) and presence (blue and red) of copolymer nanoparticles of different hydrophobicity. Blue curves show the reduction in the lag time in the presence of hydrophilic copolymer nanoparticles, and red curves show the effect of more hydrophobic copolymer nanoparticles. The concentration of β2m is 0.5 mg/mL in buffer containing 20 mM phosphate and 50 mM NaCl at pH2.5. The particle concentration in all cases was 0.01 mg/mL. Right: Distribution of lag times showing the statistical distribution of the lag times.

3.3. Comet Assay Alignment

Application of the round-robin process to the Comet assay was quiet a challenge, and involved several iterations, including alignment of the protocol between the two laboratories, use of same Comet assay reagents (including cells, serum, *etc.*), inter-lab visits and sample exchange, comparison of analysis

software and hardware and elimination of variation due to different analysis software and hardware, and application of strict random sampling rules. Table 2 shows a set of data from the two laboratories, showing that quantitative agreement can be obtained.
Once the protocol had been refined, and quantitatively similar results were obtained at both test sites, this careful, systematic and multi-site study demonstrated that amorphous silica nanoparticles from different sources (Sigma and Glantreo) do not cause genotoxicity detectable by the Comet assay in the 3T3 cell line under the conditions described, as shown in Figure 4. The results were found to be reproducible in separate tests in two independent laboratories and with different personnel analyzing the Comet results. This study does not detect genotoxicity from amorphous silica nanoparticles in the size range from 20 nm to below 400 nm, with both positive and negative surface characteristics.

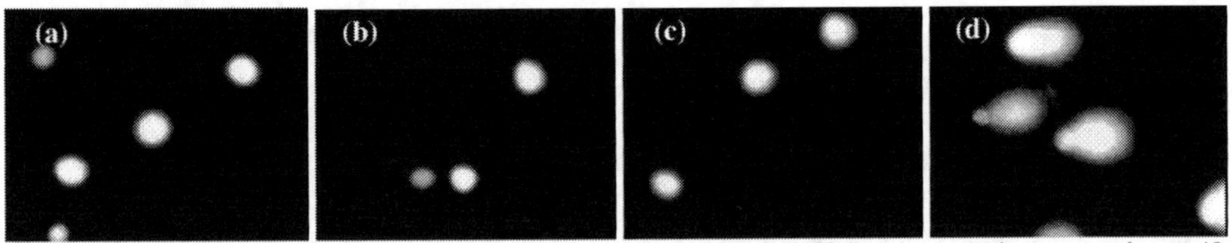

Figure 4. Examples of Comet assay images, recorded with Nikon Eclipse E400 fluorescence microscope using a x40 dry lens (NA 0.75), a) 3T3-fibroblasts incubated for 24 h with 40 µg/ml 30 nm silica nanoparticles from Glantreo, b) incubated for 24 h with the dialysate, c) negative control, d) positive control. All images are from the experiments at UU.

Table 2. Comparison of Comet assay results of Sigma Ludox CL mean and standard deviation of Tail%DNA, conducted at two different NanoInteract partner locations.

Sigma Ludox CL 420883	Partner X		Partner Y	
	Tail%DNA		Tail%DNA	
	6h		6h	
	Mean	SD	Mean	SD
40 µg/ml silica NP	6.24	5.12	7.44	7.50
Dialysate	8.46	8.86	6.42	8.56
Negative control (DMEM)	7.26	5.99	6.12	7.31
Positive control (H_2O_2)	34.30	18.44	52.34	17.99

4. Conclusions

NanoInteract is a European research project investigating the interactions between nanoparticles with living systems, using a range of approaches from classical toxicology and ecotoxicology, through physical chemistry and biophysics, and advanced biological techniques such as transcriptomics and proteomics. The key elements of NanoInteract are control of the system at all stages, from nanoparticle purity, dispersion in biological media, control of cell confluency and the exposure dose, etc., and application of round robin approaches to ensure that quantitatively reproducible results leave no room for doubt or uncertainty regarding the results, and ensures that the results can be trusted by all of the relevant stakeholders, from governments to NGOs to the general public.

An important aspect of the work is the application of a combination of existing biophysical and physical characterisation techniques to the issue of protein conformation upon binding. We have found that size and surface charge affect the nature of the proteins that bind to the nanoparticles in physiological solutions, thereby conferring different biological identities onto the nanoparticles. This may have significant impact in terms of regulation of nanoparticles in the future, and potentially for development of quantitative structure-activity relationships (QSARs). We also note the surprising feature that nanoparticles are sometimes able to induce dramatic effects on protein interactions, such as the case for protein fibrillation, and of course this observation, when combined with the potential for nanoparticles to transport to the brain implies the need to study this whole arena in more depth.

Having understood the nature of the expressed biomolecules the challenge now is to study where the particles go (using imaging techniques) and what effects they have on cells / organisms when they reach there. Significant progress has been made here within NanoInteract, and will be the subject of another article. A key achievement of the field would then be to connect this biological identity (corona) to the observed biological impact. This has not been achieved yet, but progress is being made within NanoInteract.

Perhaps the most significant outcome of the NanoInteract project to date is the suggestion that, contrary to often stated beliefs, *in vitro* toxicity test may evolve to become entirely quantitatively reproducible across multiple laboratories. This is a hopeful message as nano-EHS studies are cited broadly, not just in the scientific community, but also in the public media.

Acknowledgements

The authors wish to thank the entire NanoInteract consortium. This work is funded by the EU FP6 STReP project NanoInteract 033231. Many postgraduate students and postdoctoral researchers have been involved in this work, and their names can be found in the original articles. Clearly this article represents a very limited selection of the findings from the NanoInteract project. Additional summary articles from the NanoInteract consortium are envisaged in these other areas.

References

[1] European Commission, Nanotechnologies: A Preliminary Risk Analysis on the Basis of a Workshop Organized in Brussels on 1-2 March 2004 by the Health and Consumer Directorate General of the European Commission. 2005.

[2] Cedervall T, Lynch I, Lindman S, Nilsson H, Thulin E, Linse S and Dawson K A 2007 *PNAS* **104** 2050

[3] Cedervall T, Lynch I, Foy M, Berggård T, Donelly S, Cagney G, Linse S and Dawson K A 2007 *Angewandte Chemie Int. Ed.* **46** 5754

[4] Brant J A, Labille J, Bottero J Y and Wiesner M 2007 *Environmental Nanotechnology: Applications and Impacts of Nanomaterials* McGraw Hill NY ed. 231

[5] Linse S, Cabaleiro-Lago C, Xue W-F, Lynch I, Lindman S, Thulin E, Radford S, Dawson K A 2007 *PNAS* **104** 8691

[6] Cabaleiro-Lago C, Quinlan-Pluck F, Lynch I, Lindman S, Minogue A M, Thulin E, Walsh D W, Dawson K A and Linse S 2008 *JACS ASAP published*

[7] Barnes C A, Elsaesser A, Arkusz J, Smok A, Palus J, Lesniak A, Salvati A, Hanrahan J P, de Jong W H, Dziubałtowska E, Stępnik M, Rydzyński K, McKerr G, Lynch I, Dawson K A and Howard C V 2008 *Nano Letters, ASAP Article* DOI: 10.1021/nl801661w

[8] Yates J R, Eng J K, McCormack A L and Schieltz D 1995 *Anal. Chem.* **67** 1426

[9] Pacheco S E, Mashayekhi H, Jiang W, Xing B and Arcaro K F *Abstracts of the 2007 Annual Meeting of the American Association for Cancer Research*

[10] Lundqvist M, Stigler J, Cedervall T, Elia G, Lynch I and Dawson K 2008 *PNAS* **105** 14265

[11] Berggard T, Arrigoni G, Olsson O, Fex M, Linse S and James P 2006 *J Proteome Res.* **5** 669

[12] Mahley R W, Innerarity T L, Rall Jr S C and Weisgraber K H 1984 *J. Lipid Res.* **25** 1277

[13] Michaelis K, Hoffmann M M, Dreis S, Herbert E, Alyautdin R N, Michaelis M, Kreuter J and Langer K 2006 *J Pharmacol Exp Ther.* **317** 1246

Institute of Physics
Dirac House, Temple Back
Bristol BS1 6BE UK